HANDEDNESS & SPEECH

BRAIN PLASTICITY & EVOLUTION

HANDEDNESS & SPEECH
BRAIN PLASTICITY & EVOLUTION

by

K.A.Provins, PhD., MA.

National Library of Australia Cataloguing-in-Publication entry

Author: Provins, K. A.

Title: Handedness and speech : brain plasticity and evolution / K.A. Provins, PhD., MA.

ISBN: 9780980815931 (pbk.)

Notes: Includes bibliographical references.

Subjects: Cerebral dominance.
            Brain--Localization of functions.
            Brain--Evolution.
            Human evolution.

Dewey Number: 612.825

Dedication

This book is dedicated to the memory of my late wife Lorna and our partnership of sixty years of love, life and happiness beyond measure.

Acknowledgements

I should like to thank the University of Queensland, especially the Librarian and staff for their generous help during my tenure of an Honorary research appointment with the School of Psychology that made this review possible. I also wish to express a special thank you to Kay Provins whose editorial and management skills enabled this work to be published.

Preface

Attempts to discover a basis for the enigmatic relationship between handedness and the cerebral control of speech has a history of well over one hundred years, while the apparently universal predominance of right-handedness in the human population has puzzled investigators for even longer. For a number of reasons, the amount of research interest in the subject during the last fifty years or so has increased considerably, so that a newcomer to the field could be excused for being more than somewhat overwhelmed by the breadth as well as the depth of the literature, and the often apparently conflicting nature of the published findings. In this regard, the present review is not intended as a survey of the theories and speculations of past investigators, but an attempt to consider a re-orientation to, or re-interpretation, of the evidence available.

Clearly, presenting an examination of the overall area must inevitably be limited and less than exhaustive. However, as far as is possible, an attempt is made here to provide a guide through the maze of accumulated evidence, and to critically evaluate the major lines of research, with particular reference to the methods used and results achieved. Certainly, there have been great advances in surgical and non-invasive techniques over the years to help in the search for answers. Behavioural measures have also been refined. But the basic assumptions underlying the studies undertaken, and the approach to interpreting the available data have changed little in spite of developments in thinking that have taken place in associated basic sciences.

Ever since the finding that loss of speech was nearly always the result of damage to the left side of the brain in right-handed patients, there has been a widespread belief that the usual association between a cerebral lateralization of speech and handedness has some inborn, genetic predetermination. However, in the various genetic explanations offered so far, little consideration has been given to a key feature of the normal processes of natural selection in evolution. This requires that for a mutational change to survive and be successfully transmitted to succeeding generations, it must endow its possessor with a clear advantage over others in the competitive struggle to adapt to the demands of the environment. No such advantage can be advanced to support the notion of an inborn lateral asymmetry in *Homo sapiens*: indeed, it would more likely be a severe disadvantage. Rather, a review of the evidence suggests that, with the evolution of a genetic determination of the potential for learning or intelligence in primates and man, it is each individual's experience and interaction with his environment that shapes the form of any resulting lateral bias in manual behaviour and asymmetric development of the cerebral control of motor activity.

My own research work in the field began with the study of muscle and joint function and the role of sensory feedback in the control of skilled movements. Because of the bilateral structure of the human form, any measurements I made on one side of the body I then repeated on the other. In due course, my results showed that sometimes the findings on the two sides were the same and sometimes they were significantly different, stimulating me to enquire why.

Needless to say, the ideas set forth in the following pages are not only the ultimate outcome of my fifty years or so of extensive reading and experimental work in the field, but the result of my good fortune in having the benefit of discussions with many knowledgeable people over this period of time. In this respect, I am particularly indebted to Oliver Zangwill, who, as my tutor at Oxford, first aroused my interest in the cerebral control of speech and handedness, and to Carolus Oldfield, my PhD supervisor. But probably the strongest influence on the development of my approach to the problem was the subsequent ten years I spent as a member of the Medical Research Council Unit in the Oxford University Department of Human Anatomy. In this highly stimulating environment, with the weekly staff seminars chaired by Professor Sir Wilfrid Le Gros Clark, I was inevitably exposed to discussions about the latest neuroanatomical research on the brain and work in progress. Furthermore, since Joe Weiner (as Reader in Physical Anthropology and Head of the MRC Unit) and Le Gros' special interests were also directed towards the evaluation of fossil evidence on the history of the primates and hominid evolution, these lines of investigation also undoubtedly affected the orientation of my own line of thinking.

However, it was not until I came to Australia and took up an appointment with Malcolm Jeeves in the Psychology Department at the University of Adelaide that I attempted to pull together in an integrated form the information available on lateral asymmetry and my particular orientation to it. Again, I found myself in a very agreeable environment, enhanced by Malcolm Jeeves growing interest in the functions of the corpus callosum and acallosal patients. With the need to provide students with an up to date picture of the information available, my teaching duties naturally provided a further stimulus to my attempts to make sense of the many apparently irreconcilable findings in the literature, and to search for a more satisfactory explanation for them than was then available.

While an earlier report of these investigations appeared in two journal articles (Provins, 1997a; 1997b), the present monograph provides a much more comprehensive account of the sources consulted and research examined. The final chapter provides an overview and some introspections on the findings.

Ken Provins, PhD., MA.
Emeritus Professor, University of Adelaide

Contents

DEDICATION V

ACKNOWLEDGEMENTS... VII

PREFACE IX

Chapter 1: Handedness and Speech: Placing the Problem in
            Perspective 1

            Evolution and adaptation.. 4

            Tools, skill and evolution 5

            Brain and adaptability 9

            Manual asymmetries in pre-history............................... 15

            Handedness and Brainedness 20

            Cerebral dominance and hemisphere
            specialization 23

Chapter 2: Brain Damage, Speech and Handedness 27

            Adult aphasia.. 29

            Childhood aphasia.. 35

            Language and the corpus callosum............................... 39

            Electrocortical mapping and cerebral asymmetry 45

            Apraxia and associated motor disorders...................... 48

            Experimental lesions in non-human species................ 52

Chapter 3: Cerebral Asymmetries in Normal Subjects 57

            Neuroanatomical asymmetries 58

Neurophysiological asymmetries 65

(a) The electroencephalogram *65*

(b) Event-related potentials and manual activity *68*

(c) Event-related potentials and speech *73*

(d) Regional cerebral blood flow *76*

Behavioral (dual-task) studies 81

Chapter 4: Behavioural Studies in Non-Human Species 91

Mice 93

Rats 100

Cats 108

Monkeys and apes 112

Chapter 5: Heredity and Handedness Assessment 131

Heredity and handedness 132

Handedness assessment 140

(a) Assessing preference *141*

(b) Assessing proficiency *149*

(c) Preference and proficiency compared *153*

Chapter 6: Ontogenesis of Handedness 159

Infant motor asymmetries 160

Hand preference trends during childhood 163

Performance differences between sides 167

Consistency of asymmetry across tasks 176

Chapter 7: Skill and Handedness 179

The nature of skill 180

The effects of unilateral exercise 191

Handedness and stuttering 199

Chapter 8: Handedness and Disability............ **205**

Hand preference and intelligence 207

Asymmetry of manual proficiency and
intelligence 210

Handedness in clumsy children............212

Handedness in developmental dyslexia.........216

Autism and handedness............219

Schizophrenia and handedness 222

Twinning and handedness............225

Prematurity, extremely low birth weight (ELBW)
and handedness............228

Epilepsy and handedness 230

Handedness in immune disorders............232

Chapter 9: Cultural Values and Social Pressure **235**

Handedness and delinquency 236

Symbolic significance of right and left.........240

Handedness and social conformity............245

Chapter 10: Intracultural Differences **253**

Generational differences in handedness.........254

(a) Hand preference and age *254*

(b) Hand proficiency and age *258*

Handedness and gender 259

Socio-economic status and education 263

Profession or occupation 266

Chapter 11: Inferences and Implications **275**

Compatibility of the Social and Biological
Evidence 275

Symbolism and Speech.. 276

Facilitation of Brain Lateralization of Function.......... 279

BIBLIOGRAPHY **287**

CHAPTER 1
Handedness and Speech: Placing the Problem in Perspective

Chapter outline

Two particular characteristics are said to distinguish human from non-human behaviour. One is speech and the other is manual skill. During evolution, while speech has left no recoverable trace of its beginnings, substantial archaeological and fossil evidence has been found to provide a documentary record of the association between the appearance of manual skills and emergence of the genus *Homo*. Information relating to the increasing sophistication of the design and use of bone and stone tools in particular, together with later evidence from Palaeolithic drawings and paintings, suggest that there has been a predominant use of the right hand for skilled manual work throughout human history.

Fossil evidence of hominid evolution, and comparative studies of the anatomy and physiology of extant species of primates, has shown that another distinctively human characteristic is the relatively great increase in size and complexity of the brain. The importance of this development during the history of the primates is considered in relation to the selective advantage of a parallel advancement in behavioural adaptability and learning, and its implications for the lateralization of manual skills in response to the demands of the environment.

The relationship between handedness and speech is considered in the light of two historically significant explanatory approaches, one biological and one social. In the first, clinical evidence for the predominance of the left cerebral hemisphere for speech was used by Broca to suggest that "brainedness" determined handedness. In the other, socio-anthropological evidence was cited by Hertz to suggest that a cerebral asymmetry favouring the left hemisphere for motor control functions is the result and not the cause of a more consistent use of the right hand.

Subsequently, other investigators suggested that each cerebral hemisphere may have a differentially predominant role in a range of motor, sensory/perceptual, and/or cognitive functions, and a number of attempts have been made to assign each of these to the right or left hemisphere according to some arbitrary system of classification. However, such generalizations are considered to be unjustified oversimplifications, and the present review limits itself to seeking an explanation of the nature and origin of handedness and its relationship with speech in a critical examination of the available evidence specifically concerned with manual skills and speech production.

In everyday terms, handedness is generally understood to refer to the hand most often used by someone in the normal daily routines of life. A person who favours the right hand is said to be right-handed and one who favours the left hand is regarded as left-handed, while an individual who sometimes prefers to use one hand and sometimes the other is often called ambidextrous. But what are the reasons for these individual differences? Are they due to some inborn determinant or to the influence of environmental pressures such as custom or convention? Or is handedness the result of both nature and nurture? And anyway, why are most people right- and not left-handed?

Apart from a purely academic research interest in pursuing answers to these questions, there are also very good practical reasons for exploring the underlying basis of manual asymmetries, since deviations from the normal population bias favouring the right hand have been associated (either directly or indirectly) with a wide variety of behavioural disturbances or disorders. For example, a higher than normal incidence of left-handedness, has been reported to be found amongst delinquents or criminals, mental retardates, autistic children, cases of clumsiness, and in a variety of immune disorders. And one of the most researched associations has been the relationship between disorders of speech and handedness, and the extent to which the central control mechanisms for these functions share a common or overlapping neural organization in the brain. But as well as the clinical and educational importance of these associations (see e.g. Bishop, 1990a; Burt, 1937), left- and right-handedness may also have occupational implications with respect to the efficiency and safety of people employed in a wide variety of situations. In this respect, left- and/or mixed-handers have been reported to be at greater risk of sustaining accidental injury than right-handers (Coren, 1989, 1992; Hicks, Pass, et al., 1993).

In the following pages an attempt is made to provide a critical evaluation of the available information bearing on these associations, and to propose an explanation that most successfully accommodates the evidence. In putting forward such an explanation, consideration is also given to what is known about how, when and where manual skills, speech, and lateral bias emerged in human evolution, and the means by which such characteristics may have been transmitted down through time and successive generations.

It is a common belief that the basis of human handedness is rooted in our evolutionary origins and is a product of our genes. The fact that most people are right-handed and have extreme difficulty in undertaking certain activities such as handwriting with their other side, usually convinces them that they were "born that way". Certainly, there has been a considerable amount of research effort devoted to finding a genetic model that would substantiate the claim, but in spite of this, the matter remains unresolved. However, most of this work has focussed on the biological mechanisms of inheritance and Mendelian laws of genetics, with little or no attention being given to the vital role of natural selection in Darwin's theory of evolution and the struggle for survival. Consequently, the present review begins with a reminder of these principles.

Evolution and adaptation

The basis for the evolution of species is adaptation to the environment by natural selection. It depends upon the fact that members of any given species demonstrate individual differences, some of which give the possessor a selective advantage to survive the struggle for existence until maturity. The better adapted and more fertile survivors are then able to reproduce and leave more offspring than the others. When the characteristics which enabled them to survive are passed on to their progeny genetically, the inheritance of such benefits leads to the evolution of a population that is better adapted to its environment. Hence, all species have originated from a common ancestor at some time in the past by the same process of natural selection and heredity, and are consequently related to one another, however distantly. The wide range of characteristics peculiar to each species and their inter-relationships have been extensively researched and are now well established (see e.g. Harrison et al., 1988; Le Gros Clark, 1971).

Whereas most animals have evolved to be bilaterally symmetrical in both structure and function, there are many well-known exceptions. For example, in invertebrates, the conical helix of mollusc shells may twist to the left or to the right, and both types are common. Since mating between snails that have shells which coil in opposite directions is difficult, they produce fewer offspring, resulting in monomorphism within populations (Johnson, Clarke & Murray, 1990). In crustaceans, the paired claws of the American lobster have a quite different appearance. The major or 'crusher' claw is short, and slow but forceful in its action, while the minor claw, called the 'cutter', is long and slender and can close its pincers within 20 msecs (Govind, 1989). The crusher claw occurs with equal probability on the right or left side of the body, with its laterality being innately programmed but dependent on experience during a critical period of development. Amongst the lower vertebrates, one of the most remarkable exceptions to bilateral symmetry occurs in flatfishes (Policansky, 1982).

Although these fish have a symmetrical form when they hatch from the egg, the position of some structures changes during development. In particular, one eye migrates across the top of the head, so enabling the fish to lie on the sea bed with both eyes facing upwards. It appears that in some species, nearly all fish have both eyes on their right side while other species are mostly left eyed. Other examples of asymmetry in fishes, reptiles, amphibians, and birds have been reviewed by Bradshaw & Rogers (1993), Bisazza, Rogers & Vallortigara (1998). While there seem to be good grounds for believing that the formation of such asymmetries may be under some degree of genetic control in these animals, the development of direction of the lateral differences between right and left sides is a different question. The predominance of one particular side in a population may be due to practical problems associated with mating behaviour, as has been suggested in the case of snails and flounders. In other species, the distribution of asymmetries between right and left in different populations may vary from equal numbers of each, to one that is extremely biased in one direction or the other, without any obvious reason. However, there is no doubt that in some species, such an asymmetry may be passed on to their offspring genetically. And natural selection for an asymmetry which provides some adaptive advantage in a specific environmental setting is clearly to be expected for this characteristic as much as for any other structural or functional feature of survival value. But, as will be seen from the ensuing discussion, there are good grounds for questioning whether the evidence from asymmetries in invertebrates and non-mammalian vertebrates has any relevance for the origin of human laterality.

Tools, skill and evolution

Two particular features which have often been described as distinguishing human from non-human animal behavior are (a) speech (see e.g. Critchley, 1958; Lenneberg, 1967), and (b) manual skill (see e.g. Napier & Tuttle, 1993; Oakley, 1972). However, a human species specification based on behavioural characteristics is extremely difficult, since, as Lieberman (1975) bluntly points out, all of the immediate predecessors of *Homo sapiens* are dead. Hence, it is necessary to be very cautious about claims of uniquely human characteristics inferred from sources such as archaeological remains, fossil evidence, and the comparative behaviour of living primates and our other more distant relatives, as apparent discontinuities in form or function may simply be a result of gaps in the currently available evidence. In this respect, information relating to the evolution of speech presents even greater difficulties than for manual skill, as before the present technological age of sound recording, spoken utterances left no recoverable trace of their existence. And the advent of graphic or written forms of language was necessarily dependent on the prior development of manual skills in the use of tools.

Consequently, the search for a manifestly human behavioural criterion has centred on the evidence relating to manual skill and the proposal discussed by Oakley (1958, 1972) that tool-making as distinct from tool-using provides the vital clue. Tool-using has usually been seen as opportunistic behaviour in that readily available materials or objects may be taken and manipulated as necessary to achieve an immediate and purposeful solution to a present and pressing problem. In contrast, tool-making has often been viewed as demonstrating forward planning and conceptualization of the solution to a problem that exists at some distant time or place. However, such a distinction necessarily involves consideration of the materials or objects used since there have been numerous observations of extant nonhuman species using and making tools with leaves, wood, or bone, both in captivity and in the wild (see e.g. Goodall, 1964; Hall, 1963; Ward & Hopkins, 1993). By contrast, it is only possible to obtain a satisfactory record of the human evolutionary transition from tool-using to tool-making when the artefacts have been deliberately shaped or made from relatively enduring materials such as bone or stone. Much of this crucial evidence comes from comparing the archaeological record of these developments with the fossil evidence of a progressive change in human morphology through a continuous and gradual sequence over the same period (Leakey, 1981, 1994; Schick & Toth, 1993).

In this context, there appears to be two particular stages of hominid development that are of special significance. The first relates to the evolution (some 7 million years ago in Africa) of the first distinguishable hominid species (Australopithecus) which, from the structure of the pelvis and leg bones, habitually employed a bipedal or upright manner of walking. Although apelike in many other respects, the hands of these individuals were thus freed from their involvement in locomotion and hence available to be used in a wide range of other activities including tool-using. The second significant stage of hominid development that occurred between about 3 million and 2 million years ago, was a notable increase in the size of the brain which marked the beginning of the genus *Homo*, at about the same time as the first stone artefacts appeared (Leakey, 1994). However, another important feature in the emergence of manual skills was undoubtedly the structure and functional flexibility of the hand itself - particularly with respect to the ability of the thumb to oppose the flexion of the fingers in a finger grip. Indeed, Napier and Tuttle (1993, p.55) even suggested that it was probably the single most crucial adaptation in our evolutionary history, and that "through natural selection, it promoted the adoption of the upright posture and bipedal walking, tool-using and tool-making that, in turn, led to enlargement of the brain through a positive feed-back mechanism". Certainly, it seems to be generally agreed that the development of tool-using and tool-making was facilitated by the protohominid change to ground-dwelling from the previously predominantly arboreal habitat common to nearly all other primates. For whereas monkeys and apes have also developed a certain degree of manual dexterity, their hands have become more specialized or adapted to climbing and brachiating activities. The safe and sure use of the forelimbs and hands of most primates undoubtedly bestows on them a highly successful, mobile, and agile existence amongst the trees.

But as Napier (1961) has pointed out, the body size of animals is a highly relevant feature in the structural and functional evolution of the hand. For whereas small primates are able to live at any level from the forest floor to the top of the highest trees due to their sharp claws and light body weight, once the ratio of body size to branch diameter exceeds a certain value, larger creatures have to rely on progression by suspension (brachiation) from the thicker (and lower) boughs, or be limited to locomotion on the ground.

However, exactly when tool-using developed into tool-making, and its timing in relation to brain development and other morphological differentiation is uncertain, since these changes occurred over a period of millions of years and the evidence is all too fragmentary. Suffice it to say that the fossil record shows that the size (volume) of the brain increased from approximately 435-600 mls in Australopithecus to about 1000-2000 mls in the present-day human brain. This has been accompanied by an increasing sophistication of bone, stone, and metal artefacts to bear testimony to the great development of manual skill in hominid evolution. Both Oakley (1958) and Schick and Toth (1993) have reviewed the Palaeolithic evidence from the first crudely chipped pebbles of lava or quartz through various stages of tool development to the creation of specific implements and weapons vital to the hunter-gatherer's struggle for survival. And Toth, Clark and Ligabue (1992) have provided complementary evidence from observing the methods and skills of a present-day group of isolated people who still make and use stone tools. It appears that the simplest and earliest stone tools may have been made by breaking a stone through hitting it with another and directing the blow so as to leave a sharp edge on the core - thus producing a "core-tool", with the flakes struck from the core also being used (if suitable) as "flake-tools". Alternatively, Toth (1987) has suggested that the flakes may have been the primary objective and cores the by-product. Other more advanced techniques of manufacture by indirect percussion and pressure flaking have been reported, and the fashioning of an artefact for a specific purpose by repeated working of the surfaces and sometimes by grinding on a sandstone have also been detailed (see e.g. Semenov, 1964).

Brain and adaptability

Of particular interest in relation to the evolution of human manual skills, is the almost threefold increase in size of the hominid brain over the last 2 million years or so, unaccompanied by any similar change in body size (Passingham, 1975a; Jerison, 1991). Unfortunately, the interpretation of this increase with respect to likely changes in the relative development of different parts of the cerebral cortex is difficult, since the use of endocasts of fossil skulls to obtain such information has serious limitations (see e.g. von Bonin, 1963; Holloway, 1974; Jerison, 1973). However, using a quite different approach, some worthwhile information can be gained by making a comparison of the differences in complexity of the cortical structure and size of the specialized somatic sensori-motor areas of the brains of different living primate species. The use of such anatomical and physiological data from existing non-human primates to infer trends associated with human evolutionary development is a well-established and acceptable practice, although it also has its limitations (see e.g. Le Gros Clark, 1971; Martin, 1973).

An impressive change in the development of the primate brain relates to the enlargement and elaboration of the neocortex (Passingham, 1981), with an increasingly detailed "mapping" on it of the inputs received from the specialized sense organs of the body, and an increasingly complex mapping of the motor outputs to the musculo-skeletal system of movement effectors (Humphrey, 1986; Woolsey, 1958). In this respect, the degree of cerebral representation of these functions in each type of animal appears to reflect the relative behavioural importance to each species of the structures concerned. And in the precentral region of the human cerebral cortex, the amount of brain tissue devoted to the motor functions of the hand (especially the thumb and fingers) and face (especially for vocalization) is many times greater than for the whole of the rest of the body. As Passingham (1981) has pointed out, the motor cortex has its most direct association with manual dexterity by means of the neural pathway from the brain to the effectors known as the pyramidal tract. If body weight is taken into account, he has shown that in land mammals, the mobility of the digits is directly related to the number of fibres in the tract, and that the pyramidal system is larger and motor control of the effectors is most discrete in the higher primates and man.

However, it would be wrong or misleading to confine this brief discussion of human behavioural evolution to purely motor development. As Le Gros Clark (1956) observed, since the most conspicuous feature in the history of the primates is the steady expansion in the size of the brain, this would have been associated with changes in a whole range of behaviours. Whereas many lowlier animals ensured their survival by evolving specialized means of defence and attack (e.g. sharp teeth and claws), he suggested that the primates adapted to their environment through the use of wile and cunning, evident in higher primates and especially hominids as intelligence and/or an enhanced ability to benefit from experience through learning. And as Sperry (1958) concluded, when in evolution an opportunity existed for a given favourable behavioural pattern to be acquired by learning instead of some inborn instinctive activity, there would have been obvious advantages to be gained from an adaptation to circumstances through learning, since learning endows the individual flexibility to adjust to a multiplicity of different situations. Thus, in those species in which learning capacity had evolved to this level of development, further evolution would be expected to take the form of selection for plasticity of response capability and increased learning capacity rather than selection for further differentiation between specific behavioural response patterns (see also e.g. Dobzhansky, 1962; Dobzhansky & Montagu, 1947).

However, initial expectations that intelligence or some common ability to learn might increase progressively with phylogenetic status has been difficult to demonstrate and is subject to criticism (see Macphail, 1982; Warren, 1973). It has long been recognized that there is no simple standardized test or range of tests of adaptability that can be administered to all species (Munn, 1965; Warren, 1974). While most investigators have confined their attention to mammals, and to mice, rats, cats, monkeys, and apes in particular, differences between species in basic sensory and motor capacities as well as in motivation and temperament in the alien setting of a laboratory situation, present considerable difficulties for the design of experiments and for the interpretation of results. Moreover, as Hodos and Campbell (1969) and Warren (1973) have noted, the idea that current animal species can be ranked on some hierarchical scale of behaviour misinterprets the processes of evolution. Existing living animals are the result of past natural selection for survival and successful reproduction, and have been able to adapt to their particular environmental niches for a variety of behavioural and other reasons. While the evidence suggests that most species demonstrate some modifiability of behaviour with experience, how this can be assessed will clearly depend on the species concerned and their evolutionary history (Byrne, 1995; Jerison, 1985).

But in spite of such difficulties, Passingham (1981) has suggested that some comparisons may be justifiable. He compared the results from several studies where closely corresponding experimental conditions and learning criteria had been employed, and showed that the effects of previous training were much more beneficial for three species of monkeys than for four species of non-primate mammals tested. And in a comparison across different species of primates, Passingham (1975b) found a good correspondence between the recorded benefits of previous training and brain development as indicated by the relative size of the neocortex and medulla. Such an index of brain development is one of many that have been used to assess the degree of "encephalization" of different species, i.e. the extent to which the brain has developed in information processing capacity beyond that required for the control of basic bodily functions (Jerison, 1985; but see also Hodos, 1986).

A further aspect of evolutionary development that has been seen as significant in relation to differences in learning capacity between primate species is the relative duration of their growth and maturation periods. For example, Schultz (1969) has compared the average pre- and post-natal periods of growth in a number of living primates and shown that whereas the length of gestation only changes from 18 weeks in lemurs and 24 weeks in monkeys to 33 weeks for chimpanzees and 38 weeks for the human baby, post-natal differences between species are much greater. As Tobias (1981) has pointed out, belated postnatal expansion is clearly a function of the ultimate size of the human brain and natural selection, since a prenatal growth of the human skull proportional to that in apes would have led to considerable difficulties in giving birth to such large-headed babies. Thus the period of post-natal growth and maturation to adult status increases markedly between lemurs (2 to 3 years) and monkeys (7 years) to chimpanzees (11 years) and man (20 years). The importance of such a long dependent growing period in relation to the plasticity of the developing nervous system and the acquisition of culturally relevant physical and social skills has been emphasized by Gould (1977). As a simple illustration of the influence of experience during this period of dependency in higher primates, Oakley (1972) recounts the report of a South African naturalist (Eugene Marais) who reared an otter pup far from water and an infant baboon apart from its troop. Both were raised with food foreign to their normal environment - with strikingly different consequences. At maturity, they were returned to their natural habitat, but whereas the otter instinctively began diving for fish, the baboon displayed a terrified aversion to the grubs and scorpions that provide the habitual diet of the species, and started eating poisonous berries which would be completely shunned by normal adult animals.

Since Dobbing & Sands (1973) reported that the human brain at birth is only about a quarter of its ultimate size, it has been shown that at this stage the neural "wiring pattern" only roughly approximates the adult level of organization (Shatz, 1992). Whereas at one time it was generally thought that the complex system of neural connections within the brain was entirely predetermined and progressively assembled according to some biochemical master plan, neurobiologists are now convinced that such a view is too narrow and can no longer be sustained (Asanuma, 1991; Easter et al., 1985; Edelman, 1987; Shatz, 1992).

During the last thirty years or so, anatomical and physiological investigations of a wide variety of animals have revealed a significant degree of plasticity in the structural organization of the nervous system. And the role of environmental factors in contributing to the development of the organization of the mature brain has been clearly demonstrated by two contrasting approaches. One examined the effects on the brain and behaviour of animals reared in conditions of environmental impoverishment or sensory deprivation, while the other employed precisely the opposite, i.e. a highly stimulating or enriched environment. Certainly, the deleterious effects of sensori-motor deprivation and the benefits of varied experience during infancy have been known for many years (see e.g. Fiske & Maddi, 1961; Rosenzweig, 1971), but more recent research has shown that the type and extent of sensory stimulation and motor activity both during neural maturation and later, has a considerable influence on the numbers of synapses or interneuronal junctions formed, and on the pattern of these connections in particular regions of the brain (Greenough, 1986). And in animals in which the somatosensory input to the brain from a specified body part has been reduced or increased (by a reduction or increase in use and/or peripheral stimulation), the size of the corresponding cortical representation has been shown to be reduced or increased accordingly (Jenkins, Merzenich & Recanzone, 1990; Kaas, 1991; Wall, 1988). Similar effects have been demonstrated in relation to somato-motor activity. For example, Larson and Greenough (1981) and Greenough, Larson and Withers (1985) have shown that in closely controlled training experiments on reaching movements of a designated forepaw in rats, significant increases in the dendritic branching of neurones may be recorded in the corresponding part of the relevant sensori-motor area of the cerebral cortex. And in a series of similar experiments on kittens, Spinelli and his colleagues (Spinelli & Jensen, 1979, 1982; Spinelli, Jensen & Viana Di Prisco, 1980; Viana Di Prisco & Spinelli, 1981) have shown that as a result of learning to avoid a mild electric shock by flexing a designated forelimb, significant changes occurred in the sensori-motor cortex of the corresponding hemisphere. They found that the size of the cerebral somatotopic representation of the trained limb had increased considerably and was three or four times larger than for the untrained side, and that dendritic growth and branching of the relevant neurons was also enhanced in the hemisphere corresponding to the trained limb. Furthermore, Spinelli (1990) has reported that as a consequence of this training, the animals displayed a generalized and consistent preference for the use of the trained foreleg in a variety of other reaching and manipulative activities.

As Shatz (1992) has remarked, the finding that neuronal activity itself or use of a neural pathway is a significant influence in the ultimate shaping of the maturing nervous system has important implications. First, it provides adaptability by enabling experience to fine-tune the inherited neural mechanisms to be better suited to the demands placed upon them. And second, it is genetically conservative, in that it pre-empts the need for every one of the myriad neural connections in the brain to be specified by a correspondingly large number of genes. There is also the corollary that in the higher apes and man, the extended post-natal period of growth and maturation provides many more opportunities for shaping of the neural substrate of behaviour through learning and experience than in any of the lower primates. Hence, it seems likely that if one particular way of using the limbs becomes habitual for any reason, then the central neural mechanism responsible for the activity may be expected to develop accordingly. And if the habitual activity tends to occur more frequently on one side than the other, then the control mechanisms of the corresponding cerebral hemisphere might be expected to develop a more sophisticated neural substrate than those of the other hemisphere. Since the main sensory and motor neural pathways connecting the limbs and cerebral cortex change sides en route to their respective destinations, the control mechanisms of the left hemisphere would thus be expected to become more highly developed in those people who more consistently use their right hand, and the right hemisphere to become predominant for habitual left-handers.

Manual asymmetries in pre-history

As a result of the now extensive archaeological record on tool-making, it is reasonable to ask if this also provides any evidence on whether these tools were made by, or for, right- or left-handed individuals. Since the earliest stone artefacts were simple, hand-held core or flake tools that could have been used by either hand, some researchers (e.g. Napier & Tuttle, 1993; Spenneman, 1985) have expressed considerable reservations about the possibility of inferring the handedness tendencies of the user or maker from minor variations in their shape or form. In reviewing the accumulated information from a variety of sources, both of these reviews suggested that it is only in tools of unmistakably right- or left-handed design such as sickles and scythes made during the European Bronze Age (i.e. about 4,000 years ago) that convincing evidence can be found for a predominantly right-handed culture. Whereas up until that time, individuals had probably made their own tools as they were required, the implements had now become specialized items of manufacture needing particular technological skills and specialist tool-makers. Consequently, their design presumably became standardized either as a result of the handedness of the tool-maker or as a response to community demand. But whatever the cause, the available evidence shows that an overwhelming predominance of implements were made for right-handed use at that time (Spenneman, 1985).

However, recent attempts to reconstruct the likely methods of production of the type of hand-held core and flake tools used in Palaeolithic times, have proposed that right-handed tool-makers may have predominated some 1.4 to 1.9 million years ago in Kenya (Toth, 1985). An analysis of the stone flakes recovered from a radiometrically dated archaeological site in Kenya yielded highly suggestive evidence of a predominance of right-handedness in their manufacture, while stone artefacts from a Spanish site dated to be 0.3 to 0.4 million years old also indicated that the majority were produced by right-handed tool-makers. Although Toth's way of identifying the handedness of the manufacturer by the shape of the flakes produced is speculative, suggestive evidence of its validity was obtained by Snyder, Isaac and Jones (cited by Toth, 1985 as a personal communication) from the results of an experiment using 20 left-handed and 20 right-handed university students.

Microscopic analysis of the wear and tear on artefact surfaces has also been found to yield relevant evidence, not only on the likely methods of manufacture, but also on the use to which the tools have been put. In this respect, Semenov (1964) concluded from an examination of the design and wear of stone flakes called end-scrapers, that they were used primarily for cleaning and softening a skin after it had been removed from an animal. He reported that about 80% of the scrapers from the Upper Palaeolithic period in Europe were worn on the same side, which, in his recapitulation of the method of use, could only be attributed to their being held in the right hand. Other evidence from microwear analysis of flake tools taken from English archaeological sites dated at some 0.2 million years ago also suggest that the tool-users of those days were consistently right-handed (Keeley, 1977). In this instance, the validity of the method of analysis was tested by blind examination of specially prepared and selectively used modern replicas of Palaeolithic flint tools, and a high degree of agreement between the inferred and actual uses was demonstrated (Keeley & Newcomer, 1977). More recent Neolithic bone and antler implements (some 5,000 to 6,000 years old) found in southern Germany and Switzerland have also been shown (from grinding striations on their surfaces) to have been made by predominantly right-handed individuals (Spenneman, 1984a). However, this author found a considerable regional variation in the proportion of right- to left-handed tool-makers, with the Swiss data indicating about 6% were left-handed, and the southern German data suggesting as many as 19% were left-handed.

Another and quite different line of archaeological enquiry on the emergence of manual asymmetries relates to a determination of the handedness of the artist or the artist's subjects in various samples of pre-historic art such as cave drawings or murals, the earliest of which are believed to have been produced some 35,000 years ago (Scarre, 1988). One method of analysis assumed that the average right-handed individual drawing a face in profile will tend to produce one looking to the left side of the picture, and a left-hander will tend to draw one looking to the right (see e.g. Wilson, 1891). But as this latter author points out, such a rule does not necessarily apply to the skilled artist who would be expected to be able to draw profiles of any orientation at will. Thus, if the first Palaeolithic drawings were made by ordinary (i.e. unskilled) individuals, the relevant frequency of left and right facing profiles could provide a good indication of the relative incidence of right- and left-handedness during that period. But although the validity of the method has been tested by a number of investigators on a range of population samples from current cultures, the results have not been entirely reassuring. For example, Jensen (1952a, 1952b) tested American, Norwegian, Egyptian, and Japanese children (unselected for handedness), and in each group found a significant tendency for most subjects to draw left-facing profiles. He concluded that the effect must therefore be independent of directional reading and writing habits, and reported other evidence to suggest that the direction of profile orientation might only hold for right-handed subjects. Other studies (Shanon, 1979) on groups of American and Israeli children, and on American adults Crovitz (1962) obtained essentially the same results, with most of the right-handers drawing a profile facing to the left, but with left-handed subjects producing approximately equal numbers of left- and right-facing profiles. More recently, Schiers (1990) reported that in a group of Netherlands schoolchildren, there was only a slight tendency to draw left-facing profiles, and that there were no differences between left- and right-handers at all. It is not surprising to find then, that investigators of pre-historic art have reported somewhat mixed or inconsistent findings on the predominant type of profile in the drawings and paintings they have examined (see e.g. Cunningham, 1902; Perello, 1970; Spenneman, 1985; Uhrbrock, 1973).

More convincing evidence on the incidence of right- and left-handedness in past populations comes from a rather different method of examining works of art. In this type of approach, the analysis involves identifying the particular activity being portrayed in each drawing or engraving, and the role of each hand being used. Restricting the collection of such evidence to the portrayal of skilled manual activity and excluding merely holding or carrying functions of the hands, a number of investigators have surveyed selected pictorial sources to determine the relative frequency of demonstrations of right and left hand usage. Most of the material examined by this method is decorative in nature and consequently presumed to have been carried out by artists who were skilled specialists and who faithfully reflected in their work the characteristics and ways of life of the particular people and/or communities they depicted. But as Wiley (1934) pointed out, investigators using this type of approach need to be aware of the religious and/or aesthetic constraints that have often been placed upon artists in certain cultures in regard to symbolism for example, or symmetry of pattern.

Because of these and other problems (such as small sample size), some of the earlier published data are difficult to interpret, but more recently, several investigators have reported findings from carefully conducted surveys on reasonably sized samples of clearly defined archival material. For example, Dennis (1958) examined published reproductions of the murals in Egyptian tombs, one set of paintings having been carried out about 4,500 years ago, and the other being completed some 1,000 years later. The author found that in all the material analysed, the faces of the figures were shown in profile, with 58% and 60% respectively of the two samples facing towards the viewer's right. However, he found that over 90% of both sets of paintings depicted a preferential use of the right hand, and that this was independent of the profile orientation. Using the same method of analysis, Spenneman (1984b) studied photographs of the reliefs decorating the galleries of a large Javanese pyramid constructed some 1,000 years ago. He reported that while only 16.5% of the unimanual actions portrayed in the pictures could be classified as skilled, nevertheless, over 90% of these showed a preferential use of the right hand compared with just over 60% right hand use depicted in unskilled activities. And in a more wide-ranging survey, Coren and Porac (1977) published the results of an examination of photographs and reproductions of drawings and paintings etc. from European, Asian, African, and American sources which included samples of work carried out 5,000 years ago and at various subsequent intervals of time up until the present. Again, it was found that over 90% of the actions analysed depicted preferential use of the right hand, and that this was substantially the same irrespective of the age of the art works examined or the region of the world from which they came.

Thus, the accumulated weight of evidence suggests a predominance of right-handedness throughout the known world for at least the last 10,000 to 15,000 years wherever cultures have left evidence of their ways of life in clearly defined works of art. And information concerning hand preferences in stone tool use or manufacture is supportive of an even earlier origin of a universal right-sided preference dating back some 200,000 years or even further - perhaps to the beginning of tool-making itself. But such evidence still leaves unanswered the question of why - i.e. why there appears to have been a predominance of right-handedness in all cultures and at all times.

Handedness and Brainedness

It is a little difficult to see what adaptive advantage there could have been during evolution in selecting for an inborn tendency to favour the acquisition of skills by one particular hand or side rather than an equality of potential between them, and although many theories of handedness have been advanced from time to time, few have attempted to address this basic problem (for reviews see e.g. Bradshaw & Rogers, 1993; Corballis, 1991; Harris, 1980; Morgan & McManus, 1988; Spenneman, 1985; Wile, 1934). Yet unless the introduction of an inborn lateral bias imbued its possessor with some distinct advantage over others in adapting to the demands of the environment, it is most unlikely that it, or rather the line of its possessors, would have survived for very long. Indeed, due consideration of the question suggests that any such inborn lateral bias would have incurred more disadvantages than advantages. Clearly, a potential to acquire manual skills equally well by either the right side or the left provides a much greater capability to respond to environmental demands than an inherited asymmetric bias favouring one particular side. Furthermore, a loss or impairment of function through injury or disease affecting the favoured side would be far more devastating if the unaffected side did not have a comparable potential for motor development. But if there is no inborn organic basis for the lateralization of manual skills *per se,* could it be argued that a bias for one particular side might have evolved as the result of some other inherent and influential development? The obvious candidate for consideration here is the cerebral control mechanism for speech, which in the majority of people is known to be located in the left hemisphere of the brain. However, any suggestion of a selective advantage for speech control to be located in one particular hemisphere rather than the other would seem to be even more difficult to sustain. Yet if there is no known selective advantage for the asymmetrical distribution of neural control of speech or handedness, it is necessary to postulate the existence of some other biological or social influence which would either directly or indirectly affect both aspects of behaviour. Certainly, it is well-established that some form of association exists between the location of the central neural mechanism of speech and the nominal handedness of individuals, although the correspondence is now known to be far from perfect.

The first concrete evidence for such an association resulted from the clinical studies of aphasia made by a number of neurologists towards the end of the 19th Century, and the events leading up to the discovery have been described by many authors (see e.g. Brain, 1965; Critchley, 1970; Harrington, 1987). It was reported by Broca (see Berker, Berker & Smith, 1986) that although diseases of the right hemisphere are as frequent as those of the left hemisphere, he found that in patients with loss of speech following cerebral damage, in some 90% of cases the injury was confined to the left hemisphere. He reasoned that since nearly everyone is right-handed, the central and internal association between the external and behavioural phenomena of speech and handedness was not coincidental, and that most people may therefore be more aptly called "left-brained", with the left-handed exceptions being "right-brained". The pre-eminence of the left side of the brain he attributed to an organic predisposition and precocious development of the left hemisphere in most individuals, with a contrary predisposition in left-handers.

An alternative and contrasting social anthropological view of the origins and causes of handedness was postulated by Hertz (1909) in a paper described by Evans-Pritchard (1973) as one of the finest essays ever written in the history of sociological thought. While Hertz was aware of the factual foundations of Broca's ideas and even quoted him as saying that we are right-handed because we are left-brained, Hertz did question the justification for the inferences drawn from the clinical observations. Thus, he accepted the view that a regular connection may exist between the predominance of the right hand and the superior development of the left part of the brain, but nevertheless queried which was cause and which the effect. In other words, he questioned, why not turn Broca's proposition around and say we are left-brained because we are right-handed? According to Hertz, the influences bringing about the traditional preference for the right hand are social or religious, and founded in the concept of a symbolic dichotomy contrasting the sacred and the profane, which has always dominated the thoughts and ideas of people of all cultures in their many and varied explanations of natural phenomena. He provided evidence from a wide range of beliefs and customs to show that acceptance of this fundamental dualism throughout human history has been so universal and persistent that it has exerted both a positive and negative effect on every aspect of behaviour, including a close association of right with sacred and all that is good and desirable, and of left with profane and things that are evil or unclean.

While Broca and Hertz clearly examined the problem of handedness and hemispheric predominance from opposite ends of the socio-biological spectrum, with the benefit of hindsight and a considerable volume of research published in the meantime, which of these two approaches has provided us with the most likely basis for an explanation of the origins or causes of handedness? And what has been the impact of each on theory and practice in the fields of clinical, educational and occupational psychology? From even a superficial perusal of the literature, it is clear that Broca's influence has been strong, widespread, and persistent, while Hertz's evidence and views seem to have been almost entirely neglected. However, in the following pages, the literature reviewed will show that neither view by itself is justified, but that the social and biological information now available suggests that cerebral asymmetries for manual preference and speech production are more likely to be the result of an inborn potential for the acquisition of motor skill (unrelated to side), and an environmentally determined bias for the consistent use of a particular hand.

Cerebral dominance and hemisphere specialization

The origin of the term "cerebral dominance" is obscure, although for a long time it was generally accepted as describing the predominant role of the left hemisphere in relation to speech and handedness (Zangwill, 1964). However, current ideas and discussions of the respective functions of the right and left hemispheres are not so circumscribed. Following the remarkable split-brain experiments on animals and the subsequent clinical observations on human patients (Sperry, 1961, 1974), a virtual explosion of research on cerebral localization and behaviour developed using a wide range of ingenious techniques. As a result of this work, the conventional views of brain functions were rapidly revised and the concept of unilateral dominance of left over right hemisphere was generally abandoned and replaced by one of complementary specialization (Teuber, 1974). Various hypotheses were advanced concerning the functions to be ascribed to the left and right hemispheres, and such dichotomies as analytic versus holistic, verbal v non-verbal, serial v parallel, provided appealing simplifications, not only for the investigators themselves, but also for philosophers, educationalists, and of course, the popular press (Harris, 1985, 1988).

However, such generalizations are not only unjustified (see e.g. Bryden, 1986; Efron, 1990), they have the capacity to confuse rather than enlighten. For example, a few years ago, Bryden (1990) remarked that over the last 25 years, general scientific opinion had swung from an attitude of great resistance against suggestions that functional differences between the two cerebral hemispheres could be measured, to the other extreme of assuming that any behavioural asymmetry is an indication of hemispheric specialization. He then went on to emphasize the need, not only for a much more careful evaluation of the evidence from investigations of functional lateralization, but for a distinction to be made between different functions which may be lateralized through independent mechanisms.

The present author agrees, and believes that at least some of the present confusion concerning the more specific association between hemispheric dominance for speech and handedness is also self-generated. For example, in discussing functional specialization of the two hemispheres, many authors have failed to make any distinction between the cerebral lateralization of speech *per se* and language, and have tended to use these two terms interchangeably. Similarly, subject handedness has often been reported as just left or right without describing the number and type of criteria used, although both of these variables have been shown to significantly influence handedness assessment (Provins, Milner & Kerr, 1982). Consequently, the present study is primarily directed towards examining the evidence relating to the cerebral lateralization of manual (motor, rather than sensory or perceptual) asymmetries. Similarly, in considering the hemispheric lateralization of speech and its postulated association with handedness, the evidence reviewed will primarily focus on the motor component, i.e. speech production and not perception.

There are certainly very good grounds for adopting such a cautious approach. First, the functional hemispheric lateralization of speech originally identified by Broca (see Berker et al., 1986; Kann, 1950) related to a patient's inability to speak, in contrast to his unimpaired ability to understand everything that was said to him. Subsequent investigators of organically based language disorders have also established that the cerebral regions subserving the production and comprehension of speech are not co-extensive, and that unilateral brain damage anterior to the central sulcus is usually associated with disorders of speech production and a contralateral hemiplegia, whereas postcentral cerebral lesions more frequently result in deficits of speech perception or comprehension and an absence of hemiplegia (Benson, 1967, 1988; Geschwind, 1971; Rapin & Allen, 1988; Weisenburg & McBride, 1935; Zangwill, 1960a). Second, studies of, for example, commissurotomy patients, have suggested that the cerebral control of speech production is more strongly lateralized than comprehension (Gazzaniga, 1970; Gazzaniga & Le Doux, 1978; Gazzaniga & Sperry, 1967), while other workers have even suggested that in some instances an interhemispheric separation of expressive and receptive functions may occur, with one hemisphere being dominant for speech production and the other for comprehension (Kurthen et al., 1992). Third, experiments on normal individuals have shown that whereas subjects required to engage in two different tasks at the same time usually demonstrate deficits in performance compared with their proficiency when undertaking each task separately, this is not always true. In some situations involving concurrent speech perception and speech production, little or no interference has been recorded, suggesting that central neural mechanisms subserving these functions may be separated, if not separate (McLeod & Posner, 1985; Shallice, McLeod & Lewis, 1985). And fourth, other studies of non-clinical populations have also reported results suggesting that the behavioural asymmetries found in relation to sensory or perceptual processes are unrelated to motor asymmetries (Collins & Collins, 1971; Eling, 1983; Porac, Coren, Steiger & Duncan, 1980).

It has been observed that since the critical aspects of speech production and handedness may both involve the same serial organizational features that characterize motor skills, it is not unreasonable to assume some degree of neural specialization within a particular cerebral hemisphere for this type of activity (Corballis, 1991; Kimura & Vanderwolf, 1970). Alternatively, it has been suggested that since speech and handedness are motor activities which develop concurrently in ontogenesis, the relationship between them may instead be one of mutual facilitation or reinforcement in becoming established in one and the same hemisphere (Gazzaniga, 1970).

While problems relating to other aspects of cerebral and/or behavioral lateralization may be no less important, there is little doubt that historically, the concept of a hemispheric asymmetry of function for the control of speech and manual skills has provided the major impetus for research on cerebral specialization.

Accordingly, the next two chapters are oriented towards examining the clinical and non-clinical evidence bearing on the association between speech and handedness, and the attempts that have been made to identify the localization of these functions in the cerebral cortex.

CHAPTER 2
Brain Damage, Speech and Handedness

Chapter outline

Evidence from studies of aphasia following lesions in the adult brain show that for nearly all right-handers, speech loss results only from left hemisphere damage, whereas for nominal left-handers, lesions in either cerebral hemisphere may give rise to speech deficits, although these may be less persistent than for right-handers. The use of amobarbital testing for the functional localization of speech has tended to confirm this distribution pattern, although several studies have revealed that many patients have bilateral representation of speech. But in patients with brain damage sustained early in life, left hemisphere lesions appear to induce development of the relevant speech control mechanisms in the right hemisphere in many cases, together with an enhanced likelihood of left- or mixed-handedness

The results from studies of hemispherectomy and callosotomy (split-brain) patients also agree in identifying speech functions as being primarily lateralized in the left hemisphere (for right-handers) although there is evidence of some language processing capacity for the right hemisphere. And observations of callosal agenesis cases suggest that the lateralization of both speech and manual functions probably develop according to the pattern for normal individuals, although a somewhat higher than expected incidence of left-handedness has been reported amongst such patients.

Electrocortical mapping studies of cerebral regions concerned with speech functions have reported that vocalization or the arrest of speech usually results from the stimulation of areas in the left (dominant) hemisphere known to produce aphasia when damaged, although there are considerable differences between individuals in the localization of such sites. Furthermore, similar effects have been reported in some cases from stimulation of a number of sites in the right hemisphere - even in right-handed patients with a known left hemisphere dominance for speech.

Studies of apraxia in clinical patients tend to show greater effects of left-sided brain damage (in right-handers) than lesions in the right hemisphere, although the picture is less clear for left-handers. However, the interpretation of such results is beset with difficulties of discriminating between motor performance deficits due to the side of the lesion, and performance inadequacies due to testing of the non-preferred hand.

Ablation studies in nonhuman animals have shown the importance of a number of cortical regions for the production of movements of the limbs on either the same or opposite sides of the body relative to the lesion. Such evidence also suggests that lesions acquired early in life may produce less deleterious effects on limb movement and have less influence on the side preferred for limb usage than lesions sustained as adults.

Problems associated with drawing justifiable inferences from the available evidence have been described, and caution emphasized in extrapolating from the findings of stimulation and lesion studies to function in the normal population.

Exactly what Broca said, or meant by what he said, has been the subject of some considerable discussion (see e.g. Berker et al., 1986; Eling, 1984; Harrington, 1987; Harris, 1991, 1993c). As both Harrington and Harris have pointed out, it is necessary to appreciate not only the state of knowledge of the central nervous system in the 1860's, but also the climate of medical opinion in Broca's time in order to place the proper interpretation upon the reports of his work. But there is a further aspect to this problem. Irrespective of the exact nature of Broca's findings or ideas, it is clear that the course of scientific thinking and medical practice since then has been determined not so much by what he said, intended, or implied, but by what others have inferred, accepted, or adopted from his observations and writings, and in this respect, history has been somewhat selective (Henderson, 1986).

Adult aphasia

Certainly, it appears that Broca's primary impact at the time related to the recognition of a general principle concerning the localization of particular functions in the brain and their reference to specific convolutions of the cerebral surface (Boring, 1929, p.69-70). Thus, the report of Broca's work was rapidly followed by descriptions of a succession of findings by other investigators identifying cortical "centres" for a wide range of mental functions, so that, according to Luria (1965, p.690), by the end of the 19th Century, handbooks of neurophysiology and psychiatry were full of references to such structures as Exner's "writing centre". Henschen's "calculation centre", Broadbent's "ideation centre", etc. This was the era of what Head (1926, p.54-66) called the "diagram makers" who "imagined that all vital processes could be explained by some simple formula. With the help of a few carefully selected assumptions, they deduced the mechanism of speech and embodied it in a schematic form. For every mental act there was a neural element, either identical with it or in exact correspondence. From diagrams, based on *a priori* principles, they deduced in turn the defects of function which must follow destruction of each "centre" or internuncial path." (p.65).

This pre-occupation with the anatomical localization of brain "centres" responsible for various behavioral phenomena persisted until well into the 20th Century, leading Weisenburg and McBride (1935, p.444) to note that "The competition for new classifications and cerebral areas was so keen that anything in the way of opposition, no matter how soundly based, was swept aside. This is the only way to explain the treatment of Hughlings Jackson's psychological theories which did not gain wide recognition until about fifty years later. It is hard to down ideas of cerebral localization, once accepted, for it is on such fare that most neurologists have been brought up."

The adherence to rigidly psychomorphological concepts was strongest in France and Germany (Conrad, 1954), but appears to have been a firmly entrenched part of general neurological opinion until at least the end of the second world war (Luria, 1965, p.690). Indeed, even more recently, Walton (1977) pointed out that the terminology employed in the description of aphasia still reflected outdated views concerning the nature of speech and cerebral localization, and that there was a continuing failure to distinguish between the psychological, physiological, and anatomical bases of speech and its disorders - views originally and frequently espoused by Hughlings Jackson throughout his life and reiterated by Head (1926).

It is not surprising then, that with such a climate of opinion, the concept of the left cerebral hemisphere being pre-eminent (or "dominant") for speech in right-handers and the right hemisphere dominant for left-handers (i.e. Broca's rule - see Harris, 1991, p.2-3), also became firmly entrenched during the same period. Although exceptional cases were noted from time to time, their occurrence was regarded as so infrequent as not to constitute a serious challenge to the rule (Wechsler, 1958, p.320; Zangwill, 1964, p.104). They were instead usually classified as anomalous cases as in, for example, "crossed aphasia", even though the various explanations offered for their occurrence may have been incompatible or in conflict (see e.g. Goodglass & Quadfasel, 1954, p.522).

For neurologists, the focus was necessarily on the clinical manifestations of aphasia or similar behavioural dysfunctions, with handedness only of interest insofar as it assisted in identifying the organic basis for the disorder. As pointed out by Roberts (1959, p.90) and Harrington (1987, p.59), before 1865 the handedness of a patient suffering from speech loss was irrelevant, and for some 60 to 70 years thereafter there was no reason to suppose that simply classifying patients as either left- or right-handed could be inadequate. However, in many instances during this period, a number of investigators still omitted to record handedness at all (see Critchley, 1970, p.74; Roberts, 1959, p.90-91).

Chesher (1936) was probably the first to pay particular attention to the hand preferences of his aphasic patients and to note which hand was customarily used in a wide variety of motor acts. He classified his patients as being either fully right-handed, fully left-handed, or as having a mixed preference, and found that his evidence supported Broca's rule for those patients with a clearly defined handedness, but not for those with mixed hand preferences. He concluded that for such patients, the language mechanism is unlateralized so that a lesion in either the right or left hemisphere may produce aphasia.

Following Humphrey's (1951) demonstration that nominal left-handers tend to be less consistently left-handed than right-handers are right-handed, Zangwill and his colleagues (Ettlinger, Jackson & Zangwill, 1956; Humphrey & Zangwill, 1952a, 1952b) initiated a much closer examination of the relationship between handedness and cerebral dominance for speech in patients classified as left-handed. They concluded that aphasia in nominal left-handers is usually associated with a lesion in only one of the two cerebral hemispheres, and that this is more frequently to be found in the left hemisphere than the right.

However, one of the problems emphasized by Goodglass and Quadfasel (1954) in evaluating data from left-handers is the relative rarity of such individuals in the population, with the consequent difficulty of generalizing from the limited number available from a single clinical source. Roberts (1959) identified another problem - concerning the number of handedness criteria employed - and found that "the greater the number of questions asked and tests used, the fewer became the number of entirely right- or left-handed, until they were quite exceptional" (P.92). Consequently, he adopted what was essentially a system of dichotomous self-classification of handedness for his large series of patients operated upon for the treatment of focal cerebral seizures, and reported that aphasia occurred after operation on the left hemisphere in approximately the same proportion for left-handed patients as for right-handers. Similarly, there was no significant difference in the incidence of aphasia following operation on the right hemisphere for left-handers compared with right-handed patients. Hence, he concluded that the left hemisphere is usually dominant for speech irrespective of handedness, although he was at a loss to offer an explanation for those few cases in which the right hemisphere proved to be dominant.

By the end of the 1950's then, most investigators were convinced that for right-handed patients, aphasia was almost always associated with lesions of the left hemisphere and seldom with the right, while for left-handers, speech loss could be associated with damage to either hemisphere. Hecaen (1962, p.217) even went so far as to suggest that "nowadays it is impossible to question the fact that the region of language is situated in the left hemisphere with the right-handed subject, and there is no need to discuss such a point." A similar statement occurs in Hecaen and Ajuriaguerra (1964, p.39). Attention thus became focussed on the cerebral dominance characteristics of those individuals commonly regarded as left-handed but described by Bauer and Wepman (1955) as more likely to be ambidexters, or those in whom lateralization is not fully developed. But it has also been reported (Kimura, 1983, 1987; McGlone, 1977, 1978, 1980) that in right-handed men and women with unilateral damage to the left cerebral hemisphere, women less often suffer from speech loss or aphasia than men. Are women then, like nominal left-handers, also less functionally lateralized than men for both speech and handedness?

Clearly, interpreting the assessed effects of unilateral cerebral lesions depends not only on the validity and reliability of the measures of handedness and speech loss used, but also on such factors as the nature and extent of the cerebral injury, when the patient was examined and the time course of recovery, the age of the patient, and any history of brain damage in infancy. Further, it is important to know the nature of the patient sample and how and why it was selected for reporting since the under-reporting of some types of potentially significant data is likely for both handedness and organic symptoms. For example, omitting to report negative instances, such as cases with unilateral cerebral lesions in the usual speech region but with no aphasia, or those apparently unremarkable cases with speech loss and right hemiplegia which on closer enquiry might have revealed evidence of left-handedness (Goodglass & Quadfasel, 1954).

Of some importance then, are the reports of Gloning et al. (1969) and Gloning (1977) on 57 right-handed and 57 left-handed patients who were matched as closely as possible with respect to localization, size, and type of unilateral cerebral lesion (verified at post-mortem). The authors also recorded no significant differences between the two groups on such variables as educational background, general mental condition, age, and sex. They found that aphasia occurred in their right-handers only with left cerebral lesions, whereas in the non-right-handers, aphasia occurred with either left or right hemisphere damage. There were significantly more patients with aphasia among the non-righthanders, mainly due to the occurrence of aphasia with right hemisphere lesions in this group. But they also found that there were significantly more transient aphasias (with a duration of one day or less) amongst the non-right-handers compared with the right-handed patients. This is in keeping with other evidence (see e.g. Gloning & Quatember, 1966; Subirana, 1958, 1964; Zangwill, 1960b) for a more rapid and complete recovery for nominally left-handed and mixed-handed patients who, Gloning and Quatember (1966) considered, were likely to be less lateralized at the cerebral level.

Since there are now grounds for concluding that left- and right-handedness is not a dichotomous variable but is continuously distributed (see Chapter 7), the conventional division of patients into two handedness categories may be seen as unjustifiably rigid and likely to conceal rather than reveal the true nature of the association between handedness and aphasia. Bearing in mind the difficulties of classification of clinical cases and the wide variation between authors in the assessment of both language impairment and handedness, Alekoumbides (1978) attempted to integrate the findings from 29 of the most frequently cited series of lesion studies including those consisting of large numbers of patients. After a careful and detailed re-evaluation of the data, he concluded (p.137) that "the representation of language in the two hemispheres varies over a continuous spectrum, from strict left-brainedness, through various degrees of ambilaterality (favouring one or the other hemisphere), to strict right-brainedness" in approximate correspondence to handedness classification.

Of course, prior to the introduction of the method of anesthetizing one hemisphere at a time by means of the sodium-amytal technique (Wada & Rasmussen, 1960), there was no means of knowing if a particular patient suffering dysphasia as a result of damage confined to one cerebral hemisphere might not also display a similar or even greater speech loss from an identical lesion in the other hemisphere. Nevertheless, most subsequent studies using the amobarbital technique (e.g. Branch, Milner & Rasmussen, 1964; Rey et al., 1988; Rossi & Rosadini, 1967; Strauss & Wada, 1983) confirmed the trend of earlier clinical findings that in nearly all right-handers, speech was found to be localized in the left hemisphere, while for nominal left-handers speech lateralization was highly unpredictable, and in a number of cases it appeared to be bilaterally represented. However, there has been a considerable variation in findings from one investigation to another, and in some of the more recent amobarbital studies a significant incidence of speech representation in both hemispheres has been reported. For example, Oxbury & Oxbury (1984) noted that out of 23 patients they tested, 14 (10 right-handed) recorded aphasic errors after injection to each hemisphere; Powell, Polkey and Canavan (1987) reported that in their 27 cases, 6 (5 right-handed) had language in both hemispheres; and Zatorre (1989) found that 22 (12 right-handed) of his 61 epileptics had bilateral speech representation.

There is, of course, need for caution in drawing inferences from results on selected clinical cases in considering the implications for a normal population (Woods, Dodrill & Ojemann, 1988). Furthermore, Rausch (1987), Snyder, Novelly and Harris (1990), Loring et al. (1992), and Benbadis et al. (1995) have all emphasized the problems of interpreting the results from use of the amobarbital technique in reconciling findings from studies of different investigators in different laboratories. For example, in their particular survey of institutions using the sodium-amytal method, Snyder et al. (1990) found that the reported incidence of bilateral speech (or "mixed speech dominance") for different laboratories was significantly affected by differences in the criteria they employed for recording the presence of speech functions in the non-dominant hemisphere. In a more recently published amobarbital study by Loring et al. (1990), these authors took particular care concerning the criteria they used to record the extent of language impairment and found that in 103 patients (91 of whom were right-handed), 79 had exclusive left hemisphere language representation, 2 had exclusive right hemisphere representation, and 22 had some language functions represented in each hemisphere. These authors argued that insofar as handedness may be viewed as a continuous variable ranging from strongly right- to strongly left-handed with a majority right biased, then a correspondence between handedness and cerebral language dominance would require speech to be similarly lateralized between hemispheres, and that the results from their patients supported such a view. Thus, it appears that Broca's rule associating handedness with the hemisphere dominant for speech may well hold true, but not in the dichotomous form in which it has for so long been considered. Certainly, the exceptions to Broca's classical doctrine are now sufficiently numerous and varied to emphasize the need for a careful re-consideration of the contribution of both biological and environmental influences on the functional specialization of the two hemispheres (Joanette, 1989; Lecours, Basso, Moraschini & Nespoulous, 1984).

Childhood aphasia

One of the most notable and frequently acknowledged exceptions to the concept of left hemisphere pre-eminence relates to the apparent plasticity of the neural substrate for speech in infancy (Zangwill, 1975). According to Woods and Teuber (1978a, p.273), the standard doctrine concerning childhood aphasia emerged soon after Broca's initial study of the adult form, when one of Charcot's students noted that in a number of patients with infantile hemiplegia and complete atrophy of the left cerebral hemisphere, none had subsequently failed to acquire language. Other reports of rapid recovery from acquired aphasia in children then accumulated, giving rise to a general conviction that up to a certain age, linguistic functions could be successfully established in either hemisphere (Carmichael, 1966).

In most discussions of the evidence, the particular features distinguishing childhood from adult aphasias have been described as possessing two main characteristics (van Hout, Evrard & Lyon, 1985; Satz, 1991). The first relates to the strong predominance of the non-fluent (motoric) form of aphasia in children, including mutism and/or the lack of spontaneous speech. The second relates to the more rapid and complete recovery from acquired aphasia in children. Although there is now evidence that the difference between children and adults in the form of the clinical condition and the prognosis for recovery is more a matter of degree (or relative frequency) than of kind (see e.g. Martins et al., 1991; Paquier & Van Dongen, 1993), the distinction appears to be a justifiable and useful one (Satz, 1991).

The acquisition or recovery of language functions after unilateral cerebral injury may, of course, be due to healthy uninjured tissue in either the same, opposite, or both hemispheres subserving the behaviour (Hecaen, 1976, 1983). No such ambiguity however is attached to the results from hemispherectomy, although both Kinsbourne (1974) and Vargha-Khadem and Polkey (1992) have suggested other grounds for caution here also. Nevertheless, the results of hemispherectomy in cases of infantile hemiplegia provide the most convincing evidence relating to plasticity of the neural substrate for speech and demonstrate most clearly the differences between childhood and adult acquired aphasias (see e.g. French & Johnson, 1955; Smith, 1974).

Following the early work of Krynauw (1950), Gardner et al. (1955) and others who studied the effects of hemispherectomy on patients with infantile hemiplegia, Basser (1962) reported on an extensive series of cases with special reference to the development or recovery of speech. He provided details of 102 cases (48 with lesions of the left hemisphere and 54 right), with 35 cases of hemispherectomy (17 left, 18 right), in some of whom speech was achieved after the occurrence of the lesion while in others the lesion occurred after speech had been acquired. From his analysis of the evidence, Basser concluded that speech was developed and/or maintained in the undamaged cerebral hemisphere and that in this respect the left and right hemispheres were equipotential.

Using Basser's results and drawing together evidence from a wide range of other sources, Lenneberg (1967) went further and suggested that while the two hemispheres were equipotential for speech at birth, maturational changes in the nervous system provided a "critical period" for the acquisition of language that extended from about two years of age to puberty. Krashen (1973) subsequently modified Lenneberg's view to propose that the developmental lateralization of language functions may be complete by five years of age, although other authors (e.g. Annett, 1973a; Woods & Carey, 1979) in comparing cases with early and later incurred unilateral cerebral lesions, have suggested that equipotentiality of the two hemispheres for speech development may only extend to about the end of the first year of life.

Many investigators both before and since Lenneberg (1967) published his ideas have also provided evidence in substantial agreement with the concept of initial equipotentiality (e.g. Brown & Hecaen, 1976; Hecaen, 1976; Hood & Pearlstein, 1955; Reed & Reitan, 1969; Smith, 1974; Wilson, 1970), while many others have disputed this interpretation (e.g. Carter, Hohenegger & Satz, 1982; Dennis & Whitaker, 1976; Kinsbourne, 1975; Satz & Bullard-Bates, 1981; St James Roberts, 1981; Woods & Teuber, 1978a). However, the selection of material for examination and the assumptions made (whether explicit or implicit) in the design of an analysis can substantially affect the results. And in several subsequent carefully considered reviews of the available evidence, Bishop (1981, 1983, 1988a, 1988b, 1990a) suggested that much of the controversy in the literature can be accounted for by the paucity of data and the lack of objective and standardized measures of language functions. She considered that it was too early to conclude that the right hemisphere has some inherent limitation in the degree of linguistic sophistication it can achieve, and that while this remains a possibility, it is far from proven (1988a, p.218).

But how are these considerations concerning language lateralization in infancy related to the development of handedness? The clinical evidence from brain-injured patients with infantile hemiplegia can only be suggestive, since as Basser (1962, p.429) pointed out, in such cases hand preference right or left is obligatory. Furthermore, in those cases where hand preference has not been established prior to the occurrence of a unilateral cerebral lesion, there is no way of knowing how the individual's handedness might have developed in the absence of brain damage. More directly relevant data on mature adults have been reported by Rasmussen and Milner (1977) who contrasted the effects of sodium-amytal infused into each hemisphere in turn in 134 epileptic patients with early left hemisphere injury (i.e. sustained before 6 years of age), with the effects in 262 comparable patients whose lesions were incurred later. They reported that whereas 96% of those with later lesions classified as right-handed had speech lateralized in the left hemisphere, 81% of the early brain injured right-handers had left hemisphere speech. More notably however, whereas 70% of the left- or mixed-handed later lesioned group had speech lateralized in the left hemisphere, the left- or mixed-handed *early* brain damaged group had speech lateralized to the left hemisphere in only 28% of cases, to the right hemisphere in 53%, and bilaterally in a further 19%. This effect of early brain damage compared with lesions of later onset in inducing a significantly greater right or bilateral cerebral speech representation associated with an enhanced incidence of left- or mixed-handedness, has also been reported in a number of other and more recent studies (see e.g. Guerreiro et al., 1995; Roberts, 1959; Satz et al., 1988; Srauss, Wada & Goldwater, 1992).

From these data it seems reasonable to infer that in patients with early left brain damage, where the size and location of the lesion is sufficient to obstruct the lateralization of speech to the left hemisphere, it is also likely to hinder the development of right-handedness. Or put another way, a cerebral lesion which impedes the development of right-handedness is also likely to interfere with the lateralization of speech to the left hemisphere. Presumably, lesions of later onset (unless extensive), are less effective in producing atypical speech and handedness patterns because the neural organizational substrate for these functions is already too well established. However, complete removal of the dominant hemisphere in later life (due to a cerebral neoplasm) does not necessarily preclude the recovery of some degree of verbal expression (Smith, 1966; Smith & Burklund, 1966). As the myelination (and hence functional maturation) of nerve fibres in the corpus callosum continues until at least the end of the first decade of life and for much longer in the association areas of the cerebral cortex (Yakovlev & Lecours, 1967), the disruptive effect of a brain lesion of a given size and location may also be expected to vary with physical and mental development, and not to be related to a specific age level in any particular individual.

Language and the corpus callosum

Clearly, the information gained from the use of hemispherectomy and the introduction of the amobarbital technique as well as evidence from unilateral brain lesions, added considerably to our knowledge of the cerebral mechanisms underlying speech and their association with handedness. But perhaps the most remarkable work reported during the last thirty years or so has come from studies of split-brain patients. To the extent that localized excision of cortical material (Penfield & Rasmussen, 1950) or hemispherectomy (Krynauw, 1950) had previously been carried out as a radical treatment for *inter alia*, intractable epilepsy, callosotomy was re-introduced (after Van Wagenen & Herren,

1940) for the same purpose by Vogel in 1961 (Sperry, 1974). With the development of specialized testing methods for these callosotomy patients, careful comparison of the sensory, motor, and cognitive abilities of their separated hemispheres led to new insights into the lateralization of these functions (see e.g. reviews by Gazzaniga, 1970; Gazzaniga & Le Doux, 1978; Sperry, 1967, 1974).

Although such patients typically display surprisingly few effects of callosotomy in their daily lives, specific tests of motor performance have revealed significant deficits even 5 to 10 years subsequent to the operation (Zaidel & Sperry, 1977), and studies of language dependent tasks have shown even more marked effects. For example, Gazzaniga and Sperry (1967) reported that spoken responses to stimulus material presented exclusively to the left hemisphere (in their right-handed subjects) demonstrated little or no impairment, whereas similar testing of the right hemisphere showed it to be totally incapable of initiating speech. Similarly, when stimuli presented exclusively to the left hemisphere required a written response, the preferred right hand displayed no apparent difficulty in complying, whereas when similar material was confined to the right hemisphere, none of the objects could be described or named in writing using either hand. By contrast, using non-verbal responses, the comprehension of language (both written and spoken) was shown to be present in both hemispheres, although the minor (right) hemisphere appeared to be functionally less proficient. Subsequent testing programs demonstrated that simple words could also be expressed by the minor hemisphere through the use of the left hand either by writing or by selecting and arranging plastic letters (Gazzaniga, Le Doux & Wilson, 1977; Levy, Nebes & Sperry, 1971). Sidtis et al. (1981) confirmed these results with post-operative examination of one callosotomy case but found no evidence of right hemisphere writing ability in a second case. And in a follow-up study over a period of three and a half years of one particular patient who was operated upon when he was 15 years of age, Gazzaniga et al. (1979) reported that apart from an increasing sophistication in the processing of language by the right hemisphere during this time, the patient had also begun to develop speech using his previously mute side of the brain.

Although Sidtis et al. (1981) have suggested that the results of these studies indicate that the language system of the right hemisphere differs quantitatively rather than qualitatively from that of the left hemisphere, Zaidel (1985, 1990) has pointed out that the pattern of linguistic competencies of the right hemisphere demonstrated in such patients does not correspond with any recognizable stage in normal development. For example, the combination of muteness with substantial language comprehension is abnormal. Nevertheless, Zaidel suggests that most of the components of right hemisphere language are comparable to the level normally achieved by 3 to 6 year-old children. He postulates that up to that age both hemispheres probably participate in language development but that thereafter, different components are lateralized to different degrees in the two hemispheres, with speech usually being established in the left.

As a congenitally occurring - but relatively rare - anatomical anomaly, agenesis of the corpus callosum has also been reported. The incidence of such cases in the general population is uncertain (Grogono, 1968; Jeeves, 1990), although with the introduction of more sophisticated diagnostic techniques, the number of cases identified in life will no doubt increase. Certainly, the number of detailed behavioural studies has grown considerably as a result of the interest created by the work of Sperry and his colleagues on patients with surgically disconnected hemispheres. However, in contrast to the severe behavioural deficits discovered in callosotomy patients, callosal agenesis cases tested with the same specialized techniques reveal surprisingly little impairment (Milner & Jeeves, 1979). For example, in one such acallosal patient examined by Saul and Sperry (1968), verbal responses dependent on either right or left hemisphere processing appeared to be normal or near normal, although some functional deficits were noted on non-verbal and perceptual tasks. The patient was said to be left-handed or ambidextrous (Sperry, 1970) and further examination by means of the amobarbital technique revealed that the control of speech production had developed in both hemispheres (Sperry, 1974). These findings led Sperry (1968, 1974) to conclude that they reflected the plasticity of the nervous system and the compensatory capacities of the developing brain. However, according to Sauerwein, Nolin and Lassonde (1994), in eight cases of callosal agenesis tested with the amobarbital method, the cerebral control of speech was found to be lateralized to the left hemisphere in all but two, both of whom were said to be left-handed.

Evidence concerning the asymmetry of hemisphere function in relation to the execution of bimanual movements has been provided by Preilowski (1972, 1975, 1977). Using an X - Y recorder, he found that in a simple tracking task, two right-handed partial callosotomy patients (with the anterior commissure and anterior two-thirds of the callosum sectioned), showed less improvement with training and performed less proficiently than control subjects, especially when visual feedback was removed. The patients' performances were consistently poorer when the two hands were required to work at different rates, and at their worst when the left hand had to contribute the greater input. Similar results were obtained by Jeeves, Silver and Jacobsen (1988) on two right-handed acallosal patients. Jeeves, Silver and Milne (1988) also tested right-handed normal subjects in three different age groups (6 year-olds, 10 year-olds, and adults) on the same task, hypothesizing that since maturation of the corpus callosum is not complete until about 10 years of age, their youngest group of subjects might be expected to display performance characteristics more comparable to acallosal patients than normal adults. They found that whereas the performances of the 10 year-olds were essentially the same as the adults, the 6 year-olds were significantly less proficient than the older subjects - both quantitatively and qualitatively - especially when asymmetrical movements of the two hands were required. However, unlike the acallosal and partial callosotomy patients, the 6 year-olds showed no consistent asymmetry favouring one hand or the other. The authors suggested that this was understandable if no clear left hemisphere motor control had been established by that age.

If linguistic and motor functions normally develop bilaterally but asymmetrically between the two sides of the brain, either in association or independently, the question arises as to whether or not acallosal brains conform to the usual development pattern (Jeeves & Milner, 1987). Insofar as this question has implications for distinguishing between an hypothesized prewired or plastic neural basis for lateralization, it has been pointed out by Gazzaniga (1970, p.139), that in acallosal cases during development, the two half-brains are necessarily exposed to the same environmental influences as are those of normal individuals. Consequently, if asymmetries of cerebral motor function are primarily dependent on differential usage of the right and left hands, as determined by environmental pressures and opportunities, such asymmetries might be expected to develop with age and experience in much the same way for both acallosal and normal populations.

It may then be asked - what is known about the handedness characteristics of acallosals? In her review of 24 published cases of total agenesis of the corpus callosum, Levy (1985) noted that as many as 12 were described as either left- or mixed-handed - far higher than the 10% to 12% that would be expected in a normal population. However, in her review of those published reports of agenesis of the corpus callosum that met certain specified criteria relating to IQ and behavioural tests, Chiarello (1980) found that out of 29 cases, 22 could be classified as predominantly right-handed and 7 as primarily left-handed (i.e. 24%). Since then, a further 16 cases have been documented by Sauerwein et al. (1994) listing 12 right- and 4 left-handers (i.e. 25%). Thus, the evidence certainly suggests that there may be an unusually high incidence of non-right-handedness amongst acallosals which needs to be explained. But as Jeeves and Milner (1987) have pointed out, a higher than normal incidence of left- and mixed-handedness may be expected in a neurological population, although why this should be so is an open question, and one which is considered again in Chapter 8.

It has frequently been proposed that some of the successful perceptual-motor activity reported for acallosal patients in contrast to callosotomy cases may be due to the development of a variety of compensatory mechanisms such as subtle cross-cueing strategies or to the use of alternate anatomical pathways (Gazzaniga, 1970; Sauerwein et al., 1981). However, in reviewing these possibilities, Jeeves (1986, 1990) provides grounds for considering that sophisticated behavioural strategies probably contribute little to the results of most testing programs. Although most authors concerned with exploring the compensatory mechanisms employed by acallosals have concentrated on problems of interhemispheric integration, some findings of relevance to language production have also been reported in comparisons with intra-hemispheric function. For example, Sauerwein and Lassonde (1983) and Lassonde (1986) have reported that in tasks requiring comparative judgements of pairs of stimuli presented to the same or different visual half-fields, their acallosals were able to respond verbally as accurately as the matched IQ control subjects in both the inter- and intra-hemisphere testing situations. Nevertheless, since they took twice as long to do so in both these conditions, Lassonde (1986) concluded that such a deficit could not be explained in terms of extracallosal pathways, and proposed that in normal subjects therefore, the corpus callosum may have a facilitatory or excitatory function. She suggested that each of the two cerebral hemispheres could thus only reach its full potential in the presence of the other, and implied that conversely, any dysfunction of one hemisphere would impact adversely on the efficiency of the other. Certainly, there is good supporting evidence for this view in the evaluation of the effects of hemispherectomy in a number of cases where considerable cognitive improvement has been reported after complete removal of the damaged hemisphere (Glees, 1980; McFie, 1961; Smith, 1974).

Electrocortical mapping and cerebral asymmetry

Although the operative procedures of hemispherectomy and callosotomy for the relief of intractable epilepsy enabled significant contributions to be made to our knowledge of the functional roles of the right and left hemispheres, the earlier and more circumscribed surgical treatment by localized excision of cerebral tissue (Penfield & Steelman, 1947) similarly facilitated major advances in the field. Apart from the research benefits obtained from observing the effects of cerebral ablation upon behaviour, the work was also responsible for fostering the development of electrocortical stimulation of the brain in the conscious patient. As a necessary precautionary measure prior to surgery, Penfield and his colleagues (see e.g. Penfield & Rasmussen, 1950; Penfield & Roberts, 1959; Penfield & Welch, 1951) employed the procedure of exposing the cortex under local anaesthesia and functionally mapping the cerebral surface using electrical stimulation to evoke sensory, motor or other responses from the conscious patient. In particular, they showed that although there was an inter-individual consistency in the relative position of the cortical representation of somatic motor and sensory functions along the pre- and post-central gyri, the actual size and hence location of each such cortical area varied considerably for different regions of the body and from one patient to another. Similarly, they found that vocalization and the arrest of speech could be produced by stimulation in the somatic sensory-motor areas of the cortex, or speech disruption from other frontal, temporal, and parietal regions, including the supplementary motor area. They reported that the location and extent of these areas of speech disruption corresponded in general with the regions in which lesions or the surgical excision of cerebral tissue were known to produce aphasia. Subsequently, other workers (e.g. Mateer & Cameron, 1989; Ojemann & Whitaker, 1978) suggested that the effects of stimulation or a circumscribed lesion in these regions were likely to be very similar insofar as both provide a focal source of "noise" in the neural system. In this respect, the type of language disruption reported by Penfield and Roberts (1959) varied considerably, with the effects being described as either arrest, hesitation, slurring, repetition, or distortion of speech; confusion of numbers while counting; inability to name with retained ability to speak; and difficulty in reading or writing.

Using a similar technique, Ojemann and his colleagues attempted to clarify the effects of stimulation on language production by testing specific verbal and non-verbal oral motor activities during stimulation of a wide range of cortical sites on the exposed cortex of the same patient. For example, Ojemann and Whitaker (1978) examined the naming errors and abilities of patients subjected to cortical stimulation and reported that in some individuals, interference with speech occurred over a much wider expanse of the left lateral cortex than had previously been reported, although in the eight left hemisphere dominant cases they examined, only stimulation in one particular zone produced language disruption in every patient. These results were later confirmed on a larger sample of left hemisphere dominant cases by Ojemann (1983a) who suggested that this region formed part of the final common mechanism for the organization of speech. Evidence supporting this view was reported by Ojemann and Mateer (1979a, 1979b, 1979c), and in a later study by Ojemann et al. (1989) on the left (dominant) hemisphere of 117 patients, it was reported that a pattern of at least one frontal and one temporal site essential for oral naming of objects was found in about two-thirds of the patients tested by electrical stimulation. But while the majority of the patients had highly localized sites for evoking naming errors, the spatial distribution of these across the cerebral cortex varied considerably from one patient to another.

Although few cases have been reported, the evidence suggests that electrical stimulation of the non-dominant hemisphere (as determined by amobarbital) outside specifically motor and sensory cortical areas does not usually interfere with the production of either speech or non-speech vocalization (Mateer,

1983; Ojemann & Whitaker, 1978). Certainly, out of 14 right-handed patients tested for right hemisphere disruption of speech by Penfield and Roberts (1959, p.133), only one was affected by stimulation. Furthermore, out of six nominally left-handed cases, the same authors reported only one as having speech disrupted by stimulation in the right hemisphere. However, in a later study by Andy and Bhatnager (1984), three strongly right-handed patients with left hemisphere dominance for speech (determined by amobarbital) were examined for interference with speech production by carefully controlled electrical stimulation of their exposed right hemisphere. In all three patients, speech was clearly disrupted by stimulation of a number of different cortical sites, supporting the view that the hemisphere not dominant for speech nevertheless possesses some significant latent language function.

While stimulation studies have provided good evidence to show which regions of the left cerebral cortex (in left hemisphere dominant cases) may participate in language processing, they have also demonstrated a considerable variation in the degree of involvement of any specific site from one person to another. For example, in a series of 21 patients known to be left hemisphere dominant for speech, stimulation in traditional language zones produced naming errors in as few as 23% and in no more than 61% of the sites tested (Ojemann, 1983a). And in a study of 8 male and 10 female patients, Mateer, Polen and Ojemann (1982) reported a significant difference in the effects of cortical stimulation on the naming ability of the two sexes. Overall, naming errors were produced from 63% of all sites sampled in the males but from only 38% of all sites in the females. A further analysis of these data, together with those reported by Ojemann et al. (1989) provided suggestive evidence that, in addition to sex, there may be individual differences in the pattern of cortical organization for naming ability related to differences in verbal IQ level. However, there are other possible influences that may contribute to such individual differences, e.g. methodological, clinical, and/or anatomical, and these have been discussed by Ojemann (1983b, 1994).

Clearly, localized electrical stimulation within some "language area" of the cerebral cortex is a most unnatural event, and usually inhibits speech or vocalization; it does not produce a behaviourally meaningful response. Thus, there is the same need for caution in the inferences to be drawn from such evidence in relation to speech and language as there is in the interpretation of results from cortical stimulation in relation to limb movement and manual skill. Here, it has been shown that stimulation never produces a skilled movement or learned response - only basic muscle activity or an inhibition of voluntary movement (Penfield & Rasmussen, 1950). Furthermore, Penfield and Jasper (1954, p.64) reported that the effects of stimulation of the primary motor cortex are the same in a child 8 years old as in a man of 60, or in an accomplished pianist as compared with a manual labourer. Hence, as differences in motor performance between the right and left limbs in normal human behaviour are associated particularly with over-learned movements (see Chapter 7), gross differences between hemispheres relating to such basic motor responses should not therefore be expected. Nevertheless, differences in the spontaneous electrical activity on the two sides of the brain have been reported in relation to voluntary movements of the limbs and this work is discussed in some detail in Chapter 3.

Apraxia and associated motor disorders

Although the most obvious and characteristic motor effects of a lesion of the left hemisphere in right-handed patients usually relate to disturbances of speech and a right hemiplegia, interference with the execution of certain other voluntary movements described as apraxia is also frequently noted. According to a survey of relevant studies by De Renzi (1989), the incidence of apraxia in patients with left hemisphere brain damage varied from 28% to 55%, and in cases of right brain damage, from 2% to 9%. But while much has been written on the various forms of apraxia and their classification, the topic has been described by Freund (1987) as probably the most controversial issue in the field of motor control, and dogged by the vagueness of definitions.

Evidence relating to the nature or degree of association between the different forms of apraxia described in the literature has been reviewed by Square-Storer and Roy (1989) and Roy and Square-Storer (1990). In the latter paper, the authors define apraxia as a disturbance in the performance of movements or gestures which cannot be explained on the basis of motor weakness, ataxia, dementia, or poor comprehension in aphasia. Kimura and Archibald (1974) concluded from their examination of the literature that the generally accepted view of previous investigators was in terms of an inability to perform purposeful or voluntary movements not due to paralysis of the muscles. Another succinct definition by Geschwind (1975) appears to combine the essential elements of both the above concepts by suggesting that the apraxias are disorders of the execution of learned movements which cannot be accounted for either by weakness, incoordination, sensory loss, or incomprehension of or inattention to commands.

Historically, Hughlings Jackson seems to have been the first to describe the association between a speech disorder and an apractic dysfunction (Freund, 1987), with symptoms primarily relating to the inability of a patient to comply with a request to initiate a movement voluntarily (e.g. "put out your tongue"), although in the same individual, control over the same muscles appeared to be exercised without difficulty in such routine acts as eating and swallowing. The subsequent development of research on the clinical and neuro-anatomical aspects of apraxia have been reviewed by many authors (e.g. de Ajuriaguerra & Tissot, 1969; Faglioni & Basso, 1985; Freeman, 1984) who all refer to the important contributions of Liepmann (for selected papers, see Brown, 1988). According to Faglioni and Scarpa (1989), the three main tenets of Liepmann's interpretation of the symptoms of apraxia that have been shared by the great majority of subsequent authors are: (a) that purposeful movements are primarily controlled by the hemisphere dominant for hand preference, (b) that the corpus callosum is an essential feature in the control exerted by the dominant hemisphere, and (c) that the lateral parietal and frontal cortex are the cerebral regions most critically involved in praxic control. Consideration will now be given to the first two of these.

Faglioni and Basso (1985) reviewed the evidence from a number of investigations designed to assess the incidence of apraxia or to quantify the apraxic performances of left and right brain-damaged right-handed patients, and reported that a consistently greater impairment in the initiation of voluntary movements was associated with left-sided lesions across all studies. Nevertheless, it is clear that by no means all individuals within a given study showed the same effect. For instance, in the quantitative investigation of unselected right-handed brain-injured patients reported by De Renzi, Motti and Nichelli (1980), 20% of their 80 right brain-damaged cases as well as 50% of their 100 left brain-damaged patients were classified as apraxic. However, the apraxia in their left brain-damaged sample was not only more severe than in their right brain-damaged patients, but also tended to be more often associated with aphasia. If consideration is confined to severe cases, the incidence of apraxia in their patients becomes 32% of those with left hemisphere damage and 4% for the right brain-damaged group.

The picture with regard to left-handers tends to be rather more obscure, and although Faglioni and Basso (1985) reported that in the great majority of these subjects, apraxia resulted from right brain lesions, the authors consider that the evidence demonstrating a dissociation between apraxia and aphasia in such patients is sufficiently great for them to conclude that there can be no common neurological basis for these two disorders, although the debate on this issue continues (Kertesz, 1985; Poeck, 1988; Square-Storer, Roy & Hogg, 1990).

But what of the postulate that the corpus callosum is an essential feature in the dominant hemisphere's control of movement? The strongest advocate of this view has been Geschwind (1965), whose research interest in apraxia and its neurological basis began with the post-operative examination of a tumor patient reported by Geschwind and Kaplan (1962). These authors found that while this right-handed patient was able to comply with verbal requests to carry out simple acts (such as combing his hair or waving goodbye) with the right hand, he was unable to do so, or made an inappropriate response when asked to carry out the same tasks with his left hand. However, when asked to imitate the examiner's movements, he performed these correctly with either hand. Thus, in these and other tests, the patient displayed the behavioural characteristics later recognized as typically associated with callosotomy (see e.g. Gazzaniga, Bogen & Sperry, 1967), and the authors concluded that the symptoms could best be described in terms of a disconnection syndrome, i.e. separation of comprehension of verbal commands from control of their execution. Post-mortem findings later confirmed that the patient had indeed suffered from destruction of the anterior four-fifths of the corpus callosum (Geschwind, 1975).

In this paper, Geschwind (1975) reviewed the available evidence and put forward a modified theoretical model to include certain other types of apraxic disorder. The first related to those somewhat paradoxical cases of apraxia due to unilateral brain lesions with associated contralateral hemiplegia unaccompanied by aphasia (e.g. Heilman et al., 1973; Heilman, Gonyea & Geschwind, 1974). To accommodate such cases, Geschwind postulated a dissociation between the hemisphere dominant for speech and the hemisphere dominant for manual skills. Other cases of apraxia for individual limb movements despite successful voluntary motor control over certain axial or midline structures, were considered to be due to a failure to initiate appropriate activity via the pyramidal system and a consequent reliance on extrapyramidal pathways which resulted in a less than adequate behavioural outcome. And other cases of apraxia for independent limb movements in spite of their successful integrated use in quite complex sequences of whole body activities was explained by Geschwind postulating a special capacity for the minor hemisphere to comprehend axial commands and to initiate relevant motor activity through the non-pyramidal pathways.

Certainly, Geschwind's (1975) modified neurological explanation for these apraxias helps to account for many of the earlier findings previously unexplained. However, as Haaland and Yeo (1989) have pointed out, some of the difficulties of interpretation of clinical reports on limb apraxia may also be attributed to poor methodological controls and to the necessity to employ testing of the hand ipsilateral to the brain lesion (due to contralateral hemiplegia). If the right-handed patient with a left hemisphere lesion is thus tested on the left side, in the absence of an appropriate control group for comparison, what may actually be interpreted as a hemisphere-of-lesion induced deficit may actually be a non-preferred hand effect - or at least confounded with it. Certainly, Jebsen et al. (1971) have provided evidence to show that in cases of unilateral brain damage producing hemiparesis, the motor functions of the hand ipsilateral to the lesion are significantly impaired compared to the performance of matched controls irrespective of the side of the cerebral injury. And Finlayson and Reitan (1980) found that on tests of grip strength and finger tapping, neither lesions of the left nor right hemisphere were pre-eminent in producing ipsilateral or contralateral motor deficits. Rather, the significant effects they recorded appeared to be a reflection of the hand preferences of the patients.

Experimental lesions in non-human species

While work on the effects of brain lesions on motor behaviour in nonhuman primates might be expected to throw some light on the etiology of apraxia in clinical cases (see e.g. Walshe, 1935), Ettlinger (1969) reported that his investigations of the effects of cortical ablations in more than 50 monkeys had never resulted in a motor condition resembling apraxia. In explanation, he suggested that this was no doubt because the classical definition of apraxia had a strong language-dependent component. And in a detailed discussion of the relevance of work on subhuman species, Kolb and Wishaw (1985) also suggested that conventional concepts and approaches to the study of apraxia had been too restrictive, and that progress in the field would be enhanced by a consideration of the many relevant studies of motor mechanisms in subhuman primates.

Certainly, many investigators (e.g. Peterson, 1934; Warren, Ablanalp & Warren, 1967; Warren & Nonneman, 1976) have explored the possibility that the concept of cerebral dominance might be tested by studying the manual preferences, dexterity, and problem-solving behaviour of animals, and have examined the capacity of their subjects for recovery following experimental cortical ablations. In this regard, unilateral lesions of the nominally dominant hemisphere (i.e. contralateral to the preferred hand) have been shown to induce a change in manual preferences in a number of species (e.g. rats, cats and monkeys) and on a variety of manipulative tasks. Peterson and Fracarol (1938) began the search for a cerebral locus for the mechanism controlling hand preference in rats by comparing the behavioural effects of frontal versus occipital and temporal lesions. They reported that only ablations in the contralateral frontal lobe were effective in producing changes in manual preference - a conclusion which was subsequently confirmed in a long series of collaborative studies by Peterson with other colleagues (see e.g. Peterson & Barnett, 1961; Peterson & Devine, 1963; Peterson & Gucker, 1959). Similarly, in a number of experiments on cats, Warren and his co-workers (see e.g. Forward, Warren & Hara, 1962; Warren, Ablanalp & Warren, 1967; Warren et al., 1972) showed that lesions in the sensorimotor cortex contralateral to the preferred paw produced a long-lasting reversal of manual preferences, whereas corresponding ablations of the posterior cortex either in isolation or in combination with section of the corpus callosum had no effect on preferential paw usage. And in monkeys, Deuel and Dunlop (1980) showed that unilateral lesions in the periarcuate region of the cerebral cortex in all 12 animals tested induced a preferential usage of the hand ipsilateral to the brain damage, irrespective of the side of their pre-operative hand preferences or training. In contrast, out of 10 monkeys subjected to symmetrical bilateral one-stage cortical lesions, 8 of the animals maintained a preference for their pre-operative trained and most-used hand.

However, the results of experimental cortical lesions have also been shown to be related to the age or maturity of the animal concerned. For instance, Castro (1977) compared the effects of unilateral frontal cortical lesions in rats subjected to surgery when adults, with animals operated upon neonatally but tested at maturity. Limb preferences recorded in a unimanual food retrieval task showed that post-surgery, all adult operated animals preferred to use the limb ipsilateral to the lesion, whereas half the neonatally operated subjects displayed a contralateral or ambilateral limb preference. Similarly, Whishaw and Kolb (1988) reported that the deleterious effects of unilateral motor cortex ablations on a food-reaching task in rats was greater for animals operated upon as adults than for those given lesions on their day of birth. The brain damage tended to induce preferential use of the ipsilateral forepaw, and in both groups there was a bilateral decrease in performance which was proportional to the size of the lesion.

Burgess and Villablanca (1986) have also reported the effects of hemispherectomy, comparing the limb preferences of cats operated upon when adults with neonatally-lesioned subjects. These authors assessed their animals for limb usage on a variety of performance tests and found that although their neonatally hemispherectomized cats were significantly impaired in comparison with unoperated control subjects, they nevertheless demonstrated fewer motor deficits and less limb use bias than their adult-lesioned counterparts. And in an extensive series of experiments on monkeys, baboons and chimpanzees, Kennard (1936, 1938, 1940, 1942) reported marked differences in the degree and rate of recovery from brain damage when it was incurred in infancy compared with the post-operative sequelae in adults. She found that following unilateral and bilateral ablation of the motor and premotor areas of the cerebral cortex in monkeys, young and immature animals recovered their motor functions more quickly and extensively than adults. The author also reported a similar and even more marked trend in her chimpanzee evidence, which she attributed to the fact that the degree of motor deficit in the adult chimpanzee following cortical ablation is much greater than in the monkey, and more similar to that observed in human patients.

Other investigators have extended these comparisons to a wide range of sensorimotor and more complex behaviours in rats, cats and monkeys (see e.g. Burgess, Villablanca & Levine, 1986; Goldman, 1976; Harlow, Akert & Schiltz, 1964; Kolb & Tomie, 1988; Villablanca, Burgess & Olmstead, 1986) with results that also provide support for the concept of a decreased functional impairment with brain damage when it is sustained early in life. However, not all the available information is supportive (see e.g. Johnson & Almli, 1978; Kolb, 1987; Passingham, Perry & Wilkinson, 1983; Schneider, 1979; Whishaw & Kolb, 1984), so that it is necessary to recognize that the "Kennard Principle" may well need some qualification (Aram & Eisele, 1992; Finger & Almli, 1988; Kolb, 1992).

But there still remains the more general question of the extent to which it is possible to derive information concerning the cerebral mechanisms responsible for normal motor behaviour from research investigations of cortical stimulation or ablation, and clinical studies of the effects of brain damage. Problems associated with the interpretation of such evidence have been described many times before (see e.g. Gregory, 1961; Walshe, 1947, p.332) and do not need detailed repetition. Suffice it to say that lesion studies certainly help to locate some of the neural regions and pathways *necessary* for a given function, but do not provide *sufficient* evidence for us to conclude that the loss or impairment of a function associated with damage to a given cerebral site identifies the anatomical locus of that function. As Brooks (1981) has pointed out, the motor cortex or other such central structure is a summing point that is part of a system which is distributed in space and time. Where and when a movement starts has no simple answer, since movements are a product of past experience (stored as motor programs), current demands, and the ever-changing state of the organism. Furthermore, the rapidly growing body of information on the plasticity of the nervous system, with its implications for recovery of function after brain injury (see e.g. Stein & Glasier, 1992) also complicates the picture.

A general discussion of the problems associated with interpreting evidence obtained from different approaches to questions of functional localization has been provided by Kertesz (1994). Undoubtedly, there is a formidable number and range of possible anatomical, physiological and behavioural variables that can influence the findings of any given research undertaking, but in clinical studies, most of these are beyond the investigator's control. No matter how similar different cases may appear to be, nor how objective the evidence available to classify them, no two patients suffer from a cerebral lesion of exactly the same size, in exactly the same place, or from exactly the same cause, with exactly the same time course. Bearing in mind the widely different ages, experiences, and educational backgrounds of patients, as well as the many other physical and behavioural features which identify each individual as unique, the difficulties of generalization are all too obvious. However, there are many other approaches that have been used to examine the cerebral asymmetries associated with speech and handedness, and the evidence available from essentially non-clinical sources of information are considered in the next chapter.

CHAPTER 3
Cerebral Asymmetries in Normal Subjects

Chapter outline

Anatomical studies of the brains of normal human subjects have reported a number of asymmetries of shape or form. Differences between sides in the surface area of certain regions of the cerebral cortex (e.g. the planum temporale) have received special consideration and comparisons have been made between these morphological asymmetries and the reported handedness of subjects, but with questionable success. Attention has also been focussed on the size and development of the corpus callosum in left- and right-handers.

Neurophysiological investigations of cerebral asymmetries have examined differences in the electroencephalograms, event-related cortical potentials, and regional cerebral blood flow patterns of the two hemispheres with encouraging results - especially from the last two techniques. Asymmetries of cerebral activity have been related to both speech and movements of the dominant compared with the non-dominant hand. However, difficulties and limitations accompany each technique and problems of interpretation of the findings are discussed.

Behavioural studies of concurrent speech and unimanual motor activity on performances of the preferred and non-preferred hands have usually reported a greater interference between the two behaviours when the unimanual activity involved the preferred hand. This has been widely interpreted as demonstrating some common or overlapping central neural control mechanism for these motor skills - presumably located in one and the same cerebral hemisphere (i.e. the left for right-handers), although the results from experiments with nominally left-handed subjects are less easily explained. Developmental studies using the same concurrent task paradigm have reported that the greater interference with unimanual performances of the preferred hand tend to remain unchanged with age, although the adequacy and reliability of the evidence is open to question.

Neuroanatomical asymmetries

Evidence from studies of normal behavioural development tend to support the view that there is a close and increasing correspondence with age between the acquisition of linguistic and manual motor skills in infancy (Molfese & Betz, 1986, 1987). Nevertheless, there are clearly problems in obtaining relevant non-invasive evidence for the cerebral lateralization of these functions. Although various reports of anatomical and physiological asymmetries of the cerebral hemispheres of the human brain have been made from time to time during the last 100 years or so, it has usually been considered (see e.g. Galaburda, 1995; Rubens, 1977; Stein, 1988; von Bonin, 1962) that these differences between the two sides tend to be small and variable, and difficult to reconcile with the well-documented differences in function.

However, during the last twenty years or so, studies of certain morphological features of human and non-human primate brains and skulls (see reviews by Galaburda, 1984; Galaburda et al., 1978; Le May, 1976, 1977, 1984, 1985; and Le May & Geschwind, 1978) have reported that anatomical asymmetries are common in both human and nonhuman specimens (although less frequent in the latter) and are generally in the same direction. The left occipital pole is frequently wider and protrudes more posteriorly than the right, whereas the right frontal pole is frequently wider and protrudes further forward than the left. Since the shape of the skull and brain are closely associated, these features are to be seen in each structure. Several authors have attempted to relate these asymmetries to the handedness characteristics of subjects by comparing measurements of the cerebral hemispheres taken from computerized tomography (CT) scans with hand preference classifications. For example, Le May (1977), Le May and Kido (1978) and Bear et al. (1986) reported finding such an association for right-handers, but a reduction or reversal of the asymmetry in left-handers. Other authors using other criteria have also reported relevant associations. For example, Kertesz et al. (1986, 1990, 1992) have employed magnetic resonance imaging (MRI) techniques to obltain a number of linear and area measurements of cerebral asymmetry in both right- and left-handed normal subjects who were given manual, auditory, and visual performance tests of laterality. The authors reported evidence supporting an association between various handedness and/or laterality measures and several of their recorded anatomical asymmetries.

However, Chui and Damasio (1980) employed CT scans of 50 right-handers and 25 left-handers in their comparison of anatomical and hand preference criteria, but detected no such association. These investigators reported the presence and lateralization of frontal and occipital petalia according to expectation, but found that the occurrence of these asymmetries was unrelated to the hand preferences of the subjects. Similarly, Koff et al. (1986) measured hemispheric asymmetries in the CT scans of 172 subjects (146 right-handed and 26 left-handed) and compared them with the degree of right- and left-handedness measured by questionnaire, but found no significant correlation. Again, a tendency was noted for an anatomical size asymmetry favouring the left occipital pole and the right frontal pole, but this was found to be similar for both right- and left-handers. And again, Faglioni and Scarpa (1989) reported that skull asymmetries in their 160 brain damaged patients also demonstrated a right frontal and left occipital predominance, but that in a subgroup of 72 who were tested for apraxia and classified as either apraxic, borderline, or normal, there was no consistent relation between hemisphere dominance for praxis and skull asymmetry.

In partial explanation for their findings on hemisphere asymmetries, Chui and Damasio (1980) referred to the possible importance of outside forces acting on the skull in the recumbent posture - especially in infancy (see e.g. Konishi et al., 1987), while, as a result of finding a significant correlation in 15 patients between occipital length differences and the length of the planum temporale on the two sides, Pieniadze and Naeser (1984) suggested that these anatomical asymmetries may be more related to language functions than to handedness. In this regard, Burke et al. (1993) reported that in an investigation of four cerebral asymmetries measured from CT scans in 25 aphasic patients, they found a significantly positive correlation between the occipital width asymmetry and rate of recovery of language skills, although Kertesz (1988) and Kertesz and Naeser (1994) reported no such significant correlation between any of their CT scan asymmetries and recovery of spontaneous speech. Indeed, they suggested that occipital asymmetries were more likely to be related to other aspects of language function than to motor speech output.

Certainly, Galaburda (1991, 1995) has concluded from his review of the relevant neuroanatomical evidence, that there is no cortical area or architectonic pattern, histological feature, ultrastructural characteristic, or system of neuronal connections which is seen in one hemisphere and not in the other. He suggested that the only anatomical feature which could have significance for a functional distinction between the hemispheres is a difference in size of homologous areas of cerebral cortex on the two sides. It is of some relevance then, that in a detailed investigation of the surface features of the temporal language region of the cortex, Geschwind and Levitsky (1968) reported that the planum temporale tended to be larger in the left hemisphere than in the right in 65% of the 100 adult brains they examined at post mortem. This was confirmed by similar findings in 14 neonatal and 16 adult brains reported by Witelson and Pallie (1973) and in a further study of 100 adult and 100 infant brains by Wada, Clarke and Hamm (1975). Many other investigators have published concordant results (including data for the fetus - see e.g. Chi, Dooling & Gilles, 1977), and in a re-examination of the anatomical material originally studied by Geschwind and Levitsky, Galaburda et al. (1987) found that asymmetries of the planum temporale are graded, and occur along a continuum that approaches a skewed (to the left) normal distribution. It has also been reported that these asymmetries are due entirely to the variation in size of the smaller side, since ignoring the direction of asymmetry, there is a highly significant correlation (r = -0.85) between absolute asymmetry and the size of the smaller side, but no such correlation with the larger side (Rosen, Galaburda & Sherman, 1987, 1990). However, many reviews of the evidence in this field (e.g. Galaburda et al., 1987; Strauss, Kosaka & Wada, 1983; Witelson, 1977, 1980, 1983; Witelson & Kigar, 1988a) have expressed an increasing concern with the methodological difficulties encountered, and consequently the interpretation of results, since it is clear that defining the extent of anatomical regions such as the planum temporale is not a simple matter (see e.g. Galaburda, 1993, 1995; Witelson & Kigar, 1992).

While some authors (e.g. Annett, 1992a; Steinmetz et al., 1991) have offered grounds for considering that there may be a close association between asymmetries of the planum temporale and handedness, intuitively it might be expected that any correspondence between anatomical hemispheric asymmetries of the motor functions of language and manual activity would involve those areas of the cerebral cortex known to be more concerned with speech production than perception. Yet in the Wada et al. (1975) examination of adult and infant brains, these authors reported that the area of the frontal operculum was slightly but nevertheless significantly larger in the *right* hemisphere. And in their review of anatomical investigations of the frontal speech regions, Witelson and Kigar (1988a) found no consistent evidence to suggest that there is a morphological asymmetry favouring the left side comparable to that reported for the planum temporale. Furthermore, in a comparison of the extent of anterior and posterior speech areas on both sides of the same brains, Falzi et al. (1982) found no correspondence between the asymmetries they recorded in these two regions. In a more recent and extensive study using MRI brain scans on normal healthy volunteers, Habib et al. (1995) reported significantly larger areas for both the parietal operculum and planum temporale in the left hemisphere of consistent right-handers compared with non-right-handers. But even so, they found that the asymmetries of these two regions were not significantly correlated with each other, suggesting that the directional bias was not necessarily coincident for any given individual. By contrast, Witelson and Kigar (1992) in a detailed postmortem examination of the brains of people tested for handedness before death, found that while there were asymmetries of certain segments of the Sylvian fissure, the overall asymmetry was minimal. But unexpectedly, they did find an association between individual differences in hand preference and differences in the size of one segmental measurement that varied in the same direction in *both* hemispheres. However, in a somewhat different approach, Scheibel (1984), Scheibel et al. (1985), and Simonds and Scheibel (1989) reported differences between the right and left motor speech regions of the frontal lobe at the histological level. From autopsies of patients free of any known neurological pathology, these authors found evidence to suggest that an early preponderance of dendrite growth in the right hemisphere during ontogeny is followed by enhanced dendrite growth in the left hemisphere coincident with the development of conceptualization and speech.

A good deal of interest has also focussed on the identification of individual differences in the size and/or extent of the anatomical connections between the two cerebral hemispheres through the corpus callosum (Driessen & Raz, 1995). For example, Witelson (1985) made post-mortem examinations of the brains of 42 subjects on whom hand preference data had been obtained before death and found that those with mixed hand preferences had a larger midsagittal area of the corpus callosum than consistent right-handers by about 11%. In subsequent expanded studies reported by Witelson (1989) and Witelson and Goldsmith (1991) using the same procedure, variations in callosal morphology were again found to be associated with differences in handedness classification and age in males but not in females. Other investigators (Habib, 1989;

Habib et al., 1991) have found a similar association between consistency of hand preference and the size of the corpus callosum in healthy living subjects using the magnetic resonance imaging (MRI) technique, although others (e.g. Kertesz et al., 1987; O'Kusky et al., 1988) have reported no sex or handedness differences in relation to callosal size.

Relevant to these findings are the reports of Witelson and Kigar (1988b) and Clarke et al., (1989) that the corpus callosum increases in size rapidly in the fetus and then at a declining rate postnatally until adult size is reached at about five years of age. Furthermore, accompanying this increase in overall size of the corpus callosum with maturation, there is good evidence for a decrease in the actual number of callosal axons in the vertebrate brain during late pre- and early post-natal neural development (Clarke et al., 1989; Koppel & Innocenti, 1983; LaMantia & Rakic, 1984). Relating this evidence to their own findings, Witelson and Nowakowski (1991) speculated that such a naturally occurring loss of callosal axons might be related to the development of human hand preference in males, with a different mechanism underlying lateralization in females. But which is cause and which the effect? For example, the number of fibres retained in the corpus callosum during maturation may be a function of their usage rather than sex *per se*, since Yakovlev and Le Cours (1967) showed that maturation of callosal fibres extends up to ten years post-natally - corresponding to the period of rapid development and acquisition (i.e. learning) of a wide range of unilateral and bilateral manual skills.

Certainly, the importance of such regressive events as neuronal death and selective collateral axon elimination in determining the final form of the mature nervous system has gained increasing recognition over the last fifteen years or so, in comparison with the previously more widely recognized role of progressive phenomena such as proliferation of nerve cells and the formation of complex patterns of neural connections (Cowan, Fawcett, O'Leary & Stanfield, 1984). For example, Huttenlocher (1979, 1990) and Huttenlocher et al. (1982) have shown that there are considerable changes in synaptic density in the human cerebral cortex with age over the life span, and particularly during infancy and early childhood. These authors found that synapse production increased rapidly during the first one or two years of life to reach a peak value between about four months and seven years of age (depending upon the cortical region studied), and then to decrease steadily to approximate the adult level by about eleven years of age.

Evidence of similar developmental variations in cerebral metabolic rates with age have come from studies using the technique of positron emission tomography (PET). For example, Chugani, Phelps and Mazziotta (1987) and Chugani and Phelps (1991) have reported that an overall postnatal increase in cerebral metabolism reaches a maximum value by about 3 to 4 years of age and then levels off until around 9 years of age when it declines steadily to approximate the adult rate after about 15 years of age. The authors suggested that these changes in metabolism correspond to the three underlying variations in neuronal activity - the initial metabolic rise being related to the period of rapid over-production of neurons and synapses, the plateau period of metabolic activity being associated with the phase of maximum plasticity of neural function, and the third or declining stage of energy demand reflecting the selective pruning or elimination of unnecessary neuronal connections.

In discussing these functional aspects of the organizational plasticity of the developing human cerebral cortex, Huttenlocher (1994) concluded that whereas the initial overproduction of synapses is genetically determined, the selective survival of synaptic connections is dependent on use. Thus, the growth and development of the normal neurological substrate of behaviour not only defines, but is defined by the function it subserves. And in this respect, Huttenlocher suggested that the developmental time course of synaptogenesis for both the right and left cerebral hemispheres is the same. Hence, the considerable overproduction of synaptic connections during infancy not only allows for a fine tuning of the anatomical substrate to accommodate the environmental demands characteristic of normal development, but it also provides an effective inbuilt survival mechanism to minimize or reduce the effects of any brain damage sustained at a time when the infant is most vulnerable to injury or disease (Huttenlocher, 1992).

Neurophysiological asymmetries

(a) The electroencephalogram

The most prominent feature of the electroencephalogram (EEG) that is characteristic of the normal, awake, and resting subject is the alpha rhythm, and a neurologically oriented view of its nature and origin is the thalamic pacemaker model proposed by Andersen and Andersson (1968), although Lippold (1973) has provided cogent evidence for an alternative concept based on myogenic origins. Both explanations have their merits and the question remains controversial (Bell, 1972; Nunez & Katznelson, 1981).

Certainly, the normal electrical rhythms recorded from the surface of the intact human scalp are known to reflect or be affected by a wide range of variables. Momentary fluctuations in muscle activity and movement artifacts as well as such subject characteristics as age and the level of sleep or wakefulness are well-documented influences; others, including cerebral dominance have long been postulated (Margerison, St John-Loe & Binnie, 1967). Although several of the earlier investigators reported resting alpha asymmetries between the two sides of the head (e.g. Cornil & Gastaut, 1947; Raney, 1939) such findings have not been confirmed by others (e.g. Glanville & Antonitis, 1955; Provins & Cunliffe, 1972a), and in reviews of these and subsequent investigations, Donchin, Kutas and McCarthy (1977) and Donchin, McCarthy and Kutas (1977) concluded that the association between EEG asymmetries and cerebral dominance was more complex than originally conceived, and the evidence somewhat confused by the wide variation in measurement and analysis techniques employed.

But in spite of improved recording methods and greater experimental control of both wanted and unwanted variables, the evidence in respect of hemispheric asymmetries in the resting EEG record remains inconclusive. Many investigators have reported finding few or no parametric differences between the right and left sides, either in normal adults (Rugg & Dickens, 1982), in newborn infants (Peters, Varner & Ellingson, 1981; Varner et al., 1977), or in children between 8 and 10.5 years of age (Colon et al., 1979). These results have also been supported in a longitudinal study by Benninger, Matthis and Scheffner (1984) who recorded no trend for either a right or left preponderance of activity with age in children between 4 and 17 years of age followed for periods of up to 7 years. Yet others continue to report consistent differences between hemispheres, with higher levels of alpha activity in the right hemisphere for adult subjects (Autret et al., 1985; De Toffel et al, 1990), for children from 6 to 17 years of age - at least in the occipital region (Gasser et al., 1988b), and for the rate of development of EEG activity with age (Thatcher, Walker & Giudice, 1987). In considering the possible reasons for the discrepancies in the findings of different investigators, Butler and Glass (1976) suggested *inter alia*, that the recording of alpha asymmetries in nominally resting subjects (where it occurs) may well be due to individuals engaging unwittingly in some form of silent mental activity, of which the experimenters would be unaware.

Certainly, much of the interest in EEG activity as an indicator of hemisphere specialization has centred around the possibility of inducing lateral asymmetries in simultaneous bilateral recordings by experimentally engaging the subject in tasks assumed to be processed predominantly in one particular hemisphere (see e.g. Gevins, 1983). Thus, bilateral EEG records of subjects undertaking mental verbal activity have been compared with the corresponding records of the same subjects engaged in spatial imagery or a music task (e.g. Galin & Ornstein, 1972; McKee,

Humphrey & McAdam, 1973; Morgan, Macdonald & Hilgard, 1974). These investigators and many others since (see e.g. Butler & Glass, 1987; Galin et al., 1982; Stein, 1988) have found evidence for an asymmetry of EEG activity (primarily in the alpha band) in accordance with conventional wisdom which, in normal right-handed subjects assigns predominance in verbal processing to the left hemisphere and spatial processing to the right. However, these findings have not been without criticism (see e.g. Beaumont, 1983; Donchin, Kutas & McCarthy, 1973; Donchin, McCarthy & Kutas, 1973) and certainly, some investigators (e.g. Gevins et al., 1979; Rugg & Dickens, 1982; Ruoff et al., 1981) who have given special consideration to the non-cognitive aspects of the tasks used, have found little or no evidence to support the concept of an EEG correlate of hemisphere specialization for purely cognitive functions. Indeed, such findings have been interpreted by Yingling (1980) and Gevins et al. (1979) as suggesting that task induced EEG asymmetries only occur when the tasks entail a clear motor component, and Stein (1988) has suggested that the evidence supports his contention that hemisphere specialization should be considered in terms of differences in style of motor output rather than in terms of contrasting perceptual processes.

Nevertheless, other studies that have carefully controlled the motor and sensory aspects of the experimental situation have continued to show significant differences in the concurrent EEG record taken from each side of the head, depending upon the cognitive nature of the task employed (see e.g. Butler & Glass, 1985; Davidson et al., 1990). It is therefore unwarranted to attempt to generalize from the evidence available to date. As well as the nature and level of difficulty of the cognitive tasks used, EEG asymmetries appear to be influenced by a wide range of behavioral factors including attention and emotion (Ray & Cole, 1985) and sensorimotor activity (Autret et al., 1985; De Toffol et al., 1990), while they are also affected by a number of measurement and analysis variables such as the electrode locations from which recordings are taken and the frequency band examined (Gasser et al., 1988a, 1988b).

(b) Event-related potentials and manual activity

However, the bulk of work undertaken on the cortical electrical correlates of language-related behaviour during the past 25 years or so, has employed the technique of signal averaging event-related potentials (ERP's) derived from the cerebral processing of auditory and visual speech and non-speech stimuli (see e.g. Mateer & Cameron, 1989; Molfese, 1983). And as in conventional EEG studies relating to hemisphere specialization, critical surveys of the development of work in this field have emphasized the many methodological problems that provide difficulties in interpreting the experimental results (see e.g. Donchin, Kutas & McCarthy, 1977; Donchin, McCarthy & Kutas, 1977; Rugg, 1983; Rugg et al., 1986). Although much of the work has been concerned with examining the cerebral activity related to perceptual and cognitive processing of verbal compared with non-verbal material, of more direct interest to the present enquiry are those investigations which have recorded the brain potentials associated with speech production and manual movements.

Cortical potentials characteristically associated with the initiation of voluntary movements have been identified and named somewhat differently by different investigators (Tamas & Shibasaki, 1985). However, two potentials which appear to be the most robust and readily recognized features of the cortical waveform and most widely researched, have been described by Deecke, Scheid and Kornhuber (1969) and Deecke and Kornhuber (1977) as the Bereitschaftspotential (BP) or Readiness Potential (RP), and the Motor Potential (MP). These authors found that whereas the BP is characterized by a slowly increasing surface negativity beginning bilaterally about 850 msecs before a unilateral movement, the MP features a sharp increase in negativity about 50 msecs prior to the movement and is primarily restricted to the precentral area of the cerebral cortex on the contralateral side. For right-handed subjects, the authors reported that in their study of voluntary isolated unilateral movements of the index finger, both the BP and MP were significantly larger over the dominant hemisphere, although this difference was not found for left-handed individuals. Somewhat similar results were reported by Kutas and Donchin (1974, 1977) for voluntary unilateral squeezes of a hand dynamometer. These authors found that the RP was consistently larger over the hemisphere contralateral to the movement for right-handers using either hand, and to a lesser extent for left-handers when they used their right hand but not when they used their left. They also reported that increasing the force of the unilateral hand squeeze increased the overall size of the cortical potentials recorded but had no significant effect on the hemispheric asymmetry. Subsequently, other authors (e.g. Bashore et al., 1982; Papakostopoulos, 1980) confirmed that the amplitude of the BP is greater over the hemisphere contralateral to a voluntary unimanual movement irrespective of the handedness of the subject.

Comparing unilateral movements of the index finger of the right hand with simultaneous movements of the fingers of both hands in right-handed subjects, Kristeva et al. (1979) also found an amplitude asymmetry of the BP favouring the left (i.e. contralateral) hemisphere preceding unilateral movements, but surprisingly, an asymmetry favouring the right (non-dominant) hemisphere prior to bilateral movements. Kristeva and Deecke (1980) repeated this experiment with left-handed subjects and compared unilateral movements of either the right or left index finger with simultaneous movements of both. The authors found a contralateral preponderance of BP amplitude for both the left and right unilateral movements but a symmetrical BP for the bilaterally simultaneous movements. The authors concluded that whereas the initiation of finger movements in the non-dominant hand of right-handed subjects engaging in bilateral activity seem to require more effort than equivalent movements of the dominant hand, the absence of BP asymmetry for similar movements by nominal left-handers is in keeping with the view that such people are less lateralized or more ambidextrous than dextrals. Examining the MP (rather than the BP) and other components of the cortical potentials preceding unilateral finger movements in a group of right-handed subjects, Tarkka and Hallett (1990) have reported that although the amplitudes of the potentials they measured did not differ between the dominant and non-dominant hands, the latencies were significantly shorter for the dominant hand. This suggested to the authors that a significantly earlier cortical preparation time was needed for movements of the non-dominant hand.

Several investigations have been carried out to determine if any changes in movement-related potentials occur with age, and two features in particular appear to be of special interest to the present enquiry. The first relates to the monitoring of eye movements (as unwanted or associated movements) in experimental situations where subjects were required to engage in self-paced voluntary manual movements (e.g. squeezing a hand dynamometer) and to avoid blinking or eye movements during the process. The second feature relates to the shape and polarity of the cortical potentials preceding the execution of such voluntary manual movements. With regard to eye movements, Chisholm and Karrer (1983) and Chisholm, Karrer and Cone (1984) measured the number of experimental trials which had to be rejected due to contamination of cortical records by eye movement artefacts, and compared these with the number of artefact-free trials, using the ratio between them as an overall index of voluntary motor control for each subject. The authors found that for both non-dominant and bilateral hand movement trials, there was a significant correlation between the index and age of their subjects (ranging from 8 to 19 years of age), and a similar but non-significant trend with age for the index derived from trials with the dominant hand. In other words, if the eye movement index (i.e. relative rejection rate) can be accepted as an independent measure of the successful suppression of unwanted motor activity, the evidence reflects a clear increase in voluntary muscular control with increasing age. In a subsequent investigation by Chisholm and Karrer (1988) using a different manual activity (isolated movement of a designated finger) and four groups of subjects ranging from 5 or 6 years of age to between 24 and 50 years of age, a similar significant effect of age was reported for the eye movement index obtained for both the non-dominant and dominant hands.

With regard to the shape and polarity of the cortical potentials, Chisholm and Karrer (1983) and Chisholm et al. (1984) in their hand-squeeze experiments, and Warren and Karrer (1984a, 1984b) in their thumb-pressing studies, all reported age-related changes in the waveform of the event-related potentials preceding the requested voluntary manual movement. They found that high amplitude positive and negative components of the waveform were characteristic of their youngest subjects, but that this progressively changed with age to the predominantly negative waveform usually recorded in adults. The authors hypothesized that the positive components reflected (at least in part) the effort needed to inhibit the unwanted, irrelevant or extraneous motor activity associated with as yet relatively undifferentiated or unlearned manual movements. However, other investigators (e.g. Kristeva & Tchakaroff, 1986) have expressed only qualified support for such an explanation.

In subsequent reports by Chisholm and Karrer (1986, 1988), the cortical potentials related to specific discrete finger-lifting movements were compared between trials in which associated (i.e. unwanted) movements of other fingers were recorded and those trials in which no associated movements were detected. A between trials comparison of the waveforms averaged over subjects within each of four age groups again indicated a decreasing positivity with increasing age, which was confirmed by a significant interaction between age groups and trial set. Furthermore, Chisholm and Karrer (1988) reported that irrespective of age, the percentage of associated (unwanted) movements was significantly greater for designated finger lifts of the non-dominant compared with the dominant hand.

Whereas much of the earlier work was concerned with the cortical activity preceding single isolated voluntary movements, there has been an increasing recognition of the need to place the movements examined in the context of more sophisticated goal-directed activities. In one such study by Grunewald et al. (1979), the ERP's both before and during unimanual movements were recorded for two types of task using both the right and left hands of right- and left-handed subjects. From their results, the authors described a goal-directed movement potential that extended from before the movement began until the action ended and which appeared to have two functionally differentiated features. One was a lateralized component restricted to the precentral area of the hemisphere contralateral to the hand used and specifically related to the execution of the movement itself. The other was a bilaterally symmetrical component that appeared to be dependent on the relative difficulty of the task or the particular effort needed in using the non-preferred compared with the preferred hand - especially in right-handers (see also Deecke, 1987, p.249).

A somewhat complementary investigation by Schreiber et al (1983) recorded the cerebral potentials associated with right-handed subjects (a) writing their own signature, (b) drawing a pentogram, and (c) scribbling. The authors found that the Bereitschaftspotential had its earliest onset over the supplementary motor area (SMA) and that it began some 3 secs before starting to write, 2.5 secs prior to drawing, but only 1.5 secs preceding scribbling. Furthermore, they reported that for the frontal lobes, there were statistically significant hemisphere differences in the size of the BP, with the potential larger over the left hemisphere prior to writing but larger on the right side during drawing.

These results, together with those from subsequent studies of simultaneous and sequential bilateral manual movements (e.g. Benecke et al., 1985; Lang et al., 1988, 1989, 1990) suggest an important role for the SMA in the preparation, initiation and timing of serial motor activity. There are, of course, difficulties in interpreting the significance of event-related potentials preceding complex voluntary movements as described by Rohrbaugh et al. (1986, p.219-222). And Kornhuber et al. (1989) have discussed at some length the problems associated with drawing inferences from the spatio-temporal distributions of neural activity at both the cortical and sub-cortical levels in relation to the role of specific structures in the initiation and execution of volitional actions.

Certainly, much progress has been made, but as Jung (1984) has pointed out, further clarification of the basis for handedness classification itself - especially for left-handers - is a prerequisite in neurophysiological studies designed to investigate the cerebral lateralization of control of manual skills. Clearly, such clarification is essential in any attempt to understand the enigmatic relationship which appears to exist between handedness and the hemisphere dominant for speech. But what of the cortical potentials preceding speech? Are asymmetries also detectable here?

(c) Event-related potentials and speech

Ertl and Schafer (1967) were the first investigators to report evidence for a cortical ERP preceding speech although they did not record from both sides of the head. However, they later (Ertl & Schafer, 1969) published data suggesting a possible contamination of their record by muscle activity emanating from the upper lip region, and hence concluded that the earlier evidence must be considered equivocal. Shortly afterwards, McAdam and Whitaker (1971) also reported an association between ERP's and language production processes in the intact normal human brain. They found that whereas spitting and coughing gestures were correlated with bilaterally symmetrical cortical potentials, spontaneous word production was preceded by slow negative potentials which were significantly greater over Broca's area in the left hemisphere. Although Morrell and Huntington (1971) were initially critical of these results, in a later study (Morrell & Huntington, 1972) they essentially confirmed McAdam and Whitaker's findings, but they reported that in the majority of cases, hemisphere differences were small. Grabow and Elliott (1974) also reported no clear hemispheric asymmetries and found great difficulty in unambiguously distinguishing between ERP's preceding speech and a wide variety of possible sources of error. Nevertheless, examination of many such sources of contamination by Grozinger, Kornhuber and Kriebel (1975) convinced these authors that the evidence still favoured a lateralization of such preparatory cortical potentials.

Other evidence supporting an association between asymmetries in the ERP preceding language production and the hemisphere dominant for speech (determined by other means) has been presented by Low, Wada and Fox (1976) and Low and Fox (1977). These authors reported that in epileptic patients who were given bilateral amobarbital testing for hemisphere speech dominance prior to surgery, the results for 12 of the 15 patients were correctly predicted by asymmetries of the ERP's. These authors also recorded the ERP's for cued speech production in 43 normal subjects, and reported a significant difference in the temporal potentials recorded on the two sides, with greater negativity occurring opposite the subject's preferred hand. But they noted that not all recordings showed this relationship, and within subjects there was considerable variability in both direction and degree of asymmetry of the slow potentials between the right and left hemispheres.

Also of special interest here is the investigation by Levy (1977) who intensively studied 8 (ultimately reduced to 4) right-handed normal subjects on six spoken utterances of varying complexity. He reported that there were not only reliable lateral ERP differences in which the left hemisphere potentials were relatively more negative than those at homologous sites on the right, but that this was most evident in (although not completely limited to) the inferior frontal electrode location. Furthermore, he found that the sequentially complex articulations which were judged by subjects to be least like language were associated with reliable hemispheric ERP asymmetries, whereas a monosyllabic language-like utterance was not. The author suggested that the data were thus consistent with the notion that Broca's area is involved in high level motor programming ("macro-sequential assembly") of the articulatory mechanism which is not necessarily confined to a verbal versus non-verbal distinction usually associated with cerebral specialization.

Nevertheless, other studies have questioned the validity of such results. For example, Michalewski, Weinberg and Patterson (1977), Szirtes and Vaughan (1977a, 1977b) and House and Naitoh (1979) all failed to find handedness related hemispheric asymmetries in their ERP's for either speech or non-speech utterances, and Szirtes and Vaughan also provided evidence suggestive of a strong EMG contribution from cheek and other facial areas to the recorded cortical potentials. Similarly, Brooker and Donald (1980) who averaged the potentials from various muscular locations associated with speech production as well as other possible sources of artefact, also found heavy contamination of the EEG record by muscle potentials from the face and scalp. But more importantly, they reported that all their significant EEG lateralization effects had a corresponding EMG effect and that when this was removed by analysis of co-variance, the EEG asymmetry also disappeared.

In a study which overcomes some of the problems of scalp recording, Fried, Ojemann and Fetz (1981) recorded ERP's directly from the exposed cerebral cortex of the dominant hemisphere (as determined by prior amobarbital testing). Sites surrounding the Sylvian fissure or in the frontal lobe from which recordings were taken revealed two simultaneous potential changes to the cued silent naming of presented visual objects. One was a slow potential shift (occurring at motor or premotor sites) that was similar to the potential change recorded during spontaneous overt vocalization, and the other was an alteration in character of the electrical activity (occurring at sites posterior to the Sylvian fissure) that appeared as a flattening of the overall record corresponding to desynchronization of the EEG seen in behavioural states associated with arousal. Eliminating the phase-related association between respiration and the beginning of speech has also posed special difficulties (Grozinger et al., 1977, 1980), although they appear to have been successfully overcome by Deecke et al. (1986). These authors recorded the cerebral potentials preceding speech in 35 healthy young right-handed subjects who began holding their breath at irregular intervals prior to initiating each utterance. In these experiments, it was found that the Bereitschaftspotential began bilaterally approximately 2 secs before the onset of speech although it became significantly lateralized towards the left hemisphere during the last 100 to 200 msecs of the foreperiod. The authors concluded that their results were compatible with the view that both hemispheres are intimately involved in the preparatory organization of any voluntarily initiated movement, and that the late development of an asymmetry favouring the left hemisphere corresponds to the activity of some final common motor mechanism involved in the production of speech.

(d) Regional cerebral blood flow

However, recording the electrical potentials of the brain is but one of several physiological methods of assessing changes in cortical activity associated with behaviour. Employing a radio-isotope clearance technique to determine regional cerebral blood flow changes with mental activity in both normal and clinical cases, Ingvar and Risberg (1967) recorded a significant general increase in the cerebral blood flow and a significant redistribution of flow between particular cortical regions. Since then, many other studies with similar objectives have been published using similar techniques (for relevant reviews see e.g. Ingvar, 1983; Risberg, 1980, 1987). In the normal resting subject, many authors have commented upon the remarkable similarity in the regional cerebral blood flow patterns in the two hemispheres (e.g. Halsey et al., 1979; Larsen, Skinhoj & Lassen, 1978; Risberg, 1980; Roland et al., 1982), although others have reported differences between the two sides, either in blood volume (Carmon et al., 1972) or blood flow measures (Prohovnik, Hakansson & Risberg, 1980) in favour of the non-dominant hemisphere. But what are the effects of undertaking unimanual movements or initiating speech?

In the first such study on manual activity carried out by Oleson (1971), large increases in cerebral blood flow were reported to occur (primarily in the cortical hand area) during vigorous efforts to open and close the contralateral hand. Similar efforts by the ipsilateral hand appeared to have much less effect, although the measures of contralateral and ipsilateral activity were made on different subjects. Ingvar and Schwartz (1974) and Roland and Larsen (1976) have also reported significant increases in the cerebral blood flow of the motor cortex associated with rhythmic squeezes of a rubber ball by the contralateral hand, but the most pertinent evidence on the question comes from two studies by Halsey et al. (1979, 1980). These authors recorded the simultaneous bilateral hemisphere blood flows in both right- and left-handed normal subjects during a unimanual finger movement task which involved repetitively opposing the thumb to each finger in succession. For the right-handed subjects it was found that during left finger movements, highly significant increases in regional blood flow occurred in the upper Rolandic region of the contralateral hemisphere, but not for finger movements of the right hand. For left-handers, approximately half of the subjects presented a mirror image of the findings for the right-handers, i.e. finger movement of the right hand produced a greater and more localized blood flow increase in the contralateral hemisphere than did the left hand. For the other left-handers, no significant asymmetries in cerebral blood flow accompanied either left or right hand movement. The authors concluded from their data that the greatest cortical vascular effect appeared to be associated with the activity of the non-dominant hemisphere in producing movements in the more awkward hand.

Following a demonstration of the likely role of the supplementary motor area of the cerebral cortex in the programming of voluntary movements (Orgogozo & Larsen, 1979), Lauritzen, Henriksen and Lassen (1981) studied the regional cerebral blood flow during writing and unimanual finger tapping movements in 16 normal healthy subjects. They found that during these movements, there was a bilateral increase in blood flow in the supplementary motor area and an increase in the hand area of the contralateral primary sensorimotor cortex. And in a lengthy series of investigations by Roland and his colleagues (see e.g.

Roland, 1985; Roland et al., 1980, 1982), these authors examined the regional cerebral blood flow in both clinical patients and normal healthy young volunteers during the planning, mental rehearsal, and execution of a wide range of voluntary movements. From their results, they were able to distinguish between the involvement of three cortical areas in particular - the primary motor area, the premotor area, and the supplementary motor area, as well as certain sub-cortical regions. They found that whenever unimanual voluntary movements involving programming were undertaken, the supplementary motor area was active bilaterally, irrespective of whether the movements were executed or just mentally rehearsed, although the degree of participation of this cortical area was less in the absence of actual movement. By contrast, the primary motor area was not activated during mental rehearsal but only when movements were actively produced, and then only in the hemisphere contralateral to the unimanual movement. The authors also found that the premotor area tended to be active whenever the supplementary motor area was active although the reverse did not occur.

Studies of cerebral blood flow in relation to speech production have also examined regional cortical changes although the initial research was confined to measuring one hemisphere at a time. In the first such investigation, Ingvar and Schwartz (1974) found that during speech, the pattern of regional blood flow to the left hemisphere in 10 right-handed, neurologically normal patients changed markedly from that recorded during rest. These developments took the form of an increase in flow to the premotor, middle, and lower Rolandic as well as the anterior and middle Sylvian areas, approximating to the upper speech cortex identified by Penfield and Roberts (1959), the hand-arm-face-tongue area of the sensorimotor region, and the anterior (Broca's) speech cortex, as well as the mid-Sylvian area, thus presenting (on the surface of the left hemisphere) a Z-like configuration of augmentation. However, there were no significant changes in the *mean* hemisphere blood flows for the 10 subjects during speech compared with their resting values. Similarly, Larsen et al. (1977,

1978) reported that in 18 right-handed patients they examined, a non-significant 3% increase was recorded in the mean hemispheric blood flow to the left hemisphere during speech in those 9 subjects whose measurements were made on the left, although a significant 10% increase was recorded in the right hemisphere accompanying speech in the other 9 subjects who were measured only on the right. Notably however, in 7 subjects who were measured a second time (3 left and 4 right hemisphere), the mean values during rest were less on the second occasion by an average 6%, and during speech by an average 10%. Nevertheless, these authors found that the regional redistribution of the cerebral blood flow within each hemisphere during speech was similar to that reported by Ingvar and Schwartz except that no increase was detected in the anterior Sylvian region corresponding to Broca's area. The pattern of changes noted in the right hemisphere corresponded closely to those recorded on the left side although they were somewhat less well defined on the right.

Later studies using simultaneous bilateral recordings of cerebral blood flow have largely confirmed as well as extended these findings. For example, Ryding, Bradvik and Ingvar (1987) examined 15 patients during periods of (a) rest, (b) speech production, and (c) humming a tune, and recorded a highly significant increase in mean hemisphere blood flow in both hemispheres during speech and humming compared with resting values. Irrespective of condition (i.e. rest, speech or humming), there was a high degree of symmetry in the regional blood flow pattern with consistently higher values for the frontal regions in both hemispheres. Nevertheless, some asymmetries during speech were also recorded, with a notably greater flow in the precentral (motor) region and Broca's area of the left hemisphere, in contrast to an increase in the postcentral (sensory) cortex and a motor region corresponding to laryngeal activity in the right hemisphere. But overall, the authors concluded from their results that the two hemispheres are both intimately involved in the speech production process and in this respect, at variance with the classical concept of an exclusively dominant (and usually left) hemisphere. Formby,

Thomas and Halsey (1989) arrived at a similar conclusion from their investigation of 12 music (voice) students, 13 college choir members, and 12 other subjects with no formal music training. All subjects had their cerebral blood flows measured from corresponding regions simultaneously over the two cerebral hemispheres while resting and during three different experimental conditions in which they either recited, sang, or hummed the national anthem. The authors reported that across all three groups of subjects, the overall changes as well as the within and between hemisphere variations in blood flow for the various tasks typically were within 5% of resting blood flow values and that none of the differences were statistically significant.

In a series of experiments using positron emission tomography (PET), Petersen et al. (1988. 1989) monitored cerebral blood flow changes in normal subjects during visual and auditory presentation of verbal material, and during speech output with and without the immediate requirement for semantic association. The brain areas activated by each level of task were identified by direct subtraction of the PET images obtained during paired behavioural states (i.e. task minus control), enabling the authors to distinguish between those cortical regions specifically active during sensory processing and those areas active during speech output irrespective of the sensory (input) channel. They reported that during speech, focal activity was recorded in the supplementary motor area of the medial frontal cortex and in the mouth region of the sensorimotor cortex bilaterally, with further activity to be seen in regions of the left inferior frontal lobe (including a site buried in the Sylvian sulcus) close to the traditionally defined Broca's area. Additionally, in the right hemisphere, activity was also recorded in a region of the lateral Sylvian cortex which the authors suggested may be the homologue of the buried Sylvian site identified in the left hemisphere. In a subsequent PET-scan study on normal human subjects, Wise et al. (1991) found that a significant increase in cerebral blood flow could also be recorded from the left premotor and prefrontal cortex (including Broca's area) and the supplementary motor area during a non-vocalized attempt to generate words (verbs).

In reviewing the evidence from regional cerebral blood flow studies and similar metabolic investigations of cortical motor functions, most authors have emphasized the integrated appearance of the hemispheric activity they have recorded. For example, Lassen and Larsen (1980) and Roland (1985) have stressed that undertaking a motor task (manual or linguistic), involves the activation of a number of different cortical regions, the pattern of which varies both within and between hemispheres. These pattern variations appear to depend on the type and degree of sensory control of the movement as well as its programming complexity. Nevertheless, for most motor activities studied, a remarkable similarity in the pattern of regional cortical involvement has been recorded between the two hemispheres, with differences usually being a matter of degree. The principal exception to this finding relates to the primary motor area where, in unimanual movements, only contralateral activity has been reported (Roland, 1987).

Behavioral (dual-task) studies

Non-invasive, non-clinical behavioural evidence relating to some common or shared asymmetric cerebral mechanism for the production of speech and manual movements is based on the concept of a limited information processing capacity in the human decision-making system. This originated from observations on the intermittent nature of responses in continuous behavioural tracking tasks and speculations as to their cause (Craik, 1947, 1948), which stimulated a wealth of experimentation and theoretical modelling in a variety of perceptual-motor contexts (see e.g. Broadbent, 1957, 1958; Welford, 1952, 1968). Kimura (1961a, 1961b, 1967) later employed the Broadbent dichotic listening technique as a possible means of assessing the cerebral lateralization of perceptual verbal processing, and Kinsbourne and

Cooke (1971) subsequently developed the dual-task paradigm as a similar behavioural method to examine the likelihood of some shared hemispheric process in the execution of linguistic and manual skills. It was hypothesized by these latter authors that if right-handed subjects have speech production processes primarily localized in the left cerebral hemisphere, the degree to which manual motor activity and speech share a common organizing mechanism may be tested behaviourally by comparing the extent to which each activity interferes with the other when both are carried out at the same time.

In this first experiment (Kinsbourne & Cooke, 1971), subjects were required to balance a dowel rod on either the right or left index finger for as long as possible while simultaneously repeating orally a previously heard sentence presented by the experimenter. The manual balancing performances of the subjects recorded under these conditions were compared with their corresponding results obtained during trials in which no concurrent verbal activity was required. The authors reported that in the non-verbal condition, the right hand performed better than the left but that this advantage disappeared in the concurrent verbal situation. In a subsequent series of four experiments on both right- and left-handed subjects, Hicks (1975) used the same manual task but varied the vocal demands of the situation, and found that whereas concurrent verbalization only interfered with the right hand balancing performances of right-handed subjects, it affected the performances of both hands in left-handers. Humming instead of talking also produced the same interference effects, suggesting that the verbal content *per se* of the concurrent task was not a critical feature. However, a similar investigation by Johnson and Kozma (1977) comparing the effects of concurrent humming and concurrent speech with a control (silent) condition on dowel-balancing for both male and female right-handed subjects showed no significant effect for humming. As expected, speech significantly interfered with right hand balancing performance for the males, but had no effect on the manual achievements of either hand for the female subjects.

And in another investigation on male subjects, Majeres (1975) reported no significant interference on dowel balancing with either hand due to concurrent vocal activity.

Using both a bimanual and unimanual sequential finger tapping situation with right-handed subjects, Hicks, Provenzano and Rybstein (1975) found that concurrent vocalization had significantly greater effect on manual performance than silent rehearsal of the same verbal tasks, and that it tended to interfere more with the performance of the right hand than the left. However, they further reported that increasing the difficulty of the verbal task (lists of letters) in terms of a decreasing approximation to English, increased the degree of interference recorded on the manual task independently of the hand used. A similar but complementary study was reported by Hicks et al. (1978) in which the effects of concurrent speech and concurrent humming were compared with a control (silent) condition on the performance of a unimanual finger sequencing task employing three different levels of fingering difficulty. Concurrent speech or humming significantly interfered with the manual performances of both hands, but this was greater for the right hand than for the left and the effect increased linearly with increasing finger sequencing difficulty.

In a series of three experiments designed to investigate the effect of further variations in the vocal and manual skills employed, Lomas and Kimura (1976) also provided qualified support for the earlier evidence. For example, in a repetition of the dowel-balancing situation in which assessments were made of the interference produced on balancing times by either orally repeating a well-known nursery rhyme or by uttering a non-speech vocalization of a presumably equally well-known song in comparison to a control (silent) condition, the authors found that non-speech vocalization significantly interfered with balancing performance by both left and right hands compared with the control situation, but not in comparison with speaking. In an attempt to control more rigorously the form of the manual movements, Lomas and Kimura examined the effect of the same three concurrent linguistic task conditions (i.e. speaking, non-speech vocalization, and silence) on the sequential finger tapping performances of each hand for both right- and left-handed subjects. They found that for right-handers, speaking significantly interfered with finger sequencing of the right hand only, and that there was no effect of non-speech vocalization. However, for left-handers, both conditions significantly interfered with the performances of both hands. To examine other specific movement variables, the authors' third experiment tested the effects of the same three linguistic conditions on gross arm sequencing performances and the single (index) finger tapping rates of each hand in a group of right-handed subjects. The results showed that while speaking significantly interfered with performances on the arm sequencing task, and more so for the right side, it only interfered with single finger tapping bimanually. As Lomas and Kimura themselves observed, their study raised rather more questions than it answered, although they suggested that their results did support the concept of an overlap of the neural output mechanisms for the production of speech and manual movement sequences of the right hand in right-handers.

Certainly, the accumulating evidence indicated that the problem was rather more complex than originally envisaged. McFarland and Ashton (1975a) for example, suggested that more adequate or suitable controls were needed to avoid ambiguities in the interpretation of findings, and that laterality research required performance measures applicable to the presumed specialized processing capacities of both the left and right cerebral hemispheres. They subsequently published the results of a series of experiments in which the effects of concurrently engaging in a wide variety of verbal and non-verbal cognitive tasks were tested on the speed or consistency of unimanual finger tapping performances with the right and left hands in right-handed subjects (McFarland & Ashton, 1978a, 1978b, 1978c). These authors also introduced measures of the effects of concurrent manual activity on the performances of the verbal and non-verbal tasks. In general, their results showed significant interference with right hand performances by verbal tasks and significant interference with left hand performances by non-verbal tasks. Similar disruptive effects were reported for the verbal tasks by right hand activity and for the non-verbal tasks by concurrent use of the left hand. Thus, the authors cautiously concluded that their evidence supported the viability of an interpretation favouring either an extended Lomas and Kimura (1976) overlapping neural mechanism explanation, or an attentional bias concept such as that proposed by Kinsbourne (1973b).

Shortly afterwards, Kinsbourne and Hicks (1978) attempted to pull together the rapidly growing body of evidence from dual-task experiments into a more comprehensive explanatory model by employing the concept of "functional cerebral distance". They proposed that the degree of interference recorded between the concurrent performances of two specific tasks was inversely related to the "functional distance" between their respective cerebral control centres. They suggested that the effective functional distance was dependent not only on the degree of spatial separation between the two foci of cerebral activity corresponding to each control centre, but also on the extent of the neural connections between them - which could be either intra- or inter-hemispheric. And in a later elaboration of the model by Kinsbourne (1981) and Kinsbourne and Hiscock (1983), the idea of cortical inhibitory barriers was introduced to provide for the ability of the individual to counter interference effects by learning to segregate the controlling processes. Kinsbourne and Hiscock pointed out that the application of the model had three possible outcomes: (a) if the timing of the critical events in the two concurrent tasks were predictable, then sequential rather than simultaneous cerebral processing would reduce or avoid interference; (b) if the individual learned to segregate the tasks by establishing an inhibitory barrier between the two processing foci, then again interference between task performances would be minimized or eliminated; or (c) if (a) and (b) were inapplicable or impossible, then interference would occur, the magnitude of which would depend on the particular values of a wide range of task and other variables.

Since then, attention has been drawn to the problem of whether dual-task effects are more related to hemisphere speech or manual dominance by the findings of Orsini et al. (1985) in their study of a large sample (257) of left-handed subjects as well as 215 right-handers. These authors found that concurrent performance of finger tapping with either a verbal fluency task or silent reading produced a greater decrement in finger tapping with the dominant hand compared with the non-dominant hand regardless of subject handedness. While this conforms to expectation for right-handers, the results are somewhat surprising for left-handers, since, as Simon and Sussman (1987) point out, clinical studies suggest that only a minority of left-handers have unambiguous right hemisphere control of speech. Accordingly, these latter authors also tested a relatively large number (120) of left-handed subjects in addition to 140 right-handers on the effects of concurrently performing each of three different speech tasks on the speed of unimanual finger tapping by the right and left hands. Again, it was shown that the performance decrement in tapping speed recorded for the dominant hand was greater than that displayed by the non-dominant hand regardless of subject handedness. Similar results have also been reported by van Strien and Bouma (1988) in their dual task investigation using 24 right- and 60 left-handed subjects. They found that the effects of concurrent speech on speed of unimanual finger tapping with the index finger produced a greater decrement in performance with the preferred compared with the non-preferred hand for both right- and left-handers. However, concurrent speech with sequential tapping of the fingers of one hand produced the same amount of performance decrement for the preferred and non-preferred sides irrespective of handedness. More recently, Kosaka et al. (1993) studied 22 epileptic patients whose hemisphere speech dominance had been determined by amobarbital (18 left, 4 right), and whose handedness assessments (by inventory) classified only 2 of the subjects as left-handed (both with left hemisphere dominance for speech). The authors found that patients with left hemisphere speech recorded a greater decrement in repetitive tapping performances with the index finger of the right hand than with the left during a concurrent reading task, whereas the right hemisphere speech group displayed greater interference with tapping performances of the left hand compared with the right. Since all of the right hemisphere speech subjects were right-handed, Kosaka et al. inferred that the interference effects are more closely related to hemisphere dominance for speech than for manual functions. However, the authors suggested that the findings be treated with caution since the work was conducted on clinical patients and the sample size was small - particularly with respect to left-handers.

Certainly, many relevant variables and methodological difficulties with the dual task paradigm have been identified (Green & Vaid, 1986), and the importance of output timing for the successful integration of concurrent vocal and/or manual activities has also been stressed by Peters (1990) and Summers (1990). Peters has further highlighted the difficulty in obtaining unequivocal evidence from dual task experiments on the cerebral lateralization of speech and handedness due to both the lack of an agreed handedness classification system, and to the influence of learning in the "automatization" of complex motor sequences. Certainly, an understanding of the nature of handedness and of the role of learning in the development of motor skills is critical, and a recurring issue, since it affects the interpretation of evidence regarding handedness and motor skills from whatever source. But of more immediate relevance here is the attention given to questions of handedness and learning in studies of developmental trends in dual task performances.

Initial experiments on five year old children by Kinsbourne and McMurray (1975) showed that rate of finger tapping by the right hand was reduced more than that for finger tapping by the left hand when unimanual tapping was carried out concurrently with orally reciting or reproducing verbal material. Hiscock and Kinsbourne (1978) repeated the experiment on children between 3 and 12 years of age, and Obrzut et al. (1980) carried out a corresponding investigation of normal and learning disabled children between 7 and 11 years of age. Both studies reported similar laterally asymmetrical interference effects on tapping with concurrent performance of the vocal task and no developmentally different trends for the right and left sides. But whereas Hiscock and Kinsbourne found a reduction in the interference effects on tapping rates for both hands with age, Obrzut et al reported no such reduction over their more restricted age range, while the normal subjects performed better than the learning disabled group under all conditions. Evidence from similar experiments by McFarland and Ashton (1975b), Piazza (1977), and White and Kinsbourne (1980) in which the effects of concurrent speech were compared with the effects of concurrent non-speech activity on the rate of finger tapping in children, also produced results similar to those found for adults, with again no developmental trend for differences between the two sides. However, McFarland and Ashton noted that in their results there was evidence to suggest that variability rather than rate of tapping may be more sensitive to age effects.

In another investigation, Hiscock and Kinsbourne (1980) again tested right-handed children between 3 and 12 years of age on the same tapping and oral recitation task as reported previously (Hiscock & Kinsbourne, 1978) and obtained much the same results, namely, that concurrent vocalization interfered with right hand tapping more than left hand tapping and that the performance difference between the hands did not vary with age. However, since most of the children had participated in the previous investigation one year earlier, a direct comparison between their performances on the two occasions provided suitable data for an assessment of test-retest reliability. The authors found that although the decrement ratios for both hands on the tapping task due to concurrent vocalization yielded low but significant reliabilities, the decrement ratios for the differences between hands on the two occasions were not significantly correlated. Since there must be considerable doubt about the possibility of obtaining reliable finger tapping scores (especially for the differences between hands) in very young children anyway (see Chapter 6), Hiscock and Kinsbourne's results in this respect are not surprising. In a further experiment by Hiscock (1982), right-handed children from grades 3, 4 and 5 with mean ages of 8.8, 9.9 and 11.1 years respectively, were tested on a finger tapping task while repeatedly reciting aloud a well-known tongue-twister. The author measured both tapping rates and verbal production and found that whereas speaking interfered with right hand tapping more than left hand tapping, both hands produced similar disturbances of speech. But more notably, a significant between-hands concurrent interference was reported which increased with increasing grade level. And in a subsequent investigation of the effects of verbal and non-verbal tasks undertaken concurrently with a unimanual test of tapping speed in young (5 to 6 year-old) and older (11 to 12 year-old) groups of right-handed deaf and hearing children, Ashton and Beazley (1982) reported a significant effect of age on the manual asymmetries recorded. They found that the right hand tapping rates were significantly more adversely affected by concurrent verbalization than those for the left hand, and more particularly so for the older children in both deaf and hearing groups.

However, in a further study of right-handed children drawn from grades 1 to 4, Hiscock, Kinsbourne et al (1985) found that whereas vocalization produced a relatively constant asymmetrical disturbance of rate of finger tapping with age, concurrent activity interfered significantly more with right than left hand manual performance in grade 1 children, but not at any other age level when tapping variability was used as the criterion. A later study by Hiscock et al. (1987) on children from grades 1 to 4 confirmed these results with respect to a differential side effect of concurrent speech on rate of finger tapping and age, and provided evidence to show that speech interfered with manual performance more than concurrent participation in memory encoding activity.

But not all investigators using children as subjects have reported laterally asymmetrical interference effects on speed of finger tapping due to concurrent vocalization. Hughes and

Sussman (1983) tested a language disordered and a matched normal group of right-handed children between 4 and 7 years of age and reported that although the normal children tapped faster than the language-disordered subjects, and both groups were better with their preferred hand, there were no significant main effects for type of child or hand in respect of the percentage reduction in tapping rate due to concurrent verbal activity. This contrasts with the findings of a study by Hiscock, Antonuik et al (1985) on children between 8 and 10 years of age in which subjects with good and poor reading were compared for interference effects due to concurrent reading and unimanual finger tapping. These authors reported a significantly greater effect of concurrent reading on tapping rate with the right hand relative to the left for both the subject groups irrespective of age, and a significantly greater asymmetry of interference for the poor compared with the good readers.

But are these findings necessarily irreconcilable? As Hughes and Sussman (1983) and Sussman (1984) remarked, whereas adult studies typically find interference effects of concurrent speech to be confined mostly, if not entirely to right hand tapping performance, the effects for children are much greater overall and quite marked for the left as well as the right hand. It must also be pointed out that results obtained about developmental trends derived from group data in cross-sectional studies can be very misleading (see Chapter 6). Furthermore, although investigators have concentrated their efforts on simple repetitive finger tapping as the most readily controllable and measurable motor test of handedness, the range of manual laterality criteria employed clearly needs to be extended for the dual task evidence to have any convincing generality. But which manual task(s) may be considered as valid measure(s) of an individual's handedness is a fundamental question, and one, which is addressed at length in Chapter 5.

CHAPTER 4

Behavioural Studies in Non-Human Species

Chapter outline

In spite of the very large number of controlled and uncontrolled variables contributing to the results of investigations of animal handedness, there are several features which appear to be well established. Certainly, evidence taken from the species reviewed here indicate that individual animals tend to use the same paw or hand consistently on any given task, although this may vary considerably in the expressed strength of preference. Such individual preferences in an environmentally unbiased situation are usually distributed equally between the two sides so that half the animals in a population prefer the right paw and the other half the left. The percentage of animals demonstrating weak or inconsistent preferences (frequently classified as ambidextrous) varies considerably from one investigation to another, often (but not always) depending upon the arbitrary criterion used to identify right- and left-handers. A plot of the frequency distribution of right and left preferent animals shows that this may vary from U shaped, through square, to inverted U form, although the variables responsible for these differences have yet to be determined.

Side preferences displayed by animals tend to be specific to a particular task or situation inasmuch as inter-task correlations are often positive but low and not statistically significant. However, practice or experience on a particular task or in a given situation improves the consistency of degree and direction of side preference both within and between tasks.

From family data on primates and extensive breeding experiments on both mice and rats, no evidence has been found favouring a genetic determination for the direction (i.e. right or left) of paw preference, although the results of other breeding studies suggest that the degree of side preference (irrespective of direction) may well have some genetic basis. Data from investigations on several different species in which the animal's relationship to its environment was incidentally or deliberately biased to favour the use of one particular side, show that the development of paw or hand preference in a given situation is strongly influenced by such environmental factors.

Asymmetries in a number of both morphological and behavioural characteristics have been described for a wide range of animal species (see e.g. Bradshaw, 1991; Bradshaw & Rogers, 1993; Galaburda, Sherman & Geschwind, 1985; Walker, 1980). However, since most of the work relevant to an understanding of the lateralization of human cerebral mechanisms concerned with motor control have primarily been carried out on mammals and particularly (i) mice, (ii) rats, (iii) cats, and (iv) monkeys and apes, attention here will be focussed on these. Such animals clearly demonstrate purposeful manipulative activity of their forelimbs in a variety of situations involving reaching, eating and climbing behaviour, and in unilateral movements they may also display a preference for the use of one particular side.

Certainly, in controlled laboratory conditions where experimental constraints allow only one forelimb to be used in gaining access to food for example, a consistent preference for the use of the same limb on repeated trials is the usually noted characteristic in respect of any given animal. But there are clear differences between animals in the direction and degree of their limb preferences, and although it is often stated that population distributions of animal preferences are equally divided between right and left, Walker (1980) has been more cautious in his interpretation of the literature. After reviewing the available evidence, he simply concluded that none of the mammalian species in which forepaw preferences had been systematically studied had shown a species-wide preference for one forelimb or the other. Certainly, the proportions of animals displaying right, left, or mixed preferences often vary appreciably from one study to another and a single gross summary of their findings could be misleading. Apart from obvious differences in their manipulative ability and the use of non-comparable test situations and consistency criteria in testing them, such factors as age, experience, and physical condition of the animals have also been shown to affect results.

Mice

There is no doubt that Collins' studies of paw preferences in mice (reviewed in Collins, 1970, 1977b, 1985) have made one of the most significant contributions to the experimental evidence available on animal handedness, so it is clearly appropriate to begin this part of the review with due consideration of his work. In his first series of experiments, 370 mice from seven different highly inbred strains and 221 animals from two populations of hybrids were tested in a food retrieval situation in which equal access to the food was available for either the right or left forepaw. Each trial comprised observations of 50 paw entries to the feeding tube, and animals were classified as right or left preferent according to the recorded number of right paw entries (i.e. above 25 = right paw preferent, below 25 = left paw preferent). The results indicated that approximately the same number of animals expressed a consistent preference for the right as for the left forepaw. Although the degree of consistency varied between animals, the results for those preferring the left side mirrored those preferring the right to the extent that while the majority showed a relatively high degree of consistency in their preference, some animals were much less consistent, so that the overall population distribution assumed a U shape. Collins (1968, p.11) concluded that since both right and left preferent animals were evident in each set of genetically uniform mice, the heritability of paw preference must be close to zero.

This report was followed (Collins, 1969) by an account of the results of a selective breeding program in which 22 parental pairs of a highly inbred strain of mice were taken from a population classified as right and left preferent according to the procedure already described. Four types of matings were then made in each generation, i.e. R-R, R-L, L-R, and L-L for three successive generations with all progeny tested for paw preference. The results showed that the proportion of right/left preferent offspring was not associated with the type of parental combination, and that the distribution of animals displaying a consistent right compared with left paw preference remained qualitatively and quantitatively the same. In other words, three generations of selective breeding for right and left paw preference had failed to make any impact on the recorded distribution of paw preference characteristics. Similar results have been reported by Bianki (1981).

Collins (1991) followed up this finding with a breeding program designed to select for degree of paw preference irrespective of direction by developing a HI line of animals from matings of mice that consistently displayed strong right or left preferences and a LO line from matings of animals that showed little or no overall paw preference. Selection was continued for eleven generations, and a significantly successful separation of the lines became evident at the third generation with divergence continuing thereafter. Although selective breeding was relaxed at generation 12, random mating within lines was continued until generation 28 when selection was again introduced for 3 generations. On testing at the completion of the program, results showed that the between-line separation of the paw preference behaviour had remained high during the period of relaxation and that paw preferences of the HI line mice were significantly more strongly lateralized, and those of the LO line animals were more weakly lateralized than the paw preferences of an unselected (control) group. In a subsequent study by Collins, Sargent and Neumann (1993), it was shown that the size and form of the differences between the HI and LO lines are unlikely to be due to any differences in heterozygosity of the two strains as suggested by McManus (1992).

In contrast to the negative evidence relating to a genetic influence on the direction of paw preference, Collins (1975, 1977b) showed that when an environmental bias is introduced into the test situation, this has a very marked effect on the population distribution of lateral preferences. The environmental bias was achieved by positioning the feeding tube flush with either of the two side walls of the testing chamber, so that when it was against the right wall it was difficult (but not impossible) for the animal to reach into it with the left paw, and similarly, when the feeding tube was against the left wall, the right paw was disadvantaged. He found that in these circumstances, when the tube was on the right the recorded distribution of paw preferences was strongly biased in favour of the right paw, and when the tube was on the left a mirror-image distribution was recorded in favour of the left forepaw. Thus, the distribution was converted by use of a biased world from a U shaped distribution to a J shaped curve (approximating that seen in human populations) in which only 8.7% to 14.7% of the subjects exhibited resistance to the bias. Collins (1977b) then conducted further experiments in which mice were exposed sequentially to test situations in which the first imposed bias was reversed and the animals subdivided into two groups according to their displayed resistance to adapt to this bias reversal. The adaptability of the two groups was then shown to be significantly associated with the direction recorded for the animals' original paw preferences as displayed in the unbiased world. From this the author concluded that the origin of the paw preferences was not necessarily learned but could be due to a seemingly random genetic process which he proposed in terms of an "asymmetry lottery".

Unfortunately, he presented no data for trends within each series of trials, but statistically significant increases in strength of lateralization between test runs for the same animals were reported from the first 50 trials (test 1) to the second 50 trials (test 2). In the unbiased world, no change in the overall distribution of paw preferences between right and left was recorded for the two runs - out of 709 mice tested 47.6% used the right paw more on test 1, and 47.7% used this paw on test 2. However, if the degree of consistency is considered by examining the proportion of animals using the same paw for at least 48 of their 50 reaches in each test run, then 41,8% of the 709 animals achieved this criterion in test 1, and 62.2% in test 2 (Collins, 1975). Furthermore, the author also presented data to show that in the left-biased test situation, although 12.4% of 218 mice recorded an overall right paw preference in test 1, this was reduced to 8.7% in test 2. Similarly, in the right-biased test situation, whereas 14.7% of 217 mice recorded a left paw preference in test 1, this dropped to 10.6% in test 2. Thus, continuing exposure to the same biased world continued to induce more and more of the animals to change their paw preference and to conform to the direction of the environmental bias. In other words, in an unbiased world, it appears that the direction of lateral preference is equally distributed between right and left, but that continued exposure to the same situation increases the degree of laterality demonstrated. In contrast, exposure to a biased environment induces the majority of animals to exhibit a paw preference favouring conformity with the direction of the bias, and continued exposure to the same situation continues to induce more and more animals to demonstrate conformity. But whether the original paw preferences expressed in an unbiased world were due to some genetically determined asymmetry lottery or to a purely random distribution based on chance and/or any pretest experiences of the animals, must remain an open question. Certainly, the laterality of mice exposed to the unbiased world demonstrated good consistency over time, with statistically significant intercorrelations between the paw preference scores recorded for animals tested daily over a four day period or at monthly intervals.

But to what extent have Collins' findings been confirmed by subsequent investigators? Papaioannou (1972) tested 47 inbred mice in a Collins type unbiased testing chamber on two consecutive days and obtained a similar U shaped distribution of paw preferences, with equal numbers of right- and left-handers, and a correlation of 0.99 between the preference scores recorded on the two occasions. Similarly, Signore et al. (1991a) studied 713 inbred mice (from many different strains) under the same conditions and again reported results conforming to expectation.

They also found a test-retest correlation of 0.89 for mice tested twice for paw preferences at a one week interval.

However, later studies by Signore et al. (1991b) on 450 mice from 11 inbred strains, and Betancur, Neven and Le Moal (1991) on over 1500 mice from three different strains tested in an unbiased world, reported population distribution results for paw preference that did not always conform to the usual U shape. Although their results yielded equal numbers of right- and left-handed mice in each population tested, the distributions varied between strains and sometimes took the shape of a normal curve (with the majority of animals only weakly lateralized), or a square distribution. Takeda and Endo (1993) have also reported similar variations between strains in the degree of lateralization, with the same number of animals favouring their right and left paws. Furthermore, the authors reported that for one of the strains they used (which was the same as one of those tested by Collins, 1968, 1969), their distribution of paw preferences was bell shaped, not U shaped. Although Takeda and Endo could find no simple explanation for their findings, they suggested that the results may be influenced by variations in the incentive value of the food used in the test situation, or the breeding source of the animals. Certainly, some investigators (including Collins) have used flaked or granulated cereal food strewn along the floor of the feeding tube, whereas others (including Takeda & Endo) placed a single food pellet in the tube - a methodological difference that could have practical implications concerning ease of food retrieval, apart from any question of differences in the taste appeal or incentive value of the reward offered. But variations in the findings between studies may also be due to differences in the detail of their paw entry criteria or data recording. For example, Takedo and Endo (1993) reported that the initial behaviour of a mouse on discovering food in the test chamber was to thrust its mouth to the tube opening, and after about 10 to 20 minutes when it found that this was unsuccessful in reaching the food, it started to use one paw at a time. Indeed, Betancur et al. (1991) noted that there were sometimes quite large numbers of animals that were simply nonresponders (as many as 17.8% and 20.2% for two particular strains they tested) - even after four to five sessions, with each session lasting as long as an hour.

But what constitutes a recorded paw entry? Gruber et al. (1991) specify that a paw entry in their study was counted only "when the entire paw was inserted entirely into the tube", and since the tube was *filled* with crunched cereals, there can be no doubt that each recorded reach was one that successfully obtained food. But most authors refer to the use of the right and left paws in making reaches *for* food as the performance criterion, and the reader is left to assume that all such reaches successfully obtained food. Since the rewarding of responses leads to learning, it is clear that the paw used for the very first rewarded reach is more likely to be used again than its opposite number. Consequently, it is of some special relevance to note that in a study of within-session reaches of 306 mice, Signore et al (1991a) found that for 81% of the females and 75% of the males, the paw used for the first recorded paw entry corresponded to the paw ultimately preferred overall. And in the study by Takeda and Endo (1993), progressive individual results presented for 15 randomly selected mice from each of two different strains showed that although considerable fluctuations occurred between right and left paw use during the first ten or so reaches, this soon stabilized to a consistent degree as well as direction of paw preference for each animal for the rest of the session (of 50 recorded reaches). Such trends are highly suggestive of a period of trial and error learning, with the development of consistent use of a particular paw being dependent on the relative frequency of success of the two sides in obtaining food during the first few reaches.

But to what extent is the acquisition of a particular paw preference specific to a particular task or situation? Collins (1970) reported that when mice assessed for paw preference in the usual unbiased test situation were observed reaching for food placed beneath a hole in the floor, there was a correlation of right paw entry scores between the two tests of 0.96. He also found that mice tested for grip strength of each forepaw recorded a significantly stronger grip for the preferred side (as assessed in the unbiased world), suggesting that paw preference may be a stable, generalizable behavioural trait and not confined to a particular test situation. However, other investigators have reported results somewhat at odds with these findings. For example, Waters and Denenberg (1991) tested two different strains of mice in a modified form of the Collins unbiased world and obtained quite different results. In this situation, access to food was provided *ad lib* through two oval holes at the front of a feeding unit designed to monitor the use of the right and left paws respectively over extended periods of time. Although the animals established clear and stable paw preferences in this situation, a comparison of their laterality scores on this test with their paw preference scores on the standard Collins test produced non-significant correlation values of 0.322 and 0.127 for the two subject groups studied. Similarly, when mice that had demonstrated consistent turning preferences in a Y maze (in the absence of any differential cues or reinforcement) were tested for their paw preferences in the Collins unbiased world, Papaioannou (1972) found that the preferences recorded on the two tests were unrelated. Interestingly, Collins (1988) examined the effects of learning in a similar test situation in which observer mice were allowed to watch a trained mouse obtain food by pushing through a door opening to the left or right to gain access to a food compartment. When the observer mice were themselves placed in the test unit, those previously exposed to a left-trained animal also tended to turn left, and those exposed to a right-trained demonstrator turned right. Furthermore, observer animals were shown to perform the initial test trials more quickly than control animals not given the benefit of pretest demonstrations. The author concluded that a right or left behavioural bias could be acquired by mice through observational learning as well as by constrained motor habits.

Specificity of degree of lateral bias has also been tested by Collins (1985) who examined the behaviour of mice taken from his HI and LO bred lines by placing them in a water tank and counting the number of clockwise and anticlockwise swimming rotations they made. He found that the HI line mice displayed a more strongly developed degree of preference for swimming in one particular direction (either left or right) than the LO line animals. Similar testing of HI and LO line mice for aggressiveness (Scott, Bradt & Collins, 1986) showed that although the males of both lines were equally aggressive, the HI line females showed a greater tendency to be aggressive than their LO line counterparts both in terms of number and intensity of incidents.

Thus, in mice, the evidence suggests that the *direction* of paw preference is not subject to genetic control but can be influenced very considerably by an imposed environmental bias. In contrast, the *degree* of paw preference demonstrated in an unbiased world (irrespective of direction) does appear to be influenced by some genetic factor which may also be associated with sex differences and aggressive behaviour. But to what extent are these findings peculiar to mice? Unfortunately, different investigators have tended to work with different species and often with different objectives in mind, so that direct cross-species comparisons are rare. Nevertheless, there is surprisingly good agreement in the general trends shown by the evidence that is available.

Rats

One of the investigators to have made an early and major contribution to knowledge concerning handedness in the laboratory rat was Peterson (1931, 1934) who, together with various collaborators, continued publishing reports on the topic for some thirty years. In his first lengthy series of experiments, Peterson (1931, 1934) used a food-reaching situation which may be described as the prototype of Collins' unbiased world. After classifying animals for handedness on the basis of observation of their paw preferences in this situation, the author examined the possible influence of heredity on the preferred direction of laterality by selective breeding experiments which included mating right-handers with right-handers and left-handers with left-handers for seven generations or more. He found that neither right- nor left-handedness bred true and reported data showing that eight litters of two right-handed parents yielded 11 right-handed, 19 left-handed, and 4 ambidextrous offspring, while ten litters from two left-handed parents produced 32 right-handed, 28 left-handed, and 5 ambidextrous animals. Furthermore, of a total of 153 offspring from the breeding experiments, he found that 46% turned out to be right-handed, 43% left-handed, and 11% ambidextrous, and concluded that a much more comprehensive survey of large numbers of animals would indicate that left- and right-handedness were equally distributed in the general population. The criterion he used for classifying the handedness of animals in this experiment was not based on the counting of individual reaches, but on the observation of consistency of paw use displayed by the animals during two periods of five minutes exposure to the test situation on successive days.

Other early investigators (e.g. Tsai & Maurer, 1930; Yoshioka, 1930a, 1930b) used somewhat different (although also unbiased) food-reaching situations and handedness criteria that tended to yield similar but inconsistent asymmetries in their reported distributions, which Peterson (1934, p.7) concluded could be attributed to the effects of random sampling and the relatively small numbers of animals tested. Wentworth (1942) later combined the results of Yoshioka, Tsai and Maurer, and Peterson with his own data to give information on 584 rats with a distribution of 43.8% right-handed, 44.5% left-handed, and 11.7% ambidextrous in accordance with expectation.

In his own extensive series of experiments, Wentworth (1942) followed up these earlier experiments by attempting to examine the effects of unilateral training. He used testing and training situations that were similar except that access to food in the test situation was centrally positioned in the end wall of a narrow passage so that it was equally accessible by either paw, while in the training situation the food opening was placed in either of the right or left corners of the front and side walls. He reported that in the testing situation, only those reaches that were successful in obtaining food and taking it to the mouth were counted.

In a preliminary experiment, Wentworth (1938) placed 19 rats in the training apparatus (9 when 23 days old, and 10 when 35 days old), where they remained for 14 days (younger group) or 21 days (older group), and obtained all their food by right paw reaches. They were then returned to their normal living cages for 90 days, at the end of which they were given 25 tests per day for 7 days in the test situation. All rats proved to be right-handed, with 15 of the 19 animals making no left-handed reaches at all, and only 2 showing any tendency towards ambidexterity - with 12 and 11 left-handed reaches respectively out of the total of 175 test trials given to each. In Wentworth's (1942) main experiment, 129 rats (which were the offspring of the 19 rats used in his 1938 study) were subdivided into nine groups and the effects recorded for three age levels (25, 50 or 90 days old), of three amounts of enforced training for left paw usage (1, 2, or 7 days), at three intervals of time between completion of training and testing for paw preference (90, 180, or 270 days). The results showed that after training, 112 out of the 129 animals (i.e. 86.8%) displayed a left paw preference in the test situation with little or no effect of age, length of training, or delay in testing. Even 80% of those animals given only one day's training demonstrated a consistent left paw preference. On completion of this experiment, a retraining program was attempted for 68 of the animals which had displayed a consistent left paw preference. These rats were subjected to enforced right paw training for either 1, 2, or 7 days and then tested for paw preference after 60 days post-training. Somewhat surprisingly, 52 of the 68 still retained a left paw preference, 9 displayed a mixed preference and only 7 had changed to consistent right paw use. Hence, it appeared that the original training was a much more powerful influence than the later training.

Conscious of the problem this finding posed for any experimental design in which the initial handedness test itself could produce a directional bias, Wentworth (1942) examined the effect of classifying animals for paw preference on a test of only 25 reaches. In a group of 34 experimentally naive animals, he used a criterion of 24 reaches with the same paw to classify 24 rats as consistently right- or left-handed. These were then given one day of forced training with the non-preferred paw and then retested after 60 days. He found that only 10 (41.7%) had shifted to the other hand, 2 had become ambidextrous and 12 had been unaffected by the intervening training. In comparison with the results from his main experiment where no testing or handedness classification of the animals had taken place before the forced training periods, it appeared that the initial testing session had indeed had a clear training effect. Consequently, Wentworth tested a further group of 105 untrained animals over a series of only 10 reaches, and using a criterion of 9 out of 10 with the same paw, he classified 74 as consistently left- or right-handed. He then subdivided these into an experimental group of 56 which were given enforced training with the non-preferred side for 200 reaches, and a control group of 18 animals that were given no intervening training before being retested 60 days later. Results of the retest showed that 72.2% of the control subjects had retained their initial paw preference, whereas only 26.6% of the experimental animals had retained theirs, with 41.1% displaying a complete reversal of handedness and 32.1% demonstrating ambidexterity.

Peterson (1951) later confirmed the general thrust of Wentworth's findings by examining the effects of forced training on 108 animals using the same training apparatus. Initial tests requiring consistent use of the same paw for 10 out of 10 reaches established initial paw preference and the forced training of the non-preferred side at the rate of 50 forced reaches per day in the training cage was imposed for between 50 and 1,000 forced reaches for different subgroups of animals. Retests were carried out in the unbiased world situation the day after forced practice ended. The results showed that a gradually increasing proportion of animals made a complete reversal of handedness due to the intervening training, ranging from 18% after 50 forced reaches to

67% after 1,000, while changes away from the original preference (i.e. including shifts to ambidexterity) ranged from 22% after 50 forced reaches to 86% after 1,000. In these experiments, Peterson noted that in the training cage, some animals still occasionally attempted to get at the food with the "wrong" paw, and in each case this had been prevented by the observer inserting a spatula over the food opening to interrupt the reach. The author also noted that while the percentages presented in the results gave a good indication of general trends, some individual animals did not conform to the overall picture, leading him to conclude that some influence other than practice was also operating to affect the outcome.

Peterson (1951) further reported the effects of forced training employing a very different method. In this experiment, he used 35 previously untrained animals and assessed their handedness in terms of the percentage use of the same paw during test sessions of 50 reaches per day for 4 to 10 days in the unbiased world. Adhesive tape was then bound around the preferred arm and the rat's thorax so that the arm was held firmly against the body. The binding was kept in place throughout a period lasting from 10 to 29 days, and the animals subjected to forced training of the non-preferred side for a total of between 200 and 950 reaches. A control group of 11 rats also had the preferred side bound for comparable periods of time but without any forced training (presumably these animals had unhindered and direct access to food by mouth). The results were similar to those of the previous experiment in showing a general change in handedness due to the enforced training program, and an increasing proportion of rats shifting away from their original handedness with increasing amounts of training, although there were notable individual exceptions. In contrast, none of the control group changed preferences. Hence, Peterson concluded that it was undoubtedly the enforced training that produced the change and not the binding of the preferred side *per se*.

The experiments of Peterson (1951) and Wentworth (1942) provide a very clear and convincing demonstration of the potential strength of environmental influences and learning in the production of paw preferences in rats. But to what extent are paw preferences determined in one set of conditions specific to those conditions, or are they generalizable to other environmental situations? Peterson (1934) attempted to examine this question by first assessing the handedness of seven rats in the previously described standard unbiased world, and then studying their behaviour in a latchbox test situation. This required the pressing down of a lever above a door in order to open it and gain access to food. Subjects were first given training with a long lever that projected well into the cage, and after mastering this problem, a middle-sized lever was substituted which ended flush with the wire mesh of the wall of the cage. Later, a short lever was used which required the rat to reach through the wire mesh with one paw in order to depress it. This proved to be a much more difficult task, but successful performances by the subjects were recorded six times per day for at least six days. The author reported that the tests clearly differentiated rats into right- and left-handed animals and that in no case was an inconsistent response given. He found that five were right-handed and two left-handed. Of the five right-handers, two were also right-handed on the standard test of paw preference and three left-handed in the standard situation. Of the two left-handers, one was right-handed on the standard test and one ambidextrous. In other words, four of the seven subjects did not show the same paw preference in the two situations. However, in a later study, Peterson (1938) repeated this comparison between latchbox and standard situations, and reported that of 17 animals tested, 9 were right- and 8 left-handed on the latchbox, with two right- and three left-handers demonstrating opposite handedness on the standard test, while the remaining twelve animals were consistent in their paw preferences between the two tests.

Wentworth (1942) also compared the performances of rats in his standard test situation with their behaviour on a latchbox test similar to the one used by Peterson (1934). After a week or so of familiarization and training to operate the shortest lever manually, 49 rats were given daily tests of three depressions per day for a week, although animals displaying a tendency towards ambidexterity were given 10 per day or more. The results showed that 29 of the subjects displayed the same handedness preferences as in the standard test situation and that 15 now used the opposite hand, with five ambidextrous in one situation but displaying a consistent paw preference in the other. The three rats classified as ambidextrous in the latchbox test were then given extended testing for up to 51 days with at least 10 lever presses per day. All three ultimately adopted a convincing right paw preference in this situation, although intermittent testing on the standard test during this time confirmed that all three remained consistently left-handed on this task.

Wentworth (1942) then pursued the question of the specificity of paw preference for a particular test situation by examining the behaviour of a group of 49 naive rats on four different pieces of apparatus. The first was his standard test situation; the second, a circular cage with a food well in the centre of the floor; the third was a modified Skinner box with a lever which could only be operated by one paw at a time; and finally, a piece of equipment which he called the moat apparatus. This latter required the subject to fully extend one forearm through the side of a wire cage to retrieve food placed on the other side of a moat in order to record a successful reach. Thus, all of these test situations were similar insofar as they each required the rat to make a unilateral reach with one forepaw or the other, but different inasmuch as they enforced different supporting body postures and orientations on the animal to be successful with each apparatus. To avoid unduly influencing later comparisons employing all four types of equipment, the initial assessment of paw preference was made using the standard test situation and a criterion of 9 reaches out of 10 with the same paw. The animals were then kept in their living cages without further testing for 60 days, after which the comparison trials began. These comprised five test reaches in one apparatus, followed on the second day by five reaches in the second apparatus and so on until 15 reaches had been recorded for each of the four test situations. Of the 49 rats given the initial paw preference test, only 18 met the criterion of consistent use of one paw (9 right, 9 left), so that the rest were classified as ambidextrous. In the inter-test comparisons, a less strict criterion of 12 same-paw reaches out of 15 was used, and results for the 18 identified left- and right-handers showed that compared with their original paw preferences, the same paw was used on the standard test by 13 rats, on the circular cage test by 14, on the moat test by 11, and on the Skinner box by 8. The others either did not meet the criterion (and were classified as ambidextrous), or displayed a reversal of handedness. Examining the data for correspondence in handedness across the final test results, 16 out of the 18 animals were consistent in paw preference between the standard and circular tests, 11 out of the 18 were consistent between the circular and moat tests, and as few as 6 between the moat and Skinner box situations, while only 4 were consistent in using the same paw in all four situations. Of the 31 animals originally classified as ambidextrous, 18 met the criterion for right- or left-handedness in the standard situation on the final inter-test comparisons, 28 displayed a consistent preference on the moat test, 20 in the circular apparatus, and 21 in the Skinner box. Furthermore, none remained ambidextrous in all of these later tests and 9 were found to be consistent in their paw preferences across all four tasks. Thus, out of the total of 49 subjects, only 13 used the same paw (whether left or right) in each of the final test situations, and the author concluded that rat handedness appeared to be fairly specific to the type of reach or task involved. In this respect, Wentworth suggested that the degree of correspondence in the direction of handedness displayed between the four types of apparatus appeared to follow a commonsense order of similarity in bodily orientation and reaching movements required for each. And Miklyaeva, Ioffe and Kulikov (1991) came to much the same conclusion from their study of the paw preferences of rats tested on 9 different tasks. They found that the correlation between tasks was a maximum for those situations that required the same or similar limb movements and/or body postures.

Cats

In experiments on cats, similar studies have employed a variety of unbiased test situations to assess the relative frequency of use of the right and left forepaws in unimanual reaching activities (see e.g. Cole, 1955; Fabre-Thorpe et al., 1993; Tan & Kutlu, 1991; Tan, Yaprak & Kutlu, 1990; Warren, 1977a). These investigators have also adopted a variety of laterality criteria - ranging from a greater than 50% recorded use of the same side to a required 90% consistency of demonstrated paw preference. While this has resulted in considerable differences in the proportion of animals reported as displaying a paw preference for either side - both within and between studies - the overall trends are the same as for mice and rats. For example, in his investigation of 60 cats, Cole (1955) used a criterion of 75% consistent use of the same paw to classify his animals, and reported 20% to be right-handed, 38.8% to be left-handed, and 41% to be ambidextrous. Using a similar 75% criterion, Tan et al. (1990) reported that for their sample of 66 cats, 27.3% were right-handed, 24.2% were left-handed, and 48.5% ambidextrous, while Warren (1977a) recorded mean values of 21% right-handed, 21% left-handed, and 58% ambidextrous for his sample of 83 cats tested in seven different situations using a criterion of 80% consistency. In the study by Tan and Kutlu (1991) on 109 cats using an undisclosed criterion, 49.5% of subjects were reported as right-handed, 40.4% as left preferent, and 10.1% as ambidextrous. And Fabre-Thorpe et al. (1993) using a 50% criterion, classified 52.3% of their sample of 44 cats as right-handed and 47.7% as left-handed, whereas with a criterion of 90% they identified 13.64% as right-handed, 38.64% as left-handed, and 47.72% as ambidextrous.

However, in addition to the effects of small samples and of differences in the criterion on the recorded proportions of left-, right-, and mixed-preferent animals, other variables could also have influenced the inter-study differences in the above results. For example, Cole's (1955) cats were a motley collection of animals of unknown age and undisclosed origin. Similarly, Tan et al. (1990) described their subjects as adult cats collected at random from different villages, and Tan and Kutlu (1991) simply stated that their cats were mongrels, and were presumably obtained in the same way. Fabre-Thorpe et al. (1993) indicated that they used adult cats from different litters but gave no further details. Warren (1977a) described his subjects as Siamese and mongrel cats 180 days old at the start of the experiment, which suggests that they were laboratory-reared - in keeping with an explicit statement to this effect concerning the animals employed in an earlier investigation (Warren, Ablanalp & Warren, 1967). Consequently, the genetic background and pre-experimental experiences of the subjects appear to be unknown quantities in these experiments, so that either or both of these influences could have affected the outcome. To what extent each of these variables are important remains to be seen, although the evidence already presented in relation to mice and rats is suggestive.

In experiments using the Wisconsin General Test Apparatus (WGTA), Warren (1958) studied the paw preferences of 31 cats by recording the number of times each animal used each paw to carry food to its mouth in undertaking two particular tasks: (a) when the subject was simply required to pick up food from the test tray, and (b) when the subject first needed to uncover the food by removing a wooden block. Tests on both tasks consisted of at least 100 trials, and the position of the food on the test tray was varied randomly between trials to control for possible positional biases. The first series of tests was given during the cats' first week in the laboratory when they were naive as subjects, and a second series some six months later, with a third series five months later still. Between each series, the animals were used in laboratory learning experiments unconnected with paw preference, but which nevertheless involved testing in the WGTA and practice at retrieving food from the test tray. The author reported that for task (a), the 80% criterion employed was met by 40% of cats in the first series, 61% in the second series, and 81% of animals in the third series. And for task (b), the corresponding figures were 42%, 80% and 90% respectively. Furthermore, Warren reported that the paw preference reliability correlations between the first and second series for task (a) was 0.59 and for task (b) 0.72, which increased to 0.87 and 0.95 respectively between the results for the second and third series. In examining the consistency of paw preference across tasks, the author also found an increase in the magnitude of the correlation between tasks from 0.45 for the first series, to 0.91 for the second, and 0.94 for the third series of tests. Thus, it is clear that with increasing practice on similar tasks over the intervening period between the three series of tests, consistency in use of the same paw increased markedly both within and between the two tasks.

In a later study, Warren, Ablanalp and Warren (1967) reported the effects of practice on 34 laboratory-reared kittens, first tested for paw preference when they were 60 days old and then again at 180, 210 and 360 days old. The test apparatus consisted of a box with a glass front which ended two inches above the floor, giving paw access to a platform outside on which a piece of meat was presented for each trial. A test comprised 50 reaches, and on the basis of the first test, animals were assigned to practice or control groups. The subjects in the practice group were given two to four 25 trial practice sessions per week at food retrieval in the same apparatus, but only between the first and second tests. After testing in the handedness box at 210 days, all animals were tested in the WGTA using five different retrieval situations, each of which involved the subject reaching through the bars of the apparatus to obtain food from the test tray either directly or by mastering some interposing obstruction or restriction. As the results for the early practice group appeared to show no systematic effect on the development of paw preferences, the data for all animals were pooled and inter-correlations carried out between the preferences displayed in both the handedness box at each stage of the experiment and the five WGTA test situations.

All inter-test correlations were positive and 34 out of the 36 comparisons were statistically significant. The lowest correlations (including the only two non-significant ones) all involved the results from the first test at 60 days, suggesting that early preferences were less stable or reliable than later ones.

However, once a preference had been established in the handedness box, it appeared to transfer readily to the other situations in the WGTA. Why should this be so when the paw preferences for rats seemed to be fairly specific to each task? It is difficult to answer this question with any confidence since the test situations used for the two species are not comparable, but certainly, it appears that the gross bodily orientation and hence, the type of reach employed by the animal for each test varied much more between tests for the rats than for the cats. From pictures presented in the Warren et al. (1967) report, it appears that the cat's body postures and orientation to the food retrieval problem in the WGTA varied very little from one test situation to another. The main variable appears to be the type and degree of manipulative activity required by the paw making the reach, which clearly places a premium on the development of unilateral skill or dexterity of the distal component of the movement.

It is of some interest then that in a subsequent experiment, Warren et al. (1967) reported contrasting results between the preferences developed with the handedness box and the WGTA. In this investigation, 18 of the animals employed in the previous developmental study were given 600 trials in the handedness box over a period of about one month with the previously preferred paw confined in a cuffed glove. This prevented the preferred paw from being extended outside the box or from manipulating the meat inside it. The subjects were tested with 50 free-choice trials (i.e. without the glove) in the handedness box after 200, 400 and 600 forced practice trials followed by 100 free-choice trials in each of the five WGTA test situations, and a further 200 free-choice trials in the handedness box. The results showed that the forced training produced a significant change in paw preference in the handedness box which appeared to be permanent since the effect was still evident on retest 15 months later. This was confirmed by a comparison of the paw preferences before and after training in the handedness box which produced a correlation of - 0.22, but the effect had not transferred to the WGTA test situation as the before and after tests on this apparatus yielded significantly positive correlations for all five test conditions.

Monkeys and apes

Although many studies of handedness in nonhuman primates have been reported, relatively few have been carried out on sufficiently large numbers of animals of the same species by the same investigator under controlled conditions to provide valid and reliable data on population distributions. However, there is a considerable amount of interest in the field as may be judged by the controversial paper by MacNeilage, Studdert-Kennedy and Lindblom (1987), and the subsequent flood of responses from other investigators published in the same and later issues of the journal concerned.

One of the earliest well-designed studies on primates was carried out on 84 adolescent and mature rhesus monkeys by Warren (1953), who reported the hand employed for food retrieval in each of 24 trials per day for five days using the WGTA. The results showed that hand preferences were distributed in a U shaped curve, with 46 of the 84 monkeys using the same hand (23 right and 23 left) in more than 90% of the trials, and 53 using the same hand (27 right and 26 left) in more than 80% of the 120 trials.

Later, Lehman (1978) recorded the hand preferences of 171 rhesus monkeys in a carefully controlled food-retrieval situation in which each animal was given a total of 600 trials at the rate of 100 trials in two sessions per day for three days. The frequency distribution of preferred hand use for the population assumed a U shape with no statistically significant difference between the number of animals displaying a preference for the right hand (80) and those preferring the left (91). Similarly, Hamilton and Vermeire (1988) reported that as a result of laboratory testing of over 50 rhesus monkeys for hand preferences on three tests of varying complexity, the population distributions of preference were either flat or slightly U shaped for each task. And using a consistency criterion of 66.6% for direction of handedness, the authors found no significant difference between the number of animals preferring the right and left sides for any of the distributions.

Another study by Brooker et al. (1981) on 67 Bonnett monkeys reported the observation (from behind a one-way screen) of hand preferences in two activities - feeding and searching during a number of laboratory recording sessions. Feeding reaches were recorded when an animal grasped a piece of food and brought it to its mouth; searching was noted when an animal used a hand to push aside hay on the floor in search of food. A minimum of 100 feeding reaches and an overall minimum of 300 feeding and searching movements were recorded for each animal. The results showed that for the feeding activity, 22 animals were significantly right-handed, 21 significantly left-handed, and 24 displayed no significant preference for either hand. And for the searching activity, 9 were significantly right-handed, 12 significantly left-handed, and 46 recorded no significant preference for either side.

The hand preferred for picking up food in three field studies of artificially-fed colonies of Japanese monkeys was reported by Itani et al. (1963), although no details were given concerning the criteria used for their classification of handedness. In the first study on 81 monkeys, 37% were identified as left-handed, 19.8% as right-handed, and 43% as ambidextrous. The second study comprised 111 animals, of which 33.3% were classified as left-handed, 34.2% as right-handed, and 32.4% as ambidextrous. And in the third study described by Itani et al. on 394 monkeys, 37.8% were recorded as left-handed, 29.9% as right-handed, and 32.3% as ambidextrous. In a fourth study by Tokuda (1969) on 41 Japanese monkeys using artificial feeding to assess hand preferences, the hand used for 80% or more of at least 10 food retrievals was adopted as the criterion. If this was not reached after 20 trials, an animal was considered to be ambidextrous. The author reported that by this classification, 41% were left-handed, 20% right-handed, and 39% ambidextrous. And in another field study on a population of 277 artificially-fed free-ranging rhesus monkeys, Hauser et al. (1991) reported the results of observing the hand used to open, and to hold open the lid of a hopper containing food. On the basis of one observation per animal, they found a significant population level tendency for the lid to be lifted with the left hand and to be held open by the right - in order for food to be retrieved with the left. And on a subsample of 130 individuals who were observed to operate the lid five times or more, the authors reported a clear and consistent repetition of roles for the two sides. Unfortunately, without knowing the location or orientation of the hopper, and whether it was approached equally often from the two sides, this evidence is difficult to interpret.

An early, well-designed and carefully controlled investigation on 30 adult and adolescent chimpanzees was reported by Finch (1941). All subjects were tested on four different unimanual food-retrieving tasks, and the hand used to obtain the food was recorded for each of 200 trials on each task. A summary of the results for each animal showed a continuous distribution of the frequency of right versus left hand use in the population, with approximately equal numbers of animals recording overall right and left hand preferences. More specifically, 18 of the subjects used the same hand (9 right, 9 left) in more than 90% of the 800 trials, and 25 used the same hand (11 right, 14 left) in more than 80% of the trials. Test-retest reliability comparisons between the first and second 100 trials on each task revealed that on each of two tasks only one animal reversed its hand preference, and on the third and fourth tasks, five and six animals respectively showed a change in hand preference. In a less controlled investigation of 40 captive chimpanzees observed while adopting two different postures in obtaining food, Hopkins (1993) recorded the hand used for a minimum of 50 trials in each contrived situation. He reported no population preferences for either side for one type of posture, but a significant overall preference for the right side in the other. And in a semi free-ranging population of chimpanzees, Steklis and Marchant (1987) reported the spontaneous hand use of 26 animals during six different manual activities. They found that although individual animals tended to show consistent hand preferences for one particular side (left or right) on any given task, the numbers showing a significant preference varied from one task to another with no overall population-level differences in the numbers preferring the right versus left sides. Of the 81 statistically significant individual preferences recorded for the various behaviours, 55% were in favour of the left hand.

The distribution of side preferences in the use of the hands for feeding by 31 captive lowland gorillas was reported by Annett and Annett (1991) as a result of their observation of animals in five different zoos. They found that on a 50% criterion, 16 animals became classified as left-handed and 15 right-handed, but on an 80% criterion, only two animals could be termed left-handed and only three right-handed. Indeed, the study showed a full range of hand preferences from strong left to strong right with most gorillas displaying some degree of inconsistency. A field study of 38 mountain gorillas by Byrne and Byrne (1991) reported similar results. The analysis of feeding behaviour of these animals in the wild was rather more difficult due to the variety of food types consumed and the considerable manual processing of the material before it was placed in the mouth. However, the authors' careful evaluation of the data collected showed that hand preferences approximated to a U shaped distribution with no significant population bias to the left or right.

Thus, as concluded by Lehman (1993), the evidence from reasonably large samples of both monkeys and apes suggest that while the hand preferences of individual animals may display a convincing degree of consistency in a given situation, there is little or no tendency for these preferences to demonstrate a population bias. However, other authors who have considered a wider range of studies and included the results from much smaller numbers of animals, have suggested that there may be population-level preferences, particularly for the great apes (see e.g. Hopkins & Bard, 1993; Hopkins & Morris, 1993). Unfortunately, the rearing conditions and experiential backgrounds of primate subjects are often unknown or unrecorded, including the extent of their contact with humans, which can vary considerably from one study to another. In this respect, it is important to note that Shafer (1987) found a population bias in his gorilla subjects that appeared to be confined to those raised in captivity. He reported that whereas eight of his wild-born animals displayed a left and nine a right hand preference, only two of his zoo-born gorillas were left-handed compared with sixteen that showed a preference for the right hand. Similarly, Hopkins (1995) reported that in his study of 51 captive-reared chimpanzees (2 to 5 years-old at the time of testing), of 18 animals that had been reared by their mothers, 10 were recorded as left-handed and 8 right-handed on a simple reaching task. However, of the 33 animals that had been nursery-reared by human care-givers, 12 were found to be left-handed and 21 right-handed, suggesting a strong influence of human conventions.

While there are obvious merits in obtaining data from animals in their natural habitat, other findings from laboratory-based studies may also be important in the interpretation of results from field observations. For example, Lehman (1970, 1978), and Deuel and Dunlop (1980) have shown that monkeys tend to use a particular hand more frequently whenever it is the one nearer to the object to be grasped. This has implications both for the observed preferences of a stationary (e.g. seated) subject, and for the hand preferences recorded for a free-moving animal depending upon its bodily orientation and/or direction of approach to the target object. Other situational factors such as a cramped body position or an enforced cooperative use of both hands have also been shown by Cronholm, Grodsky and Behar (1963) to produce a J shaped rather than the U shaped or rectangular distributions associated with unbiased situations.

However, in monkeys and apes, as with other species already discussed - particularly rats and cats, both the direction and degree of hand preferences observed appear to be dependent upon the specific test of preference employed. One of the earliest systematic laboratory investigations of this variable was reported by Kounin (1938) for seven animals (three rhesus, one spider, and three cebus monkeys) using three different tests of handedness. The first required the animal to reach with one arm through a hole in the centre of the floor of a circular cage to gain access to food placed in a container at arm's length underneath. The second test required the subject to reach through a hole in the side wall of its cage, to lift the cover of a food box, and to hold it open in order to gain access to the banana reward - all with one hand. The third task involved reaching through the side of the cage with one arm and then to grasp and manipulate a rake to obtain the food reward. The tests were administered in two series of 100 trials per test approximately one month apart at the rate of 25 trials per day for four days in each series. Intratest consistency for use of the same hand by each subject was high for the first and second handedness tests both within and between series, and all animals retained their own particular side of preference throughout. Consistency on the third (rake) test was much less both within and between series, with one subject making a clear change of hand preference from 71% right hand use to 98% left hand use between series. Examining hand preference scores across tasks revealed that three of the seven subjects were inconsistent in the direction of their observed preferences.

The first of a more extended series of studies on the question was, reported by Warren (1958) using 17 monkeys and the same procedure and apparatus (WGTA) as previously reported for this author's experiments with cats. Three tasks (or test conditions) were used - task (a) involved simple unimanual reaching to pick up food from the test tray, task (b) required the displacement of a wooden block covering a food well before picking up the food, and task (c) necessitated a similar removal of a cardboard square before gaining access to the food. Tests on each task consisted of at least 100 trials (comprising a series), and three series of tests were carried out with approximately three months separating corresponding tests in each series. Between each series, the animals were engaged in daily testing in the WGTA on learning problems unrelated to questions of hand preference, although these also involved practice in food retrieval from the test tray. With an 80% criterion of same side use to classify each monkey's hand preference for food retrieval, Warren found that for each task, the number of animals reaching the criterion increased in successive series. Test-retest correlations between series for hand use on each task also showed considerable improvement with time, as did the consistency of hand preference between tasks from one series to the next. These trends correspond very closely to the results for cats published by the author in the same paper, and clearly indicate that with continuing daily practice of food retrieval in the WGTA throughout the experimental series, the consistency of hand preference improved considerably not only on each task tested but also between tasks.

In a subsequent experiment on 19 rhesus monkeys, Brookshire and Warren (1962) reported the hand preferences displayed in performing three tasks designed to ensure unimanual responses, and three tasks designed to produce cooperative bimanual food retrieval movements in the WGTA. One unimanual task involved a simple reach to obtain food - similar to task (a) described above. The second required the monkey to pull the food off a horizontal wire, and in the third task, the animal had to reach into a vertically positioned transparent bottle to obtain the reward. Each of the bimanual tasks involved manipulation of some object containing or concealing food before being able to gain access to it. One required the subject to grasp the handle (within reach) of a box containing food (out of reach); one required the animal to lift a hinged lid covering a food well, and the third involved pushing aside a wooden block - similar to task (b) above. In each bimanual task, the hand used to manipulate the object was recorded separately from the hand used to retrieve the food, although for some tasks and some animals these were not always different. The experiment was conducted in two test series, during each of which, hand preferences for each task except one (the block test) were recorded on 150 trials spread over several days. The two series on any one task were separated by several weeks during which testing on other tasks was carried out in a counterbalanced order on different subgroups of subjects. Examining intratest consistency, the authors reported significantly positive rank order correlations between the subjects' hand preferences in the two series on all tasks except the block test which was administered in only one series. The median reliability of the tests was 0.76, and although a number of reversals of hand preference were recorded between series on all tasks, these mainly involved shifts from a weak preference for one side to a weak preference for the other. Tests for consistency of hand preference across tasks showed that out of the 36 comparisons possible for the nine measures used, 31 were positive but only 16 were significantly so. Thus, the evidence again suggests that individual animals are reasonably consistent in their hand preference on any one particular task, but not between tasks although there appears to be some degree of correspondence from one task to another.

In a further and even more extensive experimental series on 14 rhesus monkeys, Warren (1977b) employed five tasks designed to be unimanual tests of hand preference, and five designed as bimanual tasks, although he found that as the latter involved successive rather than simultaneous actions, most monkeys made the necessary movements sequentially with the same hand. Most of the tasks employed were similar to, or variants of the tasks previously used by the same author and described above. An initial pre-experimental series of tests using the simple unimanual reach task at the rate of 25 trials per day for 20 days was administered as an adaptive procedure to the WGTA and experimental situation. All tasks were then given at the rate of 24 trials per day for 8 days to provide a total of 200 responses per task, and the series was repeated two years later before and after cerebral surgery, to provide a further 100 responses per task both pre- and post-operatively. A careful examination of the results by the author revealed a number of interesting findings. First, a comparison of overall hand preference scores on the 15 handedness measures for the 14 subjects showed positive and high rank correlations of 0.82 or better between series. But an examination of the consistency of hand preferences across tasks revealed that some comparisons demonstrated a high and significant correlation, others a low and mostly non-significant association, and yet others displayed a changing and improving relationship from series 1 to series 3. The author accordingly classified the tasks in these three groups as group H, group L, and group C respectively. Between-group comparisons of hand preference scores on tasks in these three groups showed that the median rank correlations between L and H groups were consistently low from one series to another, whereas the association between groups C and H increased from series 1 to series 3. However, individual test-retest correlations carried out between series on each task suggested that these results were unlikely to be due to inherent differences in the consistency of hand preferences on each task. The author concluded that the results suggest that the hand preferences recorded for these monkeys were quite specific and could be accounted for in terms of the similarities or differences in the hand and/or arm movement patterns required in the different tasks, and the effects of practice on tasks similar to those in group H in the intervening periods between series. Lehman (1980a) reported similar findings in an investigation of hand preferences in 39 rhesus monkeys tested on a simple reaching task administered in two series - separated by testing on two other tasks - a discrimination learning problem (presented in two series one month apart), and a food retrieval test (given in one series only, after the second discrimination learning series). He found that the population average of right hand preferences on each task for each series was close to 50%, but that each animal tended to use the same hand for the same task for most of the time. For the simple reach task this gave an average preferred hand use of 79% for series 1 and 92% for series 2, with corresponding figures for the discrimination task of 82.3% for series 1 and 91.7% for series 2, and an average preferred hand use for the one series of food retrieval tests of 78.5%. The increase in use of the preferred hand from one series to the next was statistically significant for both the reaching and discrimination tasks. In contrast, only 13 of the animals preferred the same hand for all three tasks, demonstrating a correspondence in hand use between tasks which was no better than would be expected on the basis of chance. Lehman (1989) replicated this experiment using the same tasks and procedure with another 31 rhesus monkeys. This yielded essentially the same results with regard to intra-test reliability, but this time one of the three between-task comparisons yielded significant correlations for both direction and degree of hand preference. From the results of the two experiments, Lehman concluded that hand preference in the monkey is task specific but probably multifactorial in nature, so that some tasks share common factors to a greater extent than others.

In an experiment that employed a range of 17 manual tasks and 31 measures of hand use, Beck and Barton (1972) also assessed both the degree and direction of hand preferences in 10 stump-tail macaques. Some tasks were unimanual, others bimanual, and many required sequential actions. Where sequential and/or cooperative use of both hands were required, the hand used for each (manipulative or food retrieval) action was recorded. Testing occupied a total of 33 weeks, and replication trials were carried out on only five of the tasks, involving 11 measures of hand use. From an examination of this number of test-retest rank order assessments of reliability, the authors reported that all correlations were positive and six were significantly so for direction of preference, but only one for degree of preference. An evaluation of the extent to which the actions required in different tasks elicited consistently strong or weak preferences showed that of the 136 comparisons for food retrieval and 91 comparisons of the manipulative component, only 10 were found to be significant. A similar examination of the inter-test correlations for direction of hand preference (i.e. use of the same hand) for the various actions recorded, showed that of the 136 food retrieval comparisons only

10 reached significance, and of the 91 comparisons of the manipulative components, 20 were significant although two of these were negative. Thus, the evidence again suggests that there is a definite tendency to use the same hand for the same action in repeated trials on the same task although the degree of consistency clearly varies from one task to another. And once again, the general lack of significant correspondence in hand use across tasks provides further support for the view that there is a high degree of specificity in the strength and direction of hand preference each animal displays in any given task.

A similar specificity of hand preference has been reported for chimpanzees. Marchant and Steklis (1986) tested five captive chimpanzees in a well-controlled series of six unimanual and three bimanual tasks and recorded marked individual differences in hand use between animals. They found that while each chimpanzee tended to exhibit a clear preference for one particular side (some right, some left) on any given task, this was not necessarily the same across all tasks. The bimanual tasks in particular seemed to evoke very individualistic patterns of manual coordination. And in a later report, Steklis & Marchant (1987) recorded the results of observing the spontaneous hand usage of 26 semi free-ranging chimpanzees for six different manual activities - feeding, reaching, holding, carrying, throwing, and grooming. The number of animals showing a significant hand preference varied considerably from one task to another - from 21 out of the 26 for reaching (11 left and 10 right), to only 3 for grooming (2 left, 1 right). They also found that only 14 of the total subject sample were consistent in their hand preferences across two or more activities (7 left, 7 right). The more recent study by Hopkins (1993) who observed the hand preferences of 40 chimpanzees obtaining food from two different postures also presents relevant data for each animal. Although the author himself makes no direct comparison of the performances in the two situations, a rank order correlation of each subject's relative frequency of use of the two sides gives a non-significant rho value of 0.28, demonstrating yet again that when present, individual hand preferences tend to be specific to each particular task or situation.

Comparable results from the examination of hand preferences in gorillas undertaking various tasks have been presented in several studies. For example, Olson, Ellis and Nadler (1990) recorded the hand used by 12 captive gorillas in retrieving food (a) from the floor, and (b) from the side of the cage, and additionally for 8 of the same subjects in a WGTA situation. A total of 125 trials was given for each animal in tests (a) and (b) and 100 trials per animal in the WGTA task. The results showed that on test (a), 7 of the 12 gorillas demonstrated a significant hand preference (5 right and 2 left), on test (b) all 12 displayed a significant preference (10 right, 2 left), and on the WGTA task, 6 of the 8 animals showed a significant preference (3 right and 3 left). Examining the consistency of hand preference across tasks revealed that only 5 of the 12 gorillas displayed a significant and consistent preference in both tests (a) and (b) (4 right and 1 left), while a similar comparison of the 8 animals that completed all three tasks showed that only 2 displayed significant and consistent preferences in all three (1 right and 1 left). In another investigation of the hand preferences of captive gorillas, Fagot and Vauclair (1988) recorded the hand used by 10 animals in (a) retrieving food from the floor of their enclosure, and for 8 of these animals in three other situations (tests (b), (c), and (d)) involving apparatus requiring sequential manipulative activity in order to obtain food. A minimum of 100 trials per animal were recorded for test (a) and 30 trials per subject were given for each of the other three tests. The results showed that 6 of the 10 gorillas displayed a significant hand preference (3 right, 3 left) in test (a), and for test (d) all 8 of the animals recorded a significant preference (7 left, 1 right). Only 1 subject consistently performed task (c) with one particular hand (the left), and on task (b) half the subjects regularly used one hand to solve the problem (2 left, 2 right), with the other 4 subjects using both hands cooperatively to complete the task.

In a more extensive study of the hand preferences of mountain gorillas during 106 days of field observations, Byrne and Byrne (1991) collected the data from 38 animals involving at least 6 hours 40 minutes per animal. The authors made a detailed analysis of the feeding habits of the animals and the extent to which one hand or both hands were used consistently in six manual processing activities associated with different types of food. For one task, the authors found considerable variations in technique from one animal to another with many using both hands and others only one. Of the other five tasks analysed, one provided adequate data for only 13 of the 38 animals, but for the remaining four tasks, clear asymmetries of hand use were recorded for nearly all subjects. The numbers of animals showing a consistent and significant preference for the right side compared with those consistently preferring the left, were respectively: task (a) 13 right, 19 left, task (b) 13 right, 21 left, (c) 11 right, 18 left, and (d) 15 right, 18 left. However, none of these differences in the proportion of right versus left were significantly different from chance. Comparing subject preferences across tasks produced positive and significant intercorrelations for both strength and direction of hand use between three of the four tasks. These three tasks all involved the processing of leaf foods compared with the fourth which was a stem type of food. Only 14 of the 38 animals showed a consistent and significant preference for the same hand across all four tasks, indicating that the pattern of hand usage was very dependent on the demands of the task

Thus, the evidence from investigations of monkeys, chimpanzees, and gorillas is generally in agreement in suggesting that for most situations studied, approximately equal numbers of animals display a consistent preference for the right and left sides, although there is a good deal of inconsistency in these preferences from one situation to another. Some authors have presented data to suggest a population tendency favouring the left hand and others for a bias to the right, but such variations between studies could be due to any one or more of a multiplicity of factors. Some differences in the literature are almost certainly a function of the inadequacy of the sample size, and others due to the type of task examined, the measurement criteria employed and/or the subject's postural orientation and freedom to move in the test situation. The variable of practice or experience on a particular apparatus or task (where it has been measured) has also been shown to significantly influence results. A critical review of methodology by Marchant and McGrew (1991) has attempted to place some of these questions in perspective.

A further feature of primate hand preference of interest here relates to questions of family likeness. Since the gorilla population studied by Byrne and Byrne (1991) had been under observation by various workers for many years and information on family relationships was available, these authors were able to compare the hand preferences of mothers and their offspring for two of the manual tasks they had analysed. Data for the first task provided 22 familial pairs, of which 11 were the same - exactly as would be expected on the basis of chance, while the second task provided 13 pairs of which only 4 were the same. Fewer data were available for father-offspring or sibling-pair comparisons, but from the information reported, there was no suggestion of any familial association for direction of hand preference. A corresponding examination of the mother-offspring data for strength of hand preference again revealed no significant correlations. A similar analysis was carried out by Brooker et al. (1981) on the data obtained for feeding and searching activity from their captive colony of 67 Bonnett monkeys. In a comparison of the direction of hand preferences between mother and offspring, no significant correlations were found for either manual task. The same negative results were reported for mother-offspring comparisons using degree (i.e. strength) of hand preference. And Watanabe and Kawai (1993) also reported no evidence for the inheritance (i.e. genetic or social) of direction of hand preference in their observations of Japanese monkeys. However, in a brief report by Brinkman (1984) relating to the 48 offspring of 2 male and 15 female crab-eating macaques observed in two harem-type groups, there was evidence of a strong familial effect on hand preference.

Of 22 offspring of parents with like preferences, 19 of the young also showed a preference for the same side, whereas of the 26 offspring of parents with unlike preferences, 23 showed a preference for the same side as the mother. The author concluded that the results were suggestive of postnatal learning rather than any genetic predisposition.

In contrast to these conflicting findings on family resemblances for direction of hand preferences in primates, a number of studies have provided very convincing evidence that such preferences can be brought about by training or experience. Evidence of increased consistency of the side of hand usage both within and between tasks due to practice in particular test situations has already been noted in discussing the results of Warren (1958, 1977a, 1977b). But the first report of an attempt to investigate the effects of systematic training of hand usage in monkeys was published by Lashley (1917) on two subjects. One was an extremely wild, fierce, and uncooperative animal that also proved to be strongly biased in hand preference and quite unresponsive to training. The other was said to be much more gentle and cooperative, and displayed a less clearly differentiated hand preference which was readily changed with training. Shortly afterwards, Kempf (1917) reported the results of a more successfully controlled study of six monkeys whose initial hand preferences were determined by testing the animals in three food-retrieving situations before submitting them to a program of preference reversal training. In this the subjects were hand-fed by the experimenter who withdrew the food if an animal attempted to take it with the "wrong" hand. As a result, all animals were successfully and rapidly retrained within a relatively few trials to use their originally non-preferred side. Tests some three or four months later showed that four of the six subjects had clearly retained their changed preference, one had reverted to its original side of preference, and the other demonstrated inconsistency. A similar experiment was reported by McGonigle and Flook (1978) on six squirrel monkeys using the WGTA. Initial hand preferences were recorded in a free choice situation over four sessions of 40 trials per session. Training trials were then given at the same rate with only the initially non-preferred hand being rewarded. After five sessions, all monkeys had reached the required retraining criterion of 90% correct responses in two successive sessions. Retesting under free choice conditions (i.e. non-differential reinforcement of responses) at intervals of 1, 3, and 7 weeks after completion of training showed excellent retention of the habit in five of the animals and a somewhat less than perfect retention by the sixth.

In an attempt to discover the basis for an animal's initial hand preferences in the experimental situation, Lehman (1980b, 1980c) conducted two interesting studies. In the first investigation (1980b), he examined 46 previously untrained stump-tail monkeys on a simple reaching task. Each animal was tested individually for hand preferences on the presentation of food which, however, was withdrawn during the first reach and all other reaches with the same hand until the subject changed to using the other hand. The first reach with the previously non-preferred side was allowed to successfully obtain the food, and all subsequent reaches with either hand were also rewarded. Each animal was given 100 such reinforced trials per session, twice a day for three days, and the percentage of hand preferences then assessed over these six sessions. The author reported that for a significant majority of the animals, the initially chosen (unrewarded) hand was preferred most of the time during the six rewarded sessions. Furthermore, he found that some animals displayed a consistently strong preferred hand usage throughout these reinforced trials whereas others were less consistent during the first few sessions but had become equally consistent in their hand usage by the last session. In the second report, Lehman (1980c) tested 58 untrained cynomolgus monkeys for hand preference on a similar simple free choice reaching task for 100 rewarded trials per day for three days. The author reported that for the entire group of subjects, the hand preferred on the first reach was significantly associated with the hand preferred overall. However, subdividing the animals into three age groups (estimated from body weights), he found that while such a significant association was true for each of the two older groups considered separately, it did not hold for the youngest animals. He concluded that the hand preferred by individual monkeys was not randomly determined, but that past experience and developmental factors influence the initial choice of hand preference in a new situation, and that this is a function of the nature of the task and the maturity of the particular animal concerned.

Further valuable information on the effects of differential training of the two hands has been provided by Preilowski, Reger and Engele (1986). In their first experiment, these authors trained two rhesus monkeys to produce a specific pressure between two fingers of one hand at progressively higher levels of difficulty - as determined by seven increasingly stringent performance standards. Each animal had a free choice with regard to hand use, but one subject only ever used its right hand, and the other animal always used its left. Four blocks of 36 trials (maximum) were given daily and the completion of each level of difficulty depended on a success rate exceeding 90% on three successive blocks of trials. The data presented showed an S shaped learning curve, levelling off at the successful completion of level 6 after approximately 40 trial blocks. A subsequent second phase of the experiment required each animal to use one hand during the first two blocks of trials each day, and then to use the other exclusively for the remaining two blocks. Thus, one hand was being retrained while the other was undergoing training on the task for the first time. The results showed that the previously trained side rapidly relearned the task and attained level 6 after about 15 blocks of trials, whereas little or no positive effects of the animal's previous training appeared to transfer to the other, inexperienced side. In a second experiment, six rhesus monkeys were used, three of which were trained exclusively with one hand on a similar task until five levels of difficulty had been mastered, while the other three subjects were required to alternate hands from day to day through the same performance levels. The averaged results for each group of three subjects showed that the five different levels of difficulty were each completed successfully much more rapidly by those using the same hand all the time than by those alternating between right and left hands, although the differences between groups became more and more evident with increasing levels of difficulty. A similar experiment described briefly by Preilowski (1990) using arm movements to produce the required force instead of the fingers has provided evidence for a somewhat greater degree of transfer of training between sides, suggesting that transfer effects may be different for finger and arm movements.

CHAPTER 5
Heredity and Handedness Assessment

Chapter outline

Stemming from clinical observations of the importance of an intact left cerebral hemisphere for the unimpaired speech of right-handed brain-damaged patients, it has been widely assumed that an individual's handedness is a behavioural marker of some more fundamental biological asymmetry. Although no simple hereditary hypothesis has been found to satisfactorily account for the directional (i.e. right or left) component of human handedness, evidence is now accumulating from family and twin studies to suggest that it is more likely to be the strength of manual asymmetry (irrespective of direction) that has some underlying genetic basis. Questionnaire studies show that both the strength and direction of hand preferences for any given individual may vary considerably from one task to another, although nominal right-handers tend to be more consistent in their choice of side than nominal left-handers. And the assessment of an individual's overall hand preference has been shown to be very dependent on both the type and number of handedness criteria employed - especially for non-right-handers.

Similarly, differences in the proficiency of motor performances of the right and left hands of any given individual also vary considerably from one manual activity to another depending upon the task demands, giving rise to low inter-correlations between handedness performance scores on different tasks.

Comparisons between preference and performance measures of handedness, and evaluation of the published data, suggest that the contrasting results obtained from the two methods of approach are primarily due to differences in the number and type of criteria employed in each.

In concluding his review of studies of handedness in nonhuman species, Warren (1980) clearly felt frustrated to find that the experimental evidence appeared to show none of the features deemed to be characteristic of human cerebral dominance. He reported that there seemed to be no general population bias in manual activities favouring the right side (or the left), and that hand preferences were inconsistent between tasks. Such preferences also appeared to be strongly dependent on or determined by learning and experience. Furthermore, he reported that no differences in function between the two cerebral hemispheres had been found that could be related to hand preference. Consequently, he inferred that "the human pattern of handedness and cerebral laterality is species unique and that no truly homologous traits are to be found in nonhuman animals" (p.357). However, while Warren's findings in respect of non-human species are clearly well founded, there is now evidence available to show that his conclusions make unjustifiable assumptions concerning the nature of human handedness and cerebral asymmetries.

Heredity and handedness

In explanation of the predominant preference by the vast majority of people for the use of the right hand in most spontaneous manual activities, Broca (see Berker, Berker & Smith, 1986) proposed that the side of preference must be determined by some organic predisposition for the corresponding development of the left side of the brain. He reasoned that if chance and imitation were the causes, then why had no population of predominantly left-handed people ever been discovered? Furthermore, why do some individuals become left-handed despite all efforts to the contrary? He suggested that this must be due to an inversion of the organic predisposition, against which neither imitation nor education could prevail. He also drew support for his views from a belief in the inherent unequal strength of the two hands and a precocious ontogenetic development of the left cerebral hemisphere. He further suggested that the predisposition and early development of the left hemisphere also accounted for the establishment of the organizational basis for speech in the left hemisphere, although he argued that both speech and the habitual skills of the right hand were only acquired after a long and specialized training which began in early infancy.

There is clearly an appealing logic in Broca's views, and it is evident that until now, it has generally been assumed that while environmental influences may moderate the development of behavioural motor asymmetries, the overwhelming population bias towards right-handedness is primarily the result of a genetically determined functional predominance of the left cerebral hemisphere (Brain, 1945; Henschen, 1926). And in clinical cases where the usual association between handedness and the hemisphere dominant for speech has broken down, the explanation of stock-brainedness originally proposed by Foster Kennedy (1916) has enjoyed considerable favour amongst neurologists (Zangwill, 1955). However, Plomin (1991) has cautioned that inferring a genetic determinant from the study of any aspect of human behaviour is fraught with considerable difficulties, and that it is rare indeed for genetic variance to account for as much as half of the variance of the behavioural traits investigated.

Certainly, it was not long after the rediscovery of Mendel's laws of heredity at the turn of the twentieth century that researchers began to explore genetic explanations for the relative incidence of left and right handedness in families, and to consider that left-handedness behaves as a Mendelian recessive character (Jordan, 1911, 1914; Ramaley, 1913). Furthermore, it is clear from these authors' discussion of their findings that they were both familiar with Broca's rule locating speech in the hemisphere contralateral to the preferred hand (see e.g. Jordan, 1911, p.117-

118; 1922, p.382; Ramaley, 1913, p.733-735), and that they believed the primary cause of left- or right-handedness was to be found in some anatomically asymmetrical condition of the brain. In a subsequent study, Chamberlain (1928) reported on a survey of 2,177 university students concerning their own handedness and that of their parents, brothers and sisters, supplemented by reports from respondents to newspaper advertisements. Since he had difficulty in consolidating the mass of handedness information he collected this way, he adopted a strict criterion of only classifying subjects as left-handed if they wrote with the left hand. He then found that in the total student sample, 4.39% were left-handed but that in families where one parent was left-handed 9.7% and 13.77% of the children were left-handed according to whether the father or mother (respectively) were left-handed. In his supplemented sample of families with both parents left-handed, he reported that 46% of the children were left-handed. If the writing hand is a clear indication of handedness and left-handedness is the result of genetic coding as a Mendelian recessive, then left-handedness should only be evident in children of two left-handed parents. Consequently, Chamberlain concluded that while there could be little doubt that left-handedness was inherited, a more sophisticated explanation was needed to account for his results.

The same trends regarding the increased incidence of left-handedness in children from families with one left-handed parent (and especially where both parents were left-handed) compared with the incidence in families where both parents were right-handed have also been reported by Rife (1940) and Annett (1973b). These studies employed questionnaire surveys, but whereas Annett classified her respondents in the same way as Chamberlain (i.e. according to the writing hand), Rife adopted a criterion of consistency of hand preference such that his group of right-handers included only those who said they used the right hand for every one of ten inventory items, with all other subjects being arbitrarily classified as left-handed. Bryden (1982) summarized the findings from these three studies and included data from his own large survey to show that while the actual proportion of left-handed offspring varied from study to study for whatever reason, the majority of children of two left-handed parents still tended to be right-handed. Later, Bryden (1987) published the results of a family study which also included the handedness of grandparents. Since in most cases the grandparents did not raise the second generation, the author assumed that they would have had no direct social influence on the hand preferences displayed by those children. He found that the incidence of left-handed offspring of either left-handed parents or left-handed grandparents was no different from (or even less than) the incidence of left-handed offspring of right-handed parents or grandparents. The fact that the trend was slightly in the "wrong" direction he considered to be an effect of sample size. Certainly, these data provide no support for a simple genetic explanation of the directional characteristic of hand preference.

In contrast, evidence is now accruing to suggest that the strength of hand preference (irrespective of preference direction) may have a genetic component. In a questionnaire study of 459 complete family units (i.e. both parents and at least one biologically related offspring) that also included data on 434 offspring pairs, Coren and Porac (1980) measured hand, foot, eye, and ear preferences for both direction and consistency of laterality. With regard to handedness, the only significant effect of parental influence on the direction of the preference of the offspring was in respect of the mother. The authors found that 13.6% more children of left-handed mothers were left-handed than children of right-handed mothers. And analyses of family resemblances for all four laterality measures of preference direction showed that of 96 parent-offspring comparisons only 11 were statistically significant. However, similar evaluation of the data for the degree of consistency of lateral preference without regard to direction showed that of 40 correlations between parent-offspring and sibling pairs for the four laterality indices, as many as 15 were significant - i.e. 38% compared with 11% for preference direction. The strongest handedness associations were between parent and male offspring (statistically significant for both mother-son and father-son correlations).

Bryden (1982) also reported evidence from both preference and performance tests of handedness to show that there is a closer association between parents and offspring for strength of asymmetry than for direction. On the manual performance task, when the scores were evaluated by using the relative proficiency of the two hands, correlations between parent and offspring were low and non-significant except for the father-son comparison. But when the performance scores were analysed without respect to right or left, significantly positive correlations were obtained for all parent-offspring comparisons except in respect of mother and son. Similarly, in the Bryden (1987) family study of hand preferences which included grandparents, although the association between generations with regard to direction of hand preference were no different for the offspring of left-handed parents or grandparents compared with their right-handed counterparts, the results of an analysis for strength of handedness was quite different. He found that re-classifying individuals as "strongly" or "weakly" handed (irrespective of direction) produced significant results for 8 of the 11 comparisons made. There was a clear and consistent tendency for strongly-handed parents or grandparents to have a higher proportion of strongly-handed offspring than weakly-handed parents or grandparents. The evidence therefore suggests that whereas some factor determining strength of handedness may be inherited, the agent responsible for direction of handedness is not.

However, apart from the difficulties described by Bishop (1980a, 1990b) in the evaluation and use of family data concerning left-handedness, familial associations, in themselves, cannot distinguish between genetic and environmental influences (Collins, 1977a; Plomin, 1991). Other studies which have addressed this question more directly have compared the incidence of left-handedness in families where those rearing the children are the biological parents, with families in which one or both of the carers is not. For example, Hicks and Kinsbourne (1976) attempted to examine this question using data from university students concerning their own and their parent's writing hand (as a measure of handedness) and found that in step-parent families, the handedness of the biological parent and offspring were significantly related, but that the handedness of the step-parent and offspring were not. Similarly, Longstreth (1980) reported that the incidence of left-handed writers in the offspring of two natural parents (one of whom was left-handed) was significantly greater than that in step-parent families where the step-parent was left-handed. And in a third study, Carter-Saltzman (1980) recorded the hand preferences (using Oldfield's 1971 Edinburgh Inventory) of adopted children and compared these with those for biologically related children. She reported that while the incidence of left-handedness in children was significantly related to parental left-handedness in her biological sample, no such significant relationship was found for the adoptive group.

Nevertheless, Longstreth (1980) recognized that a major difficulty in the interpretation of these results was the mean age of acquiring a step-parent, which for his subjects was 7.55 years (well after the time when hand preferences first appear or writing habits are formed). Similarly, in the Hicks and Kinsbourne (1976) study, it appears that the average age of their subjects when they acquired a step-parent was even older - 13 years (see Liederman & Kinsbourne, 1980). In contrast, Carter-Saltzman (1980) confined her investigation to those families who had adopted their children in infancy (before their first birthday), so that it might be assumed that these results would be more reliable. However, the average age of her children when assessed for handedness, is not given (it varied between 4 and 22 years), and her subject samples small in crucial categories. Furthermore, the extraordinarily high overall incidence of 50% non-right-handedness in the offspring reported for both biological and adoptive groups clearly raises questions about the normality of the populations examined.

Reference should also be made here to the results of the Colorado Adoption Project published by Rice and Plomin (1983) and Rice, Plomin and De Fries (1984), in which observations were made of the handedness of 152 early-adopted and 120 non-adopted (control) infants when 12 and 24 months old. These authors found significant trends in both strength and direction of handedness with age, but reported inconsistent parent/child hand preference correlations, which were no better than could be accounted for on chance alone. A subsequent study by Sandino and McManus (1998) also using early adoption data obtained from the Colorado Adoption Project, compared children's handedness when they were 7 years old and that of their adoptive and biological parents with a matched control group of non-adoptive parents and children. But again, the authors found no significant familial trends in any of their measures of direction of lateralization.

Another type of family study has focussed on the relative incidence of left- and right-handedness in identical and fraternal twins, although various authors (e.g. Carter-Saltzman et al., 1976; Nagylaki & Levy, 1973) have argued that twin data are unsuitable for use in genetic studies of handedness. As reasons for these views, they cite the increased incidence of pathology associated with twinning, and the occasional occurrence of mirror-imaging as well as an overall higher frequency of left-handedness in both types of twins compared with the single-born. However, in a critical evaluation of the relevant literature, McManus (1980a) concluded that such objections are not warranted.

Certainly, there are problems in the use of twin data to support any simple genetic explanation of right- and left-handedness. The first relates to the observed frequencies of same and different handedness in twin pairs. Since identical twins have exactly the same genetic make-up, if handedness direction is the phenotypic expression of some genotypic determinant of asymmetry, then it would be expected that the individuals in each twin pair would both be right-handed or both left-handed, whereas the occurrence of like-handedness in fraternal twin pairs would be no different from the incidence of same-handedness in ordinary sib-pairs. To pursue this question, Coren (1992) examined the data from thirteen published studies that included information on both identical and fraternal twins with acceptable standards of handedness measurement. He found that only 76% of identical twin pairs had the same handedness, and that this was in fact identical with the incidence recorded for the fraternal twin pairs, and similar to what he suggested would be expected from chance pairings of unrelated individuals. Since then, a carefully conducted study by Derom et al. (1996) on same-handedness preferences of a large sample of identical and non-identical twins, has also reported no significant difference in the incidence of concordant/discordant-handedness between the two types of twins.

Collins (1970) earlier examined the data from four different published twin studies and found that for both types of twins (i.e. identical and fraternal) and for non-twin siblings, the proportions of RR, RL, and LL pairs conformed closely to that which would be expected from a binomial distribution, and that the frequency of handedness types within both twin and non-twin pairs can be accounted for on the basis of chance and an overall bias to the right. The same results were obtained by Collins (1977a) in a further detailed and extensive examination of the data from 12 studies of identical twins and 10 studies of fraternal twins. From a careful consideration of these results and other lines of evidence, Collins (1977a) concluded that genes are "left-right indifferent". He suggested that although children tend to resemble their parents in hand preference, there are no grounds for believing that this is dependent on the transmission of left and right genetic alleles and that the resemblance could just as well be explained in terms of cultural influences. Nevertheless, he did not preclude the possibility of a genetic influence on the expression of a directional laterality and suggested three different ways in which this could occur. The first possibility he advanced concerned the inheritance of asymmetry itself, in which the unilateral appearance of a characteristic (but not the side) may follow Mendelian principles. Second, he suggested that the right-left dimensions of asymmetry may be randomly determined although not necessarily with equal probabilities. And third, he considered that genes could influence the strength or weakness of an asymmetry.

Although a number of different approaches have been made to examining the range of evidence in the literature (see e.g. Annett, 1972, 1985; Collins, 1977b; McManus, 1984, 1985a), there is little doubt that a common focus of most investigations has been the role of a genetic influence on human manual asymmetry. Yet few have discussed the corollary that whenever a new genetic characteristic emerges in evolution, it needs to endow its possessor with some advantage in adapting to the demands of the environment for it to be successfully transmitted to succeeding generations. It is suggested that no such advantage can be advanced to support the theoretical notion of an inborn directional asymmetry and that natural selection would favour the inheritance of an equipotentiality for the two sides. A comprehensive appraisal of the evidence clearly calls for a more Darwinian consideration of the problem.

But apart from the necessity for a fresh research orientation, there are other very good reasons for a re-evaluation of the evidence. For example, there are many problems of interpretation of findings published in the literature that are due to the way handedness has been defined and measured or studies designed and conducted. Clearly, if valid conclusions are to be drawn from the information available, then much more attention must be given in the first instance to the methods used to obtain the information.

Handedness assessment

In pre-literate or non-literate populations (e.g. human infants or nonhuman species), the observational recording of preferential limb usage in either natural or contrived situations is often the only practical means of obtaining relevant data. However, in human adult populations, the two conventional methods of assessing handedness characteristics are: (a) by questioning subjects concerning their hand preferences, or (b) by comparing the performance achievements of their right and left sides. Although a considerable amount of data have been published using both preference and performance criteria, studies of performance differences between the two sides have mostly been confined to experimental laboratory investigations for reasons relating to the availability of apparatus and the use of controlled conditions. Furthermore, in clinical situations it is usually impractical or inappropriate to assess subject performances, and questioning of the patient or a close relative concerning hand preferences is the only possible option.

(a) Assessing preference

The single most obvious question to ask a subject is: "are you right- or left-handed"? The response, of course, indicates nothing about what the subject understands by the question unless it is followed up by further questioning. The problem then becomes a matter of deciding what further questions should be posed. In the clinical field it may be a matter of considering the special symptoms or condition of the patient as to what might be an appropriate line of enquiry. No two patients are the same no matter how similar their symptoms may be or how localized the site of an organic dysfunction. Yet unless the same range of questions is put to all patients, then the possibility of discovering a pattern of handedness characteristics to correspond with any given clinical syndrome becomes remote. With normal subjects the problem would seem to be much less difficult. By definition, they possess the usual complement of healthy right and left body parts and are fully capable of understanding and responding to the questions asked. Yet here again, each subject is unique with a genetic make-up and history of opportunity and experience peculiar to each individual.

As both Lauterbach (1933a) and Hildreth (1949) noted, people differ considerably in their consistency of hand usage, with invariability being the *exception*, not the rule. Individuals are not uniformly right- or left-handed, but vary in the strength and/or direction of their hand preferences according to the task being performed, with both authors stressing that no one except a one-armed person would be exclusively right- or left-handed. Furthermore, Lauterbach provided good evidence to show that so-called left-handers tend to be much less consistently unilateral than normal right-handers. In a survey of 1,061 subjects (approximately half of whom were said to be left-handed and half right-handed), the author used an inventory of 50 items to assess both direction and degree of hand preference. The distribution of handedness indices (H.I.) derived from this subject sample produced a strongly skewed U shaped distribution, with a much sharper gradient for the right-handers (median H.I. = 93.3) compared with the slope of the curve for the left-handers (median H.I. = 16.0; where an H.I. of 0.0 reflects a consistent left hand preference and an H.I. of 100 indicates a completely consistent right hand preference). Similar findings were reported by Humphrey (1951) who used a questionnaire of 20 items in studying the handedness characteristics of 70 left-handed adults (half of whom wrote with the right hand), and a control group of 35 right-handers matched for age and education. The results showed that the right-handers were generally very consistent, with 93.1% of all their preferences being for the right hand, 4.2% for the left, and 2.7% for the use of either hand. The left-handers who wrote with their left hand were appreciably less consistent from task to task, with 71.4% of all preferences being for the left hand, 20.6% for the right, and 8% for either hand, while the left-handers who wrote with their right hand were least consistent, with only 49.9% of all their preferences being for the left hand, 37.7% for the right, and 12.4% for the use of either hand.

Unfortunately, it is not entirely clear how Humphrey made up his two groups of left-handers. It appears that they were selected on the basis of their reputation for being left-handed or ambidextrous, and it seems more than coincidental that exactly half of them wrote with the right hand. Consequently, it is difficult to know to what extent either of these groups of left-handers can be said to be representative of left-handers in the population. Nevertheless, in a follow-up of Humphrey's report, Benton, Myers and Polder (1962) questioned 106 people (who were employees of the Iowa University Hospital or relatives of patients) concerning their hand usage. Subjects were first asked if they considered themselves to be right-handed or left-handed. They were then asked if they considered themselves to be "strongly" right- or left-handed (as appropriate), and finally, which hand they employed in writing, cutting with scissors, and in using a screwdriver. The authors found that whereas 94% of the 66 self-confessed right-handed subjects were completely consistent in using the right hand for all three tasks, only 50% of the 40 self-classified left-handers were similarly consistent. Furthermore, Benton et al. pointed out that the inconsistency of the left-handers could not simply be ascribed to the preferential use of the right hand for writing since only 4 of the left-handers wrote with the right hand.

In a study which supports these conclusions, Borod, Caron and Koff (1984) gave a seven item questionnaire to 146 neurologically normal adults who were classified as right- or left-handed according to which hand they used for each of the following tasks: throwing, hammering, cutting, writing, using scissors, turning a doorknob, and using an eraser. According to their responses, points were allotted to each right and left preference expressed by each subject on each question and a "dominance ratio" was computed from the scores summarizing their overall consistency of preference. In evaluating the results, the authors found a significant difference between the dominance ratios of the right- and left-handers, with left-handers not only proving less consistent on the average than the right-handers, but also displaying a greater range of inconsistency from one subject to another.

But again, neither Benton et al. (1962) nor Borod et al. (1984) indicate how they recruited their subjects so that their results, like those of Humphrey (1951), are not necessarily representative of left- and right-handers in general. Estimates from a variety of sources over the years have quoted figures for the incidence of left-handedness ranging from 1% to 30% (Wile, 1934 p.68) depending upon the method of measurement and the population studied, with 11% to 14% being a frequently reported level for many Western societies today (see Chapter 10).

Recognizing the need for a standardized method of assessing handedness preferences, Oldfield (1971) attempted to provide a short, sensitive inventory of questions, which were easy to administer in the practical situation. In doing so, he realized that both the number and type of questions could be important, but in the absence of any firmly based knowledge of the underlying mechanism of handedness, he decided to use a modified version of Humphrey's (1951) questionnaire as a starting point. From the results of a preliminary survey of 1,128 university students and an item analysis of their responses, he devised the 10 item "Edinburgh Inventory" which has been used widely ever since.

Several other questionnaires have also been devised (Fennell, 1985) varying in length but often using many of the same items. The test-retest reliabilities of all the more frequently employed questionnaires tend to be good with some of the common test items such as writing, drawing, and throwing being more reliable than others, especially two-handed activities such as using a broom or opening a box. The reliability of a questionnaire item is undoubtedly dependent on the specific demands of a task; for example, in a questionnaire study of an unselected sample of 934 male university students by Provins, Milner and Kerr (1982), 95% of the subjects stated that they always used the same hand for writing while 4% said they usually did, but for carrying a suitcase, only 9% reported that they always used the same hand with 44% expressing no preference. And in a two-handed task such as washing dishes, 43% of the males always held the dish in the same hand, 35% usually did so, 12% expressed a slight preference for the same hand and the remaining subjects gave no preference at all. Similar trends were recorded for an unselected group of 1,032 female students reported in the same study.

However, the reliability of a test item or instrument indicates nothing about its validity and may even be misleading. For example, the writing hand has frequently been found to be one of the most reliable items for which a hand preference is expressed (Hull, 1936; Raczkowski, Kalat & Nebes, 1974), yet it is evident from the work of Benton et al. (1962) and Borod et al. (1984) that for left-handers in particular this is not a good guide to their general handedness tendencies. In fact, Chapman and Chapman

(1987) found that of 1,306 male students who wrote with the right hand, 89.7% were classified as right-handed on a 13 item questionnaire and 10.2% as ambilateral, yet of the 169 male students who wrote with the left hand only 61.5% were classified by questionnaire as left-handed and 38.5% ambilateral. Similar results were reported for 1,278 right-handed and 149 left-handed female university students. Clearly, the primary concern for any handedness measure should be to provide a true reflection of each person's handedness characteristics - in other words, to ensure that it is a valid measure. Some investigators have attempted to validate their questionnaires by presenting their subjects with the same test items in a second situation requiring a demonstration of use to check the degree of agreement between verbal and actual responses (Hull, 1936; Raczkowski et al., 1974). But again, such a procedure can be quite misleading since there is no assurance in such a process that the limited range of items selected for questioning and testing are indeed appropriate selections.

To investigate the importance of both type and number of questions used in a questionnaire, Provins et al. (1982) constructed a 75 item inventory in which was embedded the Edinburgh Inventory. To enable simultaneous comparison of both type of question and number of items, a third questionnaire of the same (10 item) length as the Edinburgh Inventory was also constructed by randomly selecting 10 of the complete range of 75 items for separate evaluation. To eliminate volunteer bias, 1,966 unselected psychology students were given the 75 item questionnaire in class and were required to hand it in before leaving, with 147 students completing the form a second time under the same conditions some six months later (establishing test-retest reliabilities ranging from 0.69 for the Edinburgh Inventory to 0.76 for the 75 item questionnaire). The responses for each subject were evaluated in terms of a "Handedness Index" calculated separately for each of the three questionnaires and their distributions compared. As might be expected, the results showed all three distributions to be heavily skewed to the right-handed end of the scale, but that the most skewed distribution of all was for the Edinburgh Inventory.

This appeared as a bi-modal distribution, truncated at the extremes but more particularly at the right hand end, whereas the 75 item results and more clearly the randomized 10 item results tended towards a normal distribution with a long negative tail. In fact, the overall effect of changing the type and number of questions asked not only produced a significant reduction in the degree and strength of handedness recorded (left or right), but also brought about a significantly greater reduction for the left-handers than for the right-handed subjects. The actual shift in the median Handedness Index value derived from assessments on the Edinburgh Inventory compared with assessments on either of the other two questionnaires was approximately two or three times greater for left-handers than for right-handers. Indeed, of the 184 subjects assessed as being left-handed by the Edinburgh Inventory, 27 became clear right-handers when evaluated by the random 10 item questionnaire, including two who could easily have been classified as strongly right-handed.

From these results, it is clear that most right-handers are sufficiently consistent in their hand usage from one task to another for their handedness assessment to be only slightly affected by whatever questionnaire is used, whereas left-handers tend to be much less consistent, with the possibility that some 15% may be more correctly classified as right-handers. The problem then arises, how wide-ranging should a questionnaire be and what type of question should be asked? Even 75 questions can only provide a relatively restricted insight into an individual's behaviour. However, the larger the size of a questionnaire, the more difficult it is to administer or to gain the cooperation of subjects and sustain their objective approach to each and every question. Conversely, the fewer the questions, the more critical each one becomes and the more important it becomes to establish its validity.

At one time, many investigators assumed that hand preference was determined by some single underlying factor which would express itself in the internal consistency of results obtained from a suitably constructed questionnaire. Research by a number of investigators using factor analysis of questionnaire data tended to support this view in respect of major well-practiced unimanual skilled activities (Briggs & Nebes, 1975; Bryden, 1977; Loo & Schneider, 1979; McFarland & Anderson, 1980; Richardson, 1978; Roszkowski, Snelbecker & Sacks, 1981; White & Ashton, 1976; Williams, 1986). But it is notable that these studies involved questionnaires of between 7 and 18 items and that later reports using appreciably longer questionnaires arrived at somewhat different conclusions (Beukelaar & Kroonenberg, 1983; Dean, 1982; Healey, Liederman & Geschwind, 1986; Liederman & Healey, 1986; Steenhuis & Bryden, 1989). Dean for example, used a 49 item inventory and isolated six different factors, one of which related to activities involving eye preference, one concerning ear preference and one relating to the use of the feet. Healy and her colleagues used a 55 item questionnaire which did not include any eye, ear or foot activities and isolated four specific factors which they identified in terms of characteristic differences in the programming of movements required by the activities sampled or the type of musculature involved and its innervation. However, Steenhuis and Bryden who also excluded eye, ear and foot items from their study suggested that the four handedness factors they identified were more likely to be distinguished on the basis of the skill required in sequencing complex motor activities. Again, since many of the items included in the Healey et al. questionnaire did not appear in the 60 item inventory used by Steenhuis and Bryden, it is not surprising that the conclusions from all three of these investigations were different. Indeed, the results tend to emphasize once again that both the number and type of questions asked in a handedness questionnaire are important. As Bryden (1987, p.64) bluntly put it "The problem is that questionnaires and factor analyses give you what you put into them. If you don't ask about hammering, you won't find out about hammering; if all of your questions are concerned with fine finger movements, you are unlikely to detect a gross movement factor".

Similar arguments apply to measures of foot preference although far fewer criteria have been used to examine population trends for footedness. For example, in most large-scale surveys which have included a reference to footedness, the usual, if not the sole question posed relates to the foot preferred for kicking a ball (e.g. Komai & Fukuoka, 1934; Nachson, Denno & Aurand, 1983; Teng et al., 1979). While such a criterion has been shown to have a high reliability (Coren & Porac, 1978; Coren, Porac & Duncan, 1979; Hull, 1936), the question remains as to whether the foot used for kicking is also the foot preferred for other activities. Porac et al. (1980) determined the foot preferences of 962 subjects using three different criteria (kicking, stepping up onto a chair, and picking up a pebble with the toes), and reported inter-correlations of 0.37 and 0.44 respectively between kicking and the other two tasks. And the extent to which the foot preferences for each of these three tasks correlated with the hand preferences of the same subjects on four different handedness criteria (throwing, drawing, using an eraser, and dealing a card) was found to vary between 0.24 and 0.63. Gardner (1941) used 7 different criteria for foot preference, and as a result of testing some 90 university students (approximately half of whom were male and half female), she reported that the overall findings for both sexes showed no clear preference for either foot. Averaging over the seven tests yielded a 41% preference for the left foot and a 59% preference for the right, with a good deal of variation between tasks - e.g. for kicking, 89% preferred the right foot, but for pressing down on a garden spade, only 46% used the right foot. And Chapman, Chapman and Allen (1987) tested 220 university students on 13 different foot preference criteria and found item-scale correlations varied between 0.26 and 0.79. Examining the correspondence between handedness and footedness in 311 subjects, these authors reported that 94% of their right-handers could be classified as right-footed, but only 41% of the left-handers could be considered to be left-footed. But while this finding is generally in keeping with other reports (see e.g. Dargent-Pare et al., 1992; Peters, 1988), the numbers of footedness items used is still relatively small and arbitrary in nature.

From the evidence presented, it is clear that both hand and foot preferences vary considerably from one task to another and that it is difficult to justify the common practice of simply summating preference scores across questions to arrive at an overall gross estimate of an individual's handedness characteristics. If both the number and type of criteria are important in preference assessments of laterality, to what extent have these problems been addressed in proficiency testing and what is the nature of the association if any, between preference and performance measures of handedness? Evidence relating to both these questions will be considered in the following two sections.

(b) Assessing proficiency

If subjects are asked to move (slowly and carefully) a designated finger or pair of fingers on one hand without moving any of the other fingers, it appears that the outcome may or may not be more successful for one hand than for the other, but that greater success with either hand on this task may have little or no association with overall hand preference determined by questionnaire (Jason, 1986; Kimura & Vanderwolf, 1970; Parlow, 1978). Similarly, an examination of response times to visual stimuli in a reaction time situation may produce results which are faster for one hand than for the other but are not necessarily related to the subjects' hand preferences assessed by other means (Annett & Annett, 1979; Baxter, 1942; Kerr, Mingay & Elithorn, 1963). It has also been shown that in the production of simple graded movements of the index finger, the accuracy achieved by the preferred side may not be significantly different from that recorded by the non-preferred side (Provins, 1956). Somewhat mixed results have been obtained for tests of hand steadiness (see e.g. Edwards, 1948; Simon, 1964) and for muscular strength on the right and left sides (see e.g. Schmidt & Toews, 1970; Toews, 1964), but in general, any differences reported tend to be small and of doubtful reliability (Provins & Cunliffe, 1972b).

However, differences between sides have been consistently reported in tasks which involve rapid repetitive movements. For example, the maximum rate of alternating flexion and extension movements about the metacarpo-phalangeal or elbow joints was found to be significantly faster for the preferred hand (Provins, 1956, 1958). Furthermore, the time interval recorded between successive movements was found to vary less for the preferred hand. Similarly, Peters and Durding (1979) and Todor and Kyprie (1980) have examined the maximum rate of tapping with the index finger and found that the preferred hand was not only significantly faster than the non-preferred side but was also significantly less variable in the time interval between taps. Other investigators have also reported significantly faster performances for the preferred side in the maximum speed of repetitive tapping made by various finger and arm movements (Kimura & Davidson, 1975; Todor, Kyprie & Price, 1982).

In a rather more complex repetitive task which required the coordination and timing of the flexor and extensor muscles acting about both the elbow and shoulder joints to produce a rotary motion of the hand in turning a crank handle, Glencross (1970) reported that the maximum speed of handle turning was significantly faster for the preferred hand. In addition, analysis of the strain-gauge records of torque applied to the handle in successive rotations showed that the consistency of cycle length was significantly and positively correlated with speed of cranking for both the preferred and non-preferred hands but that this association was significantly greater for the preferred side.

Thus, the evidence suggests that for tasks involving simple isolated (i.e. single) muscle contractions, no differences in performance between sides may be detected. However, where such contractions are serially organized in a motor program which must be run off within time constraints that preclude the benefit of correction from sensory feedback (Schmidt, Zelaznik & Frank, 1978), differences between sides do appear. Furthermore, such differences in overall (speed of) performance seem to be closely associated with differences in the recorded variability of individual components - the greater the consistency, the higher the rate of repetition. This is presumably explicable in terms of predictability of outcome of the motor program. Clearly, the more predictable the production of a movement or sequence of movements, the faster the performance may be since less time is needed for intermittent adjustments or corrective action.

However, systematically varying the level of difficulty by adjusting the accuracy requirements within tests of movement speed has also been shown to influence the performance differences recorded between the two sides. For example, Steingruber (1975) tested subjects on two tasks: (a) dotting in circles, and (b) tapping in squares, using three different circle diameters and square side lengths. He reported that as the accuracy demands and difficulty of the task increased, performance differences between sides favouring the (dominant) right hand also increased. Similarly, in an experiment by Flowers (1975) which involved subjects making rapid alternating movements between two targets which were systematically varied in size and distance of separation according to Fitts (1954) "Index of Difficulty", significant differences between sides in favour of the preferred hand were recorded at the higher levels of difficulty but not at the easier levels.

However, as Welford (1968, p.147) pointed out, in the Fitts type of task, "when the targets are wide and the distance short, the subject uses very much less than the full target width" so that for both Steingruber (1975) and Flowers (1975), the lack of proficiency differences between hands in the easier situations may have been due to their using target size and not actual response distributions (within the target) in assessing subject performances. Certainly, this interpretation gains support from the results of a study by Annett, Annett, Hudson and Turner (1979) using a pegboard task and three levels of movement amplitude and accuracy (determined by the tolerances between peg size and hole diameter). These authors not only found significant overall differences in performance between the (preferred) right and left hands on this task, but also reported that while movement accuracy had a significant effect on these performance differences, movement amplitude did not. Furthermore, in a fine-grain analysis of the performances of the preferred and non-preferred hands taken from a high speed film recording of each phase of the movements, they found that the major difference between hands was in the positioning element of the movement cycle. This was due to the primary aiming movement of the non-preferred hand being less accurate and consequently necessitating 50% more secondary or correcting movements than for the preferred hand. From this and a consideration of other relatively constant features of the performances recorded, Annett et al. concluded that the difference between sides was simply due to the motor output of the non-preferred hand being more noisy (i.e. less predictable) than that of the preferred side.

The same explanation can be applied to the performance of other aiming tasks that are not continuously repetitive but which clearly require the rapid execution of many muscle contractions in a predetermined sequence such as those involved in accuracy of throwing. In the game of darts for example, both Provins (1956) and Watson and Kimura (1989) have found significant differences in performance between the two sides in favour of the preferred hand. And in a more detailed investigation by Roy and Elliott (1986), these authors recorded the movement times and errors of subjects making single aiming movements of the preferred and non-preferred hands with and without continuous visual feedback. Like Annett, Annett, et al. (1979), they found that in movements completed in less than 200 msecs, the preferred hand was significantly more accurate than the non-preferred hand and that this difference could not be attributed to an effect of visual feedback.

Further consideration is given to the underlying nature of the relative motor proficiencies of the right and left hands in Chapter 7, but here it may be asked, to what extent are such differences between sides on one type of task consistent with the differences recorded on another? Most intercorrelational data on performance differences between the two sides over a wide variety of motor tasks in older children and adults all suggest a low or generally poor correspondence in handedness across tasks (see e.g. Borod, Caron & Koff, 1984; Buxton, 1937; Durost, 1934; Johnstone, Galin & Herron, 1979; Orlando, 1972; Rigal, 1992; Sappington, 1980). As Borod et al. concluded from their comparison of results from nine performance tests of handedness, the size of a given individual's hand dominance appears to depend on the specific performance measured. The handedness score obtained by a subject on the laterality continuum of one task bears little relationship to the handedness score achieved by that same individual on the laterality continuum of another task. In other words, the relative achievements of the right and left hands of someone undertaking any given motor activity appear to be task specific and are relatively unpredictable from one task to another. Even in four tests of manual proficiency that were designed to be closely similar in their skill demands to her standard pegboard test, Annett (1992b) found that between-task correlations of performance differences between the hands still only reached from 0.38 to 0.65, thus accounting for less than half the variance common to the pegboard test and any of the other tasks. Hence, as in animal studies and investigations of human hand preferences, although there appears to be an advantage for most subjects to use the same hand for different tasks, the degree of consistency in hand preference or proficiency tends to be low, and to be very dependent on the particular tasks being considered.

(c) Preference and proficiency compared

Although it has often been assumed that hand preference and performance differences between the hands are simply different forms of expression of some common underlying lateral determinant, Porac and Coren (1981, p.12) suggested that the evidence on the question is not convincing. These authors reviewed a number of studies in which subjects had been tested on various manual performance tasks and their hand proficiency scores compared with overall handedness ratings derived from a preference questionnaire. They found a mean agreement of 74%

between preference and skill or strength tests but with a fair degree of variation from one test to another. In contrast, some authors have reported close agreement between the two measures. For example, Annett (1985, p.211-223) concluded from tests on a number of different sample populations, that there is a good linear relationship between the results of subjects grouped according to particular combinations of their hand preferences (derived from a 12 item list of observed activities) and the performance differences between hands for the same subjects recorded on a pegboard task. A similar linear relationship was reported by Peters and Durding (1978) between the results obtained from a 7 item hand preference scale and relative hand proficiency on a finger tapping task.

In two further studies which have used a simple manual aiming test, data have been reported which contribute another interesting aspect to the problem. In the first study, Lake and Bryden (1976) found a significant positive correlation of 0.78 between the laterality scores of their manual performance task and the results of a handedness questionnaire. But when they tested the association between the two measures for right- and left-handers separately, they found a correlation of only 0.44 for the right-handers alone and 0.0 for the left-handers. Similarly, Tapley and Bryden (1985) using a group version of the same manual task, also found a high correlation of 0.75 between the overall performance test results and preference inventory data, but when the preference and performance results from the right- and left-handers were considered separately, the correlation values dropped to 0.17 and 0.20 respectively. Furthermore, consideration of the hand-difference distributions for the two handedness groups suggested to the authors that the right- and left-handers might represent two distinct sub-samples of the general population. They concluded that preference and performance measures probably assess different aspects of handedness.

In discussing the overall characteristics of the hand proficiency scores of subjects tested on several different motor performance tasks, Borod et al. (1984) remarked that the distributions for the left-handers and right-handers considered separately were normal and similar in central tendency, shape and range, and tended to be mirror-images of each other, suggesting that subjects in general tended to be faster, stronger and more accurate on their preferred side. In contrast, the distributions of preference scores based on responses to a 9 item inventory by the same subjects, were described as J shaped and significantly different for the two handedness groups. The left-handers were less lateralized than the right-handers and less consistent in their preferences so that their questionnaire scores were more widely dispersed and less skewed to the preferred side.

These findings are consonant with many other reports in the literature which focus on two particular considerations that appear to be inextricably linked, viz. (a) the characteristic shape of the population distribution of handedness bias, and (b) the relationship between preference and performance measures of handedness. In fact, Porac and Coren (1981, p.17) in their summary of the relevant literature showed the J shaped distribution was typical of preference studies and that the right-biased normal distribution was typical of performance investigations of handedness. Furthermore, these authors even suggested that hand proficiency and hand preference may be unrelated (Porac & Coren, 1981, p.12) - a view which Bishop (1989) considers does not necessarily follow from the evidence, proposing instead a statistical explanatory model in which the probability that one hand will be preferred for a given activity is directly proportional to the relative proficiency of the two sides.

But there are other good reasons for reconsidering the evidence on which these characteristic curves are based, since the data are not strictly comparable. First, hand preference assessments are typically consistency measures (i.e. each subject's averaged response scores) obtained from a list of highly selected questionnaire items considered to be "good" indicators of handedness. Such good indicators are usually the product of the investigator's intuition, although in some instances, e.g. the Edinburgh Inventory (Oldfield, 1971), the questions finally employed were derived from a longer list through the refinement of an item analysis. Not surprisingly, the use of such a highly selected series of questions tends to produce strongly lateralized responses from subjects, which results in their being placed in extreme handedness categories (right or left), thereby giving rise to a population distribution in the shape of a J curve. But everyday activities for most people do not centre around ten or twelve highly selected unimanual tasks, and it has been shown that by using a less biased or larger selection of questions, a relatively normal distribution of handedness preference scores is obtained (Provins et al., 1982).

Second, the typical normal distribution of performance differences between hands is usually derived from the results of testing subjects on a single motor task selected especially for its ease of measurement and convenience of administration, since performance testing is extremely time consuming. An examination of the distribution of hand differences on eight such motor tasks performed by the same group of subjects (Provins et al., 1982) revealed that while they varied somewhat in central tendency and dispersion from one task to another, they were all relatively normal in shape with one notable exception (handwriting) which is referred to again below. However, in spite of overall distribution similarities, as Heinlein (1929) and Borod et al. (1984) have shown, a given individual's hand difference score on one task may have little or no relationship to that same individual's score on another task.

Third, as well as the inadmissibility of comparing single task performances with averaged preference data, a further difficulty in reconciling the results from preference and performance studies stems from the very different tasks considered in each type of investigation. For example, preference inventories commonly include questions on the hand preferred when using familiar everyday objects such as a comb, toothbrush, hammer, scissors and tennis racquet or in dealing playing cards, striking a match and threading a needle. Clearly, most of these activities do not readily lend themselves to a quantifiable comparison of proficiency between hands so that performance tests have commonly used a variety of tapping, pegboard and specially devised dexterity tests which are likely to be relatively novel to the subjects (see e.g. Borod et al., 1984. p.184).

In an investigation which attempted to overcome some of these problems, Provins and Magliaro (1993) compared the performances of self-classified right- and left-handed subjects on a handwriting and grip strength task with the preferences they expressed on questions concerning handwriting and grip strength included in a 20 item questionnaire. A comparison of overall preference with differences between hands on the handwriting task for all subjects produced a significant positive correlation (0.78) which disappeared when the right- and left-handers were examined separately. A similar effect was obtained for the corresponding comparisons between preferences and performances using the grip strength data, confirming the findings of Lake and Bryden (1976) and Tapley and Bryden (1985) on their aiming tasks. However, examining the specific items on the questionnaire relating to handwriting and grip strength without regard to handedness classification proved more constructive. For handwriting, subjects' preferences were both strong and consistent from one subject to another and corresponded completely with the more proficient writing hand (whether left or right). For the grip strength task, preferences varied considerably from subject to subject in both degree and direction, and this time the correspondence between hand preference and the better performing hand (irrespective of handedness classification) was of only marginal significance. Population distributions of performance differences between sides for both these tasks published previously (Provins et al., 1982) show that for handwriting it is bi-modal and J shaped, whereas for grip strength the distribution is normal but skewed to the right. Since handwriting is probably the most favoured criterion which appears in almost all handedness questionnaires however brief, the evidence suggests that the task rather than the method of measurement *per se* is probably the main determinant of the shape of the population distribution.

Other relevant evidence on this question has been published by McManus (1985b) on over 12,000 schoolchildren (classified as right- or left-handers by writing hand) on two performance tests of handedness: (a) square-marking and (b) match-moving. The first (a) square-marking task was a paper and pencil test involving some of the same features as handwriting, i.e. holding a pencil to place marks on a piece of graph paper. In many respects it resembled Tapley and Bryden's (1985) aiming task, but whereas the latter authors reported a clear separation of the hand-difference distributions for their right- and left-handed subjects, McManus reported overlapping bi-modal distributions. But as Tapley and Bryden used university students compared with McManus' 11 year-olds, it seems very likely that the difference in age and experience of the subjects employed in the two investigations could well explain the difference in degree of separation of the respective hand proficiency distributions. The second (b) match-moving task involved picking up matches one at a time and putting them into a matchbox drawer, an activity that appears comparable with some of the tasks examined by Steenhuis and Bryden (1989) in their factorial analysis of responses to two hand preference inventories. It is noteworthy then, that for this task, the complete overlap of the hand-difference distributions for the right- and left-handers of McManus' subjects agrees remarkably well with Steenhuis and Bryden's finding of a factor specific to picking up small objects - characterized by relatively low or either-hand preferences.

Thus, the evidence suggests that hand preferences and hand proficiency differences on performance tasks may well be in agreement at both the individual and population level providing the same tasks are measured and due account taken of the age and experience of the subjects. Furthermore, it is clear that the findings are remarkably similar to those reviewed earlier with respect to animal handedness, although direct comparisons between the manual asymmetries of human and nonhuman species have been few and far between (see e.g. Seltzer, Forsythe & Ward, (1990).

CHAPTER 6
Ontogenesis of Handedness

Chapter outline

In early infancy, various motor asymmetries associated with head and arm movements and body posture have been observed and proposed as being related to, if not precursors of, later hand preferences. However, such asymmetries appear to be inconsistent and/or temporary behaviours that are more likely to be a reflection of the cumulative pre- and/or post-natal influences on the infant up to that time. An initial absence of stable hand preference during the first few months of life appears to give way to a more frequent and consistent use of one particular hand by about two to three years of age.

For the majority of children this favours the right hand in most activities, but even so, there may well be some 25% to 35% of children with mixed hand preferences at five or six years of age. By about ten years of age, most of the mixed-handers have become right-handers although the classification of individuals as right- or left-handed etc. is, of course, dependent on the particular number and breadth of handedness categories used to classify individuals as well as the number and type of hand preference criteria employed.

Using performance data, the evidence also shows that for many manual activities, proficiency differences between the sides become more clearly differentiated with age (usually in favour of the right side), although the recorded developmental trends on one task are not necessarily indicative of the trends which may occur on any other task.

Infant motor asymmetries

Systematic studies of several different motor activities observable during the first few months of life (e.g. hand and arm movements, asymmetric head or body posture, head turning and responses to stimulation) have often reported a majority preference for the right side and/or a right-sided orientation. But an extensive and detailed critical evaluation of the evidence on these lateral tendencies (Provins, 1992), has shown that there are good grounds for concluding that such behavioural asymmetries are most likely to be temporary habits that reflect the cumulative pre- and/or post-natal experiences of the infant up to that time, with little or no predictive value for later handedness development.

Certainly, observational studies of normal infant handedness using such criteria as the hand most frequently preferred for unimanual reaching and/or manipulating some presented object, have typically found that at or before about six months of age, there is no consistent preference for either side, but that by seven or eight months, an increasingly detectable difference between sides (usually in favour of the right) begins to emerge (Lippman, 1927; Ramsay, 1980), and to become more clearly evident in as many as 80% of subjects by about three years of age (Treves, Goldschmidt & Korczyn, 1983). These trends have been confirmed by Cornwell, Harris and Fitzgerald (1991) who found that in a study of three groups of infants 9, 13, and 20 months of age, of the 22 9-month-olds, 20 demonstrated no hand preference and 2 a significant preference in favour of the left hand; of the 18 13-month-olds, 5 displayed no hand preference, 10 showed a significant preference for the right hand and 3 a similar preference for the left; while of the 23 20-month-olds, 20 demonstrated a significant preference for the right hand and 3 showed no significant preference for either hand.

There is also ethological evidence that if a large number of observations are made in a free play situation over a very wide range of unimanual activities, a reliable change from no hand preference in early infancy to a majority significantly favouring the right by about the end of the first two years of life may be found in an unselected population sample (Provins, Dalziel and Higginbottom, 1987). Corresponding results have been reported by Rice, Plomin and De Fries (1984) and Plomin and De Fries (1985) in their longitudinal study of infants assessed for hand preference in free play, semi-structured and structured play situations. These authors recorded a significant change in handedness tendencies with age in their subjects, and noted that whereas only 8% of the infants exhibited a clear preference at 12 months of age, approximately 30% were lateralized at 24 months.

Undoubtedly, individual hand preferences in this age range are still very unstable. For example, in a longitudinal study by Gottfried and Bathurst (1983) on 89 children tested for the hand preferred for drawing with a crayon at five six-monthly intervals from 12 to 42 months of age, these authors reported that of the 48 males, 50% were inconsistent in hand use from one test occasion to another, and of the 41 females, 44% were inconsistent. Similarly, Archer, Campbell and Segalowitz (1988) made a longitudinal study of hand preference using an 11 item inventory (which included both unimanual and bimanual items) on 49 children tested at 18, 24 and 30 months of age, and also found that over this period, as many as 41% of the girls and 33% of the boys changed their hand preference (when classified as either right-, left- or mixed-handed).

In a longitudinal investigation of the acquisition of manual skill involving the analysis of video-taped recordings of behaviour, Connolly and Dalgleish (1989) examined the development of unrestricted self-feeding with a spoon in two groups of infants, one aged between 11 and 12 months, and the other aged between 17 and 18 months at the beginning of the study. Comparisons of each child's activity during monthly sessions over a period of six months, and between the younger and older subjects, showed that apart from systematic changes in the pattern of movements made with the development of the skill, there was an increasing consistency with age in the hand used, with most subjects displaying a preference for the right hand. In a similar detailed video-taped analysis of the unrestricted painting activities of 49 nursery school children ranging from 34 months to 58 months of age, Connolly and Elliott (1972) again found a significant tendency for hand preference to become more pronounced with age, with most subjects favouring the right hand.

Thus, in spite of the considerable differences between the experimental design of these investigations, ranging from completely unstructured situations to specific testing programs, and from multiple hand preference criteria to single item reporting, there is considerable agreement that hand preferences only become evident with age. However, the likelihood of detecting a stable pattern of hand preference or a change in observed preference with age is almost certainly dependent upon the age span being studied, the number of test occasions employed, and the coarseness (or otherwise) of each handedness category used. In addition, investigations that only provide information on group trends can give little or no indication of individual development. In this regard, it is worth recalling that in their study of reaching behaviour in 32 infants tested periodically from 24 to 52 weeks of age, Carlson and Harris (1985) showed that analyses confined to group data conceal a striking variability of hand preference not only between individuals, but for individual infants from one test period to another. They concluded that until at least 52 weeks of age, hand preference may be highly variable, and that sampling an infant's preference on any one occasion during this phase may provide a very poor basis from which to predict hand preference as little as three weeks later. Nevertheless, from the evidence so far reviewed, it is reasonable to conclude that prior to about six months of age, infants generally display little or no consistent preference for one hand or the other, but that by about two or three years of age, a significant tendency emerges for the majority of infants to show some unilateral preference - usually for the right hand. However, there may be considerable inter- and intra-individual variations from one occasion of testing to another and from one type of situation to another.

Hand preference trends during childhood

The evidence relating to changes in hand preference with age for kindergarten and grade-school children tend to confirm these general trends, although the results from different studies may sometimes appear to be incompatible. In this respect, it will again be seen that such variables as the handedness classification system employed, and/or the handedness criteria used, can seriously influence the results obtained and pose important questions for the interpretation of findings.

To take cross-sectional studies first, Harris (1957) reported that for two age groups of an unselected sample of 245 schoolchildren tested on five hand usage tasks, he found a marked decrease with age in the incidence of mixed-handedness between the seven-year-olds and the nine-year-olds, and a parallel increase in right-handedness. Similarly, Belmont and Birch (1963) examined 148 children 5 to 12 years of age for hand preference on four specific unimanual activities and classified their subjects as right- or left-handers if they performed all four tasks with the same hand, or mixed-handed if there was any inconsistency between tasks. These authors reported a significant reduction in mixed-handedness with age and suggested that preferential handedness becomes reliably established by about 9 years of age. Certainly, studies of students older than this tend to show little or no change in the direction of their general hand preferences with age (Bryden, Macrae & Steenhuis, 1991; Roszkowski, Snelbecker & Sacks, 1980).

In apparent contrast, Longoni and Orsini (1988) reported no change in hand preference across 271 4-, 5- and 6-year-old children tested on a six item inventory of unimanual activities using a simple dichotomous grouping of subjects, although for the total subject sample the mean percentages of children classified as right-, mixed- and left-handed were 65.7%, 32.6% and 1.5% respectively. These data are similar to those obtained by Coren, Porac and Duncan (1981) on 384 3-, 4- and 5-year-old (pre-school) children who were compared with 171 high school students (mean age 16 years) using similar handedness criteria. The corresponding percentages for right-, mixed-, and left-handers in their pre-school sample were 68.3%, 25.5% and 6.2%, and in the high school sample, 80.7%, 13.5% and 5.8% respectively, which, like the Belmont & Birch (1963) data, yielded a significant trend away from mixed- to consistent right-handedness with age. But when the children's preferences were simply dichotomously classified as either right or left, no significant effect with age was found. And in a study of 6,760 schoolchildren who were again dichotomously classified as right- or left-handed on the basis of three handedness criteria, Hardyck, Petrinovich and Goldman (1976) also reported no significant variation with age in the relative incidence of right- and left-handedness from grade 1 through grade 6.

Rymar et al. (1984) assessed 725 6- to 15-year-old schoolchildren on the basis of seven different handedness criteria and found a significant effect of grade level which, however, appeared to be a fluctuating rather than a systematic effect of age. Using the ten handedness criteria of Oldfield's (1971) Edinburgh Inventory, Brito et al. (1992) also found no age effect in the hand preferences of 4- to 7-year-old schoolchildren, but in a comparison of this data with the findings of their study on adults (Brito et al., 1989), they reported a higher incidence of mixed-handedness amongst the children. The percentages for male right-, mixed- and left-handers in the student sample were 57.6%, 39.0% and 3.4% respectively, with corresponding values of 63.7%, 32.9% and 3.4% for the adults. In three other studies (Curt, Meccario and Dellatolas, 1992; Curt, De Agostini, Maccario & Dellatolas, 1995; De Agostini, Pare, Goudot & Dellatolas, 1992), the hand preferences of 3 to 6 year-old children were assessed using eight handedness criteria, with results showing that while the proportion of strong right-handers increased with age, the numbers of inconsistent left- and right-handers decreased, and the proportion of strong left-handers tended to remain unchanged.

Over a somewhat wider age range, McCarthy (1970) reported on trends in the distribution of hand preferences of 1,032 children divided into ten age groups ranging from 2.5 to 8.5 years of age. She found that the proportion of subjects who displayed inconsistency in hand use across at least three of four motor activities tested, decreased from 33.3% of the youngest age group to 15.1% of the oldest. And whereas those who were completely consistent in their usage of the right hand for these various activities increased in number from 53.9% of the youngest children to 80.2% of the oldest, there was relatively little change in the proportion of left-handers with age. Furthermore, in a sub-sample of 125 children who were tested a second time approximately one month later, the author reported that whereas only 61% of children in the 3 to 3.5 years-old age group obtained the same handedness classification on each occasion, for those aged 5 to 5.5 years the corresponding figure was 70%, and for the 7.5 to 8.5 year-olds there was 76% agreement between the two assessments.

In a later study, Gudmundsson (1993) compared the hand preferences of random samples of pre-school children (aged between 3 years 0 months and 5 years 11 months) with primary school children (aged from 6 years 0 months to 9 years 11 months) assigned to five handedness categories on the basis of seven performance criteria, and found significant differences between the two groups. There was a relatively small drop (from 8.9% to 6.1%) in the incidence of consistent left-handedness between the pre-school and primary school children, but a correspondingly large increase (from 63.3% to 80.9%) in consistent right-handedness.

Similar trends have been reported in longitudinal studies of hand preference. For example, Fennell, Satz and Morris (1983) investigated 208 children using 10 unimanual hand preference items at three age levels (with means of 5.5, 7.7 and 10.9 years) and classified their subjects at each stage as either right-handed, left-handed or ambidextrous (i.e. mixed-handed) according to an overall laterality score. They found that the percentage of mixed-handers dropped from 12.5% to 4.3% between the first and last years of testing, and that most of the subjects changing from this category became right-handers. Similarly, Gaillard and Satz (1989) examined 124 children at two different ages (means of 5.75 and 9.9 years) using the Edinburgh Handedness Inventory (Oldfield, 1971), and found a significant shift in handedness from low unimanual consistency categories to greater lateral consistency with age, and again mostly towards right-handedness.

More recently, Dellatolas, Tubert-Bitter, Curt, and De Agostini (1997) published the results of a longitudinal study of the development of hand preference of 256 children aged between 3 and 6 years at the beginning of an investigation which lasted two years. Preferred hand use was assessed for eight items on five occasions during this time, and each subject classified into one of five categories on each occasion ranging from strong left (L), through weak-left (Lw), and mixed-handed (M), to weak-right (Rw), and strong right-handed (R). Left-handers (L + Lw) were deliberately over-represented (63) in the subject sample (as initially assessed) compared with right-handers (R + Rw = 181). The authors found that 54% of the children changed from one classification to another at least once over the two year period, and that overall, there was a general trend with time for numbers in the weak and mixed categories to decrease, and for numbers in the extreme categories to increase, particularly with respect to right-handedness.

A notable large-scale study by Whittington and Richards (1987) provides similar evidence. These authors reported on the hand preferred for two particular unimanual activities (writing and throwing) for over 11,000 schoolchildren, recorded at both 7 and 11 years of age. They found that the percentage of children classified as right-, mixed- and left-handed at age 7 was 79.3%, 13.2% and 7.5% respectively, which became 87.3%, 4.7% and 8.0% respectively at 11 years of age. In three further studies which produced results compatible with these findings, McManus et al. (1988) dichotomously classified their subjects (ranging from 3 to 9 years of age) as either right- or left-handed on a ten item hand preference inventory. These authors recorded a significant and clearly defined increase in degree of hand preference with age over this period but no developmental change in direction of laterality (i.e. between right and left).

Thus, the evidence from hand preference studies on young children tends to confirm and extend the earlier findings on infants of an increasing lateralization with age. But these studies also highlight the importance of the effect of type of experimental design and analysis in the actual findings recorded. For example, the likelihood of detecting a change in handedness tendencies with age is clearly dependent on both the age range considered and the number (and breadth) of categories of handedness used within the age range measured. Thus, the greater the age range, the greater the chance of finding an age effect. Similarly, no variation between categories or a reported change in the degree of hand preference within categories in one study using e.g. a dichotomous classification, may well be recorded as a change between categories in another which uses multiple hand preference classes.

Performance differences between sides

Not surprisingly then, similar considerations to those outlined above with regard to studies of hand preference also apply to the evaluation of evidence from investigations of handedness development using performance criteria. But the earliest age at which children can be tested for performance differences between the right and left hands clearly depends on their developmental maturity, interest, and cooperation in undertaking the tasks presented. For example, Briggs and Tellegen (1971) attempted to test 20 subjects at each age level from 3 to 18 years of age on a series of five manual dexterity tests. However, they found that since the 3-year-olds were either unable to understand what was required of them and/or unable to provide reliable results, the investigators were forced to exclude this age group from their findings. De Agostini et al. (1992) also described similar difficulties in testing their 3 year-old subjects.

As most performance investigations of handedness have reported on a very restricted range of motor tasks, and often on only one, it is necessary to consider the findings of different studies task by task, beginning with the most elementary. Hand grip-strength is one such task, and an increase in grip-strength with age has been well documented in the publication of data for both the right and left hands up to 32 years of age (see e.g. Fullwood, 1986; Jones, 1949; Mathiowetz, Wiemer & Federman, 1986; Woo & Pearson, 1927) before declining thereafter. For example, in their study of the grip-strength of 223 children ranging from 6 to 12 years of age, Spreen and Gaddes (1969) reported means (and standard deviations) of 8.8. Kgms (1.6) and 7.9 Kgms (1.8) for the right and left hands respectively of their 6-year-olds, which increased to corresponding values of 23.2 Kgms (3.4) and 21.7 Kgms (3.5) for their 12-year-olds. But are these differences between the two sides reliable, and if so, do they remain relatively constant with age?

In a study of 200 pre-school children, Methany (1940) reported that the proportion of individuals for whom the right hand was stronger (by 0.2 Kgms), equal to, or weaker (by 0.2 Kgms) than the left hand, was 42.5%, 22.5% and 35.0% respectively at three years of age, with corresponding values of 80.0%, 11.1% and 8.9% at six years of age. But the study by Krombholz (1989) provides evidence to suggest that group (or cross-sectional) data do not necessarily reflect individual trends. This latter author tested the grip-strength of 250 boys and 271 girls on three occasions in a longitudinal study from the beginning of the first year at school (mean age 81.5 months) to the end of the second school year (mean age 101.5 months), and found that when they first started school, the proportions of children recording the right hand as stronger, equal to, or weaker than the left were 51.1%, 8.1% and 40.9% respectively, with the corresponding figures for the end of the second year being 51.4%, 9.8% and 38.8%. Yet this apparent absence of change conceals the fact that 35% of the children who were reported to perform better with the right hand and 47% of those who did better with the left hand at the beginning of the first year at school, nevertheless performed better with the contralateral hand at the end of their second school year.

Some investigators have attempted to record the ability of children to make independent movements of particular fingers of each hand. For example, Ingram (1975) recorded the degree of success achieved by 84 right-handed subjects three to five years of age in trying to copy the finger positions of an examiner on two hand posturing tasks. She reported that the children were significantly better on both tasks with their non-preferred hand. However, in a similar task (the "Rey Test") used by Ingram that required the isolated movement of a designated finger of one hand, when unintended (i.e. "associated" or "mirror") movements were also recorded, the results tended to show that subjects discriminated movements better with their preferred hand. And in a study which paid special attention to controlling and recording the performances of 60 children (ranging from four to six years of age) on similar hand posturing skills, Ireland and Watter (1995) found no significant differences between the achievements of the preferred and non-preferred hands. Thus, the evidence suggests that in this type of manual task, and at this age level, performances are sufficiently variable for differences between sides to be somewhat unreliable.

Many other investigators have reported on a wide range of measures used to examine the ability of subjects to move a particular limb or part of a limb (e.g. a finger) without incurring unintended movements of any other body part. Excessive and/or persisting associated movements are regarded clinically as a sign of neurological immaturity or dysfunction (Fog & Fog, 1963; Stern et al., 1976), and in normal children their incidence decreases markedly with age, although there are wide individual differences (see e.g. Abercrombie, Lindon & Tyson, 1964; Cohen et al., 1967; Connolly & Stratton, 1968; Fog & Fog, 1963; Lazarus & Todor, 1987; VanSant & Williams, 1986; Wolff, Gunnoe & Cohen, 1983; Woods & Teuber, 1978b). Where quantitative measures of unintended movements of the supposedly inactive sides have been compared during active performances of the same motor task separately by the right and left hands, an age effect for each side has also been observed. For example, Edwards and Elliott (1987) compared the performances of 24 right-handed boys (mean age = 9 years, 7 months) with 24 right-handed men (mean age = 20 years, 4 months) in a unilateral rapid sequential finger-lifting task where the four fingers of each hand rested on movement-sensitive keys. The authors found that the children exhibited significantly more associated movements than the adults, although both groups of subjects demonstrated significantly more unintended movements of the supposedly inactive hand when the active hand was the non-preferred (left) hand than vice-versa. And a significant reduction in associated movements with training was recorded for the children but not for the adults whose error scores for either hand were low anyway.

Using a different manual activity, Njiokiktjien, Driessen and Itabraken (1986) tested 219 normal children between four and nine years of age on a wrist movement (pronation-supination) task and reported that overall, 49.9% of their subjects recorded unintended (i.e. mirror or associated) movements on the supposedly inactive side. But apart from the normal decrease in this activity with age, the authors also reported that at all ages, the associated movements (where they occurred) were larger on the right arm than on the left in all children except one, who showed the reverse trend. Cohen et al. (1967) also reported the same effect in their 6- to 12-year-olds, although more specifically, these latter authors noted that for subjects with definitely established hand dominance, only 8% recorded greater muscle activity on the contralateral side when the dominant hand was selectively active, compared with 68% who recorded greater muscle activity on the contralateral side when the non-dominant hand was making the deliberative movement.

Similarly, in a study of 176 right- and 19 left-handed school-children in grades 2, 4, and 6, Parlow (1990) reported that while contralateral associated movements were more common in the younger children, their incidence also varied with both task and subject handedness. Whereas right-handers more frequently demonstrated unintended activity with deliberative movements of the left hand, left-handers more often recorded involuntary associated activity when making deliberative movements with their right hand. VanSant and Williams (1986) also reported that age effects on the frequency of contralateral associated movements varied according to the task, although these authors failed to find any significant difference in the incidence of associated activity recorded for deliberative movements of the preferred compared with the non-preferred hand.

While these various investigations of associated movements show that using different tests or even different versions of the same test may give rise to somewhat different results, the evidence suggests that in general, the ability to pare-away unwanted muscle activity from wanted movements improves with age, and that this increase in discriminative ability occurs more often for selective initiation of a movement with the right hand (or fingers) in nominal right-handers than for a corresponding deliberative movement with the left.

Another frequently employed performance test of handedness is speed of finger tapping, and in most instances this has been confined to repetitive movements of the index finger. In a study designed to provide normative data on 449 children ranging from 6 to 12 years of age, Spreen and Gaddes (1969) reported a systematic increase with age in the mean number of taps made in six 10 second periods for the index finger of each hand. They reported means (and standard deviations) of 185.6 (19.6) and 169.5 (14.4) for the preferred and non-preferred hands respectively at a mean age of 6 years, increasing to corresponding values of 284.6 (31.6) and 245 (31.6) at a mean age of 12 years. Similarly, in a study of the tapping speeds of 169 children ranging from 5.7 to 14.4 years of age, Knights and Moule (1967) recorded a systematic increase in tapping speed with age for both sides (consistently favouring the preferred hand) but no significant interaction between the hand used and age. Peters and Durding (1979) also reported on finger tapping performances of the preferred and non-preferred hands of 508 children between 6 and 12 years of age and obtained much the same results.

Using two other tasks involving speed of finger tapping, Denckla (1973) recorded the results for the right and left hands separately in nominally right-handed children from 5 to 7 years of age. The first task required the subject to tap the index finger and thumb together as rapidly as possible and the time taken to make 20 repetitive movements was noted. The second task required the subject to make each finger in succession tap the thumb as fast as possible starting with the index finger, and the time taken to make 20 such successive movements was recorded. The results for the repetitive finger tapping task showed a significant increase in performance with age for both hands and a significantly better performance overall with the right (preferred) side. And the percentage of children who performed better on this task with the right hand compared with the left systematically increased with age. However, for the successive finger movement task, no significant differences were found between the right and left hands although the performance of each hand significantly improved with age. In a second and similar study using the same tasks, Denckla (1974) tested three different age groups of right-handed children 8, 9 and 10 years old. In contrast to the earlier study, apart from a significant difference between hands for repetitive finger movements, no other significant effects were found for either task or age.

In a unimanual task which involved whole hand and/or arm movements in tapping between two metal plates six inches apart for 12 seconds, Briggs and Tellegen (1971) recorded the better and poorer hand performances of 150 males and 150 females ranging from 4 to 18 years of age. These authors found that while the speed of tapping more than doubled for both hands over this age range, the differences between the hands also tended to increase, as typified by the number of taps recorded (and their standard deviations) of 26.3 (3.5) for the better hand and 24.1 (3.25) for the poorer hand of the 4-year-old males, compared with the corresponding values of 60.0 (7.59) and 50.9 (6.23) for the 18-year-old males. And in a similar task requiring subjects to tap as fast as possible between two plates 60 cms apart, Szmodis et al. (1984) recorded the performances of the right and left hands of 389 girls and 596 boys ranging from 4.4 to 14.8 years of age. The data from this investigation again show that for both sexes, while the mean tapping rate improved rapidly with age, the differences between the two sides also tended to increase steadily from a minimum for the younger children to a maximum for the older subjects. However, it is notable that in both the Szmodis et al. and Briggs and Tellegen studies, the performance distributions for each hand continued to overlap throughout the age ranges examined.

But perhaps the most often quoted study on the effects of age on performance differences between the two sides, is the investigation of 219 children ranging from 3.5 to 15 years of age by Annett (1970b). This author recorded the mean time taken by subjects in each of 24 age groups to complete a pegboard task with the right and left hands separately, and reported that while there was a marked increase in speed of performance by both sides with age, there was no systematic age effect on the difference in performance between sides. Mean right hand performances were consistently better than left hand performances for all age groups although the published standard deviations indicate very considerable overlapping of the distributions of individual differences for the two hands in the younger children that must raise serious doubts about the meaningfulness of the group means at these age levels. However, this variability rapidly decreases with age. For example, at 3.5 years, the mean times taken to complete the task (and their standard deviations) were 26.96 seconds (6.95) for the right hand and 29.53 (11.97) for the left hand which gradually declined to 9.61 seconds (0.77) and 10.58 seconds (0.93) respectively for the 15-year-olds. Similar findings on additional subject samples using the same pegboard performance test have been reported by Kilshaw and Annett (1983).

Two further studies using the Annett pegboard task on children ranging from 2.5 years to 6 years of age also reported that the relative difference in performance between the two sides was not correlated with age (Curt et al., 1992; De Agostini et al., 1992). Nevertheless, when children drawn from the same subject populations (and age range) as before were tested on a speed of dotting task, both Curt et al. and De Agostini et al. found a significant relative increase in the performance differences between the preferred and non-preferred hands with age. And in a subsequent test of the speed of dotting of a sample of 1,742 children drawn from a much wider age range (7 to 14 years of age), Carlier et al. (1993) also reported a significant effect of age on the recorded performance differences between sides.

In a study by Burge (1952) on 312 schoolchildren ranging from 5 to 13 years of age, this author tested each subject on three different tasks (grip-strength, throwing, and ball-bouncing). He subdivided his subjects into right-handers, left-handers, and mixed-handers and found that in a comparison of the means of the 6- to 9-year-olds with the means of the 10 to 13-year-olds, only two comparisons out of 18 showed significant effects (increases) in the differences between sides with age.

On a test of handwriting speed given to 249 children ranging from 6 to 12 years of age, Spreen and Gaddes (1969) recorded the time taken (in seconds) for subjects to write their own name with their right and left hands. Means (and standard deviations) of 24.6 (6.0) and 50.6 (19.3) were reported for the right and left hands respectively for the 6-year-olds, with corresponding values of 9.8 (4.6) and 19.4 (7.5) for the 12-year-olds.

Summarizing the findings of investigations on performance differences between sides clearly presents more of a difficulty in attempting to discern general developmental patterns than for the work on hand preference results. Several problems with the evidence may be identified. First, there is no doubt that most investigators have been concerned with establishing group trends or norms, and have paid little attention to changes in the relative hand proficiencies of individuals with age. However, the study by Krombholz (1989) clearly shows the misleading potential of evidence provided by group means, and reveals that considerable individual changes (and even reversals) in hand superiority with age on at least some performance criteria are compatible with no apparent changes in overall trends.

Second, most of the performance studies provide no information concerning the reliability of their findings, and when they do, it is confined to data on the right and left sides separately and not the differences between hands. But in grouped data, this latter measure can provide a very helpful indication of the stability of the relative proficiencies of the two hands of individuals. Its importance has been shown in the study by Provins and Cunliffe (1972b) who found that even in adult subjects, whereas test-retest correlations on each of seven different motor performance tasks for the preferred hand were all positive and statistically significant, those for the non-preferred hand were significant on only four of the seven tasks, and for the between-hand differences, the test-retest correlations were significant for only two.

Of the investigations reviewed here, relatively few have provided data on the reliability of between-hand performance differences, and then only in a very limited way. For example, Burge (1952) confined his test of reliability for each task to a group of 10-year-olds, while that reported by Carlier et al. (1993) related to 8- to 11-year-old children. In the study by Annett, Hudson and Turner (1974), the reliability of the Annett pegboard task was assessed on a pooled group of subjects ranging from 11 to 45 years of age, but this provides little help in evaluating data from cross-sectional studies of children ranging from 3 to 15 years of age, especially for those in the lowest age groups where the individual differences are so large.

A third point of importance relates to the task itself and the extent to which trends in performance differences between hands with age on one task may be indicative of similar trends on any other task - i.e. the generality of the findings. For example, Curt et al. (1992) found significant effects of age on performance differences between hands on one task but not on another even though the same subject sample was tested in each instance over the same age range. Such indications of task dependency for proficiency differences between hands are similar to those reported in the present review for hand preference studies, and clearly have important theoretical implications. What other evidence is available then on this particular question?

Consistency of asymmetry across tasks

The consistency (or otherwise) of childrens' manual asymmetry of performance across tasks has been recorded in a number of investigations. For example, in a study of 60 children ranging in age from 18 to 66 months, Jones (1931) divided his subjects into four different age groups and tested each of them on five different pegboard, formboard, tower-building or color-sorting tasks. The degree of use of each hand in contributing to the completion of the task was timed and compared with the total time taken by each subject on each occasion to arrive at a mean dexterity ratio for the four age groups on each of the five tasks. The author reported that there was a steady increase in dextrality with age over all tasks such that his 2-year-olds mostly used the right hand about two-thirds of the time while his 4-year-olds used it about four-fifths of the time. However, although intercorrelations of the dextrality ratios for the five tasks were all positive, they were also fairly low, ranging from 0.29 to 0.50.

It has also been shown by Bruml (1972) in an investigation of 60 children in each of three age groups (with median ages of 5:10, 8:3, and 10:4 years) on five tests of motor performance (grip-strength, pegboard, screw-turning and two tapping-speed tasks), that differences in performance between the right and left hands on different tasks varied considerably. This author reported that intercorrelations between performances on the five motor tasks based on the size of the difference between hands, ranged from 0.01 to 0.63 and were generally low and not particularly stable - especially for the lowest age group. The best correlations were (not surprisingly) between the two tapping tasks.

In another investigation using four different motor tasks (grip-strength, hand steadiness, tapping speed, and dart throwing), Heinlein (1929) made a longitudinal study over a period of three to five years (depending on the task) of 60 children varying in age from 4 to 12 years at the time of first testing. The subjects were subdivided into three handedness groups independently for each task according to their consistency in superior hand performance on successive yearly tests although not all subjects completed all four test series. The author found that for grip- strength, 61.7% were consistently right-handed, 35% inconsistent and 3.3% consistently left-handed. Comparable figures for the other three tasks were: steadiness, 68.3%, 25% and 6.5%; tapping, 73.3%, 25% and 1.7%; and throwing, 75.5%, 13.8% and 10.3% respectively. But more importantly, it was reported that of the 60 children given the four test series, only 25% were consistently superior with the right hand on all four tasks, and that 71.7% were in different handedness groups in the various test series. None of the children were in the consistently left-handed group in all four series.

Clearly, a prominent feature emerging from this examination of developmental investigations of both preference and performance differences between the hands, relates to the dependence of findings on the particular ages, tasks, and criteria employed. This is in keeping with the other evidence reviewed here that demonstrates the relative specificity of both motor skills and manual asymmetries, and the importance of learning in their acquisition. This suggests that the direction of manual asymmetry displayed in each situation by any given individual is a response to the particular range and type of environmental influences to which that person has been exposed during development, and is unlikely to be a result of some inborn determinant which would provide the same all-pervasive asymmetrical bias regardless of behavioural circumstances.

CHAPTER 7
Skill and Handedness

Chapter outline

The inter-correlations between performances on a wide range of motor activities are low, suggesting that motor skills tend to be specific to particular tasks and to become even more specific with practice. The nature of skill is shown to be a learning process, dependent initially on the discrimination and selection of the movement elements required for each particular task. The development of skill by the organization of movements in an appropriate and relevant spatio-temporal pattern is seen to be shaped by continuous and detailed sensory feedback. However, as movements become more consistent, the need for frequent corrections lessens, and the performance becomes more accomplished although never entirely predictable. The level of proficiency achieved has been shown to be closely related to the amount of practice undertaken, and improvements in performance to continue for millions of repetitions and/or years of training.

The effect of unilateral exercise and asymmetrical development of manual proficiency is seen to be a corollary of the description given for the acquisition of skill and the achievement of expertise. In the same way that the development of motor skills have been shown to be task specific, the development of hand preference or a performance difference between the right and left sides has also been shown to be task specific and dependent on experiential differences between the two sides. However, the role of motivation, social pressures, and adaptability of the individual when presented with a task requiring the development of a unilateral skill are also shown to be important.

The association between left- or mixed-handedness and a higher than normal incidence of stuttering described in early reports is less evident in more recent investigations. In this regard, there is evidence to suggest that the strictness of school discipline, and the practice of attempting to force all children to use their right hand for writing may well have been a precipitating influence in the genesis of stuttering and its past association with mixed or left-handedness in emotionally susceptible cases, particularly for intellectually retarded individuals.

The nature of skill

Investigators attempting to determine the basis for the individual differences found in tests of "manual ability" have generally recorded low but positive intercorrelations between tasks (see e.g. Fleishman & Ellison, 1962; Fleishman & Hempel, 1954; Goodman, 1946, 1947; Griffiths, 1931; Muscio, 1922; Seashore, 1930, 1951). Since the object of such testing was to find combinations of tests, which could be used for selecting those individuals most suited for a variety of occupations requiring particular manual skills, the results were disappointing, if not frustrating. Muscio concluded that there is no "motor type" and that generalizations using such terms as "motor dexterity" or "practical ability" were misleading, since a person's relative performance on any one test gave little or no indication of that individual's relative performance level on any other test. Similarly, Seashore concluded that individual differences in motor skills are largely specific to a particular type of operation or, at most, a relatively narrow group of operations, and that it might be more profitable to re-focus attention on other sources of variation affecting individual differences such as motivation, training, or health. Other studies that have attempted to find some factor, which may be responsible for motor learning have also proved inconclusive, and Martenuik (1974) suggested that, like "motor ability", there is no such factor as "motor educability".

More recently, Zelazo et al. (1993) in a controlled study of two motor behaviors (stepping and sitting) in six-week-old infants have shown that the beneficial effect of daily systematic practice on one particular motor activity is specific to that activity. This confirmed the observations of Super (1976), who found that although the pattern of development of certain motor behaviors in African children varied from that of American children, this could be attributed to inter-cultural differences of emphasis in their training. Similarly, variations in the pattern of development associated with sex or social class (Nelligan & Prudham, 1969) may also be a reflection of intra-cultural social attitudes (see Chapter 10). But of more particular relevance here is the demonstration by Buxton and Humphreys (1935) that the effect of deliberate training on a number of motor tasks was to *decrease* any intercorrelations between them, i.e. to make the skills even more specific after training than before. This suggests that specificity *per se* may be largely an artifact of differential training, a view which gains support from studies of individuals with a lowered capacity to learn, insofar as motor skills appear to be less specific in retardates compared with nonretarded subjects (Bruininks, 1974).

Certainly, the problem extends far beyond the realm of manual dexterity and psychomotor apparatus tests. In the field of sports and athletic skills involving such characteristics as strength, range, and speed of limb movements, balance, and coordination of muscle activity, Hempel and Fleishman (1955) and Fleishman (1964) have also demonstrated that individual achievements tend to be specific to a particular activity or type of activity. But as Crossman (1964) observed, since skill (defined as ability acquired through practice) enters into all aspects of human behavior, its very familiarity tends to impede progress in analyzing its mechanisms. He noted that such activities as writing, using a knife and fork, or driving a motor vehicle all seem simple and easy tasks, but really involve a range of perceptual-motor functions, the complexities of which are really only evident during the initial phase of learning to master them. In providing a concise overview of the area, Crossman emphasized the importance of the approach developed by the work of Craik (1947, 1948), Welford (1952, 1959) and others whose experimental studies led to a theoretical framework for human skilled behaviour in terms of two particular postulates: (i) that the individual behaves as an intermittent, error-actuated servo-mechanism, and (ii) that the individual behaves as a limited capacity communication system.

In attempting to provide a simplified classification of the wide range of everyday activities encompassed by this framework, Provins (1967) suggested evaluating skills according to the degree of involvement of the sensory or perceptual processes at one extreme, and the complexity of the effector or response activity at the other, with the amount of involvement of decision-making or cognitive processes in each instance having an overall modulating effect on the difficulty of the task. On such a continuum, the problems of driving a motor vehicle - particularly on a tortuous route through a city centre - are mostly associated with the ever-changing perceptual demands of the environment. In contrast, an athlete undertaking a successful pole-vault or high jump has a stationary target to aim at but a very complex series of whole body movements to organize. However, these and other key features have been viewed in a number of ways by different theorists, and in a brief discussion of the problem, Holding (1989a) has provided a useful summary of most of these ideas.

In focussing on the nature of those manual activities frequently used as performance criteria in tests of handedness (e.g. handwriting, throwing, tapping, etc.), it is necessary to remember that although the various levels of the central nervous system must somehow provide the executive and integrative substrate for voluntary movements in an appropriate hierarchical inter-relationship (Summers, 1989), it is the peripheral structures of the limbs, viz. the bones, muscles, and joints which are the effector units ultimately responsible for the production of any movement. Furthermore, it is also necessary to stress that voluntary control over any movement is learned (see e.g. Brooks, 1983; Clark, 1995; Thelen, 1995; Thelen et al. 1993; von Hofsten, 1993), a fact which is difficult to appreciate at the adult level when so many different motor skills have already been acquired and appear to demand so little conscious effort for their performance. But a consideration of the essential features of the learning process followed by some relevant examples will help to establish these points.

At the simplest level, attempting to acquire and/or perfect a particular skill usually involves (i) the identification of some desirable goal or objective, (ii) the selection and execution of an appropriately controlled goal-directed action, and (iii) the provision of sensory feed-back to the subject on the degree of success attained in attempting to achieve the objective. These features are clearly illustrated by considering someone learning to play the game of darts. Here (i) the goal is to throw each dart at a selected part of the dartboard, (ii) the appropriate action is a coordinated movement of the hand and arm in throwing the dart, and (iii) the success of the throw is seen by the direction and distance of the impact of the dart in relation to the part of the board aimed at. By repeated throws and comparisons of their outcome with the location of the target, the learner gradually reduces his error by systematic modifications of the throwing action until an acceptable standard of performance is achieved. In the absence of knowing what is required and/or the outcome of the learner's efforts, little or no progress may be made - a situation all too familiar to those concerned with the normal motor development of infants blind from birth (Adelson & Fraiberg, 1974; Troster & Brambring, 1993), or to those involved in teaching the pre-lingually deaf child to speak (Lieberth, 1982). Certainly, the critical role of knowledge of results or sensory feedback in the learning process was clearly demonstrated long ago in the classical studies of Thorndike (1927) and Trowbridge and Cason (1932), while subsequent work on the topic has been well-reviewed by Newell (1976), Salmoni, Schmidt and Walter (1984) and Winstein and Schmidt (1989).

But somewhat less generally recognized and undoubtedly less easily demonstrated are the principal features of (ii) above in the acquisition of skills. Since there are very few gross movements of the limbs or body over which a high degree of voluntary control has not already been achieved by the adult stage of physical maturity, an experiment by Bair (1901) is particularly relevant. This investigator recorded the activity of the retrahens muscle used by subjects in learning to move their ears. He showed that while artificial electrical stimulation of the muscle did not enable subjects to learn to contract the muscle themselves, it may have assisted them to recognize their success when they did eventually manage to contract it voluntarily. More importantly however, he found that his subjects gained voluntary control over their ears by first moving them as part of some gross facial activity which included raising the eyebrows, clenching the jaws, etc. Subsequently, they learned to isolate the movement of the ears from the other unwanted muscle activity by a paring down process of progressive relaxation or inhibition, and ultimately to move one ear independently of the other.

Such a process has also been demonstrated in learning to acquire voluntary control over individual motor units in producing very fine movements (Basmajian, 1963, 1973; Basmajian & Samson, 1973), and there can be little doubt that a similar process of discrimination and selection is employed whenever voluntary control is achieved over specific limb and body movements. For example, Lunderwold (1951) recorded motor unit activity in a range of arm and shoulder muscles during ordinary typewriting by trained touch-typists and untrained individuals, and found that the unwanted or irrelevant muscle activity was always greater in the untrained subjects. And Davis (1942, 1943) recorded the electrical activity of the muscles from all four extremities in both children and adult subjects who were requested to make a simple isolated movement of one of them. He found that activity could usually be recorded in all extremities, but that the amount of activity was least from those regions of the body furthest from the actively moved limb. He also reported that although this same pattern or gradient of activity held for both children and adults, the differentiation was less sharply focussed in the case of children.

This incomplete paring down process, or elimination of unwanted muscle activity, has often been described as "motor overflow", "associated movement", or "mirror movement", and has already been considered in Chapter 6 in relation to the developmental aspects of motor skills. The evidence on developmental trends indicated that while these associated movements, decrease with age, there is also a tendency for them to occur more readily when the non-preferred hand is attempting to make an isolated voluntary movement. Certainly, in tests on both children and adults, it is now clear that for normal right-handed subjects, these unintended movements are much more evident on the contralateral side when the deliberative movement is made by the left, rather than the right hand (see e.g. Armatas, Summers & Bradshaw, 1994; Liederman & Foley, 1987; Stern et al., 1976; Todor & Lazarus, 1986a). As Kinsbourne (1973a) has noted, the ability to differentiate isolated from mass movements usually develops over a period of many years, and undoubtedly involves both selective activation of the required muscular components, and the inhibition of irrelevant or unwanted activity (emphasized by Todor & Lazarus, 1986b, 1986c). While neuromaturational factors may well restrict the early emergence of selective discrimination of the wanted neuromuscular elements in infancy, it seems clear that the continuing gradual development of successful voluntary control over individual movements is learned. Since some actions are more frequently needed than others, the opportunities for gaining proficiency in control will vary from one situation to another. And insofar as the preferred hand for a given activity is used more frequently than its counterpart, the motor discrimination would be expected to develop earlier and more successfully for the preferred side.

But as well as the elimination of irrelevant muscle activity, voluntary control over a movement involves bringing together in an appropriate spatio-temporally graded pattern the activity of both prime movers and synergists. For in everyday activities, no movement is ever produced by one muscle acting alone (Sinclair, 1957). And in the sort of whole body movements employed in activities such as golf or tennis, it has been shown that a highly complex but reproducible pattern of muscle activity in many different muscles is characteristic of a skilled player (Slater-Hammel, 1948, 1949). This author recorded the activity of a variety of muscles used during the production of a golf swing and of a tennis stroke by skilled performers in each sport, and reported that while the contraction-movement relationships in repeated attempts at each swing or stroke were extremely consistent for each individual, there were considerable variations from subject to subject. Terzuolo and Viviani (1979) and Salthouse (1984) have also shown that the timing between finger movements in typing a given word by professional typists, is highly consistent when executed by the same person, although the timing pattern may vary somewhat between typists. Similar findings were reported by Glencross (1973) for a hand cranking task where the muscle activity was monitored by electromyography and by the torque applied to the hand crank throughout each successive cycle of movement. He reported that while there were notable differences between subjects in their pattern of motor activity, the action pattern recorded for each subject was characteristic of that individual. Furthermore, he found a significant correlation between the subjects' speed of performance and their consistency of timing of successive movement cycles, confirming the findings of an earlier study on the same task (Glencross, 1970). Evidence of improvement of timing consistency with practice for such components of rapidly executed movements has also been provided by studies on hand cranking and manual tapping activities (Provins, 1956; 1958) and for dart throwing (Jaegers et al., 1989).

The implications of consistency in the generation of muscle activity for the predictability and programming of voluntary movements has led to a considerable volume of theoretical discussion in the last 20 to 30 years which is beyond the scope of the present review, but surveys of the field may be found in Holding (1989b) and Jeannerod (1990). Suffice it to say here that the main thrust of these theoretical deliberations has centred around the relative importance of "open loop" programming of muscle activity in comparison with "closed loop" sensory feedback control of the execution of skilled movements, and the particular units or level of control achieved with learning (Keel, 1982; Newell, 1985; Young & Schmidt, 1990).

However, from the evidence already presented, there is little doubt that the acquisition of voluntary control over a movement is gained initially by a process of trial and error (i.e. closed loop) activity. As repetitions of the attempted movement become more effective and predictable through the correction of errors based on sensory feedback, successful programming of the motor components can be said to have been achieved and control of the movement to have become more open loop. The relative amounts of closed- or open-loop control involved in processing a given movement will thus depend on the amount of prior practice and the level of perfection (or skill) achieved at any given stage of learning (see e.g. Brooks, 1981; Brooks, Kennedy & Ross, 1983). This is, of course, assuming that the external (i.e. environmental) demands remain constant. *Ab initio* learners of a motor task must necessarily spend most of their time consciously trying to master their movements by a closed-loop system of control, whereas the movements of skilled performers are highly predictable so that the relevant muscle activity can be programmed to require only occasional corrections based on sensory check. The skilled performer is thus enabled to devote more time and attention to other (changeable) aspects of the task, which in activities such as tennis or squash may occur quickly and unpredictably.

But like the successful development of voluntary control over individual movements (i.e. the ability to differentiate wanted from unwanted muscle activity), the achievement of increasingly higher levels of skill also requires years of practice concomitant with physical maturation and changes in the perceptual and cognitive capabilities of the child (Clark, 1988). Certainly, the progressive development of motor skills with age has been well documented for both sexes (Branta, Heubenshicker & Seefeldt, 1984; Thomas & French, 1985), but how much of the improvement may be due to physical maturation and growth and how much is the result of environmental influences and practice or experience is difficult to disentangle. In his review of the relationship between various anthropometric dimensions, strength, and motor performance, Malina (1975) concluded that most intercorrelations tend to be low or moderate, and for practical purposes not meaningfully predictive.

On the other hand, with regard to the effects of practice, it has been reported by Klemmer and Lockhead (see Klemmer, 1962) that in the process of mastering keyboard skills by punchcard operators, these workers averaged between 50,000 and 90,000 keystrokes every day, but that it took a year or more for them to reach maximum speed of operation. Similarly, Crossman (1959) compared the learning curve of adults on a wide range of perceptual motor tasks and found that their performance continued to improve with practice over very long periods of time. In his own industrial study of cigar making for example, he observed a continuing improvement in performance of the operatives for at least two years before they reached their peak, representing experience of over three million work cycles. Subsequently, Newell and Rosenbloom (1981) reviewed the evidence on learning an even wider range of perceptual motor skills and suggested that irrespective of the task parameters, improvements in performance conform remarkably well to a power law of practice - viz. that when the logarithm of the time taken to perform a task is plotted against the logarithm of the trial number, the learning curve approximates to a straight line. Thus, the more difficult and time consuming the task initially, the greater the practice required to achieve a given standard of proficiency and the smaller the increment of improvement as higher levels of skill are attained, with excellence being approached asymptotically.

Thus, the learning of skills occurs over extremely long periods of time and from millions of repetitions, although a residual variability remains due to an irreducible level of inherent "noise" in the system. Such a view would find strong support from those sports and athletics competitors who daily put themselves through rigorous training schedules in order to qualify for national or world events, and from those musicians who spend countless hours at the piano or violin in seeking to attain the rewards of international recognition as a concert soloist. Certainly, Ericsson, Krampe and Tesch-Romer (1993) concluded that to attain recognized international standards of expertise in such skilled activities, an extended period of at least ten years of deliberate practice is required. From two detailed biographical studies, these authors were able to show that the ultimate level of proficiency reached for professional violinists and pianists was closely related to the amount of time they spent practicing their skills. But the same power law of practice can be seen to apply to the myriad other perceptual motor activities undertaken by everyone as part of their daily lives - whether it be learning to walk and talk in infancy, or learning to write at school, or perhaps later to typewrite or to play golf. It is a matter of common observation that the beginner's attempts at any of these activities are hesitant, effortful, and uncertain of outcome, which after training and years of subsequent experience, become smooth, effortless, and highly consistent, but never completely predictable.

Crossman (1959) attributed the development of skill in his industrial workers to be due to the elimination of inappropriate methods - i.e. a paring down process based on a progressive discrimination against unwanted or irrelevant activity. Newell and Rosenbloom (1981) however, suggested an improvement with practice was most likely to be due to the learner processing the information in progressively larger "chunks", which in a motor task would involve building up more elaborate "schemata" (Schmidt, 1976) or higher order programs of movements (Rumelhart & Norman, 1982) as the outcome of each lower order or elementary program of muscle activity became more predictable. It seems likely that both a paring down process of irrelevant activity and a building up of larger units of relevant activity would occur in such situations although probably to different extents at different stages of learning.

If such processes are responsible for the development of motor skills, it is easy to see why they are so specific to particular tasks or situations. Clearly, someone who has developed a motor program for the shot-put is likely to be quite inexpert at the javelin throw or discus, since the activities of different muscles and joints are required to be coordinated in quite different sequences etc. in these different athletic events. And a musician who has devoted ten years or more to becoming an expert performer on the piano is unlikely to find (or expect) that such proficiency generalizes to playing the violin or flute. Similarly, laboratory tests of manual performance using tasks such as grip strength, finger tapping and pegboard completion, etc. also employ quite different hand and arm movements. The student subjects usually used in these experiments are also likely to have had quite different amounts of experience with each of these types of activity and consequently to be at very different points on the learning curve. In other words, it appears that the recorded specificity of performance on manual dexterity tasks or other motor skills at any given stage of development may not be due to any specificity of inborn abilities but to the learning of specific perceptual motor strategies which primarily require the development of specialized higher-order programs of muscle activity. To the extent that any given motor skill involves such a specialized program, which utilizes subprograms common to another motor skill, then the overall performances on the two skills may be expected to be positively related.

But how is this concept of motor skill relevant to the question of development of right- and left-handedness? In considering the characteristics of animal handedness (Chapter 4) and the measurement of human handedness (Chapter 5), evidence was presented to show that the manual asymmetries of individuals vary widely and unpredictably from one task to another, i.e. they are relatively task specific. Could then the differences between sides, where they occur, simply be another example of the specificity of skills resulting from differential learning? In the following section, consideration is given to the effects of experience or practice on differences in performance between the two sides. And in the next chapter, the importance of both capacity and opportunity to acquire skill in the development of handedness is discussed.

The effects of unilateral exercise

There have been many studies of the effects of unilateral practice on the acquisition of perceptual motor skills, but in the present review attention will necessarily be focussed on those types of motor activity which have provided the criteria for performance tests of handedness. In this respect, the effects of unilateral exercise on the development of muscle strength has probably the longest history, although much of the early work is difficult to interpret due to deficiencies in experimental design such as too few subjects or the absence of suitable control groups etc. (Slater-Hammel, 1950). Many of these studies claimed to show that systematic training of one side using heavy resistance exercises not only produced gains in strength and/or endurance of the exercised limb, but also (though to a lesser extent) of the unexercised (contralateral) limb, to which the term cross-education was applied (see e.g. Hellebrandt, Parrish & Houtz, 1947). However, Gregg, Mastellone and Gersten (1957) in their examination of electromyographic activity of the right and left elbow flexors and extensors during exercise of one arm, showed that the absence or presence of associated muscle activity in the unexercised (contralateral) arm was dependent on the level of exercise and degree of muscle fatigue in the exercised limb. Thus, as suggested by Hellebrandt et al., it seems likely that the apparent cross-education reported in earlier studies was due to an unintended exercising of the contralateral limb during deliberative training of the pre-selected side.

Subsequently, Kruse and Mathews (1958) examined the effect of unilateral strength building exercises (using repeated contractions of the left elbow flexors at three-eighths of maximum voluntary strength) on four groups of subjects over a four week period, during which each group trained on either two, three, four, or five days of each week. These authors recorded the strength and endurance of elbow flexion on both the right and left sides before and after the four week unilateral training period for the elbow flexors of the left arm for the four experimental groups, and on both the right and left sides of four matched (unexercised) control groups. The effect of the training program produced significant increases in both strength and endurance of the left (exercised) limb for those groups that trained for three, four, or five times per week, but not for the twice per week group. Furthermore, there were no statistically significant increases in strength of the right (unexercised) arm for any of the experimental groups nor for either side in the case of the control groups, which suggests that where differences between sides are found in random population samples, these may well be due to differences in the extent to which the respective limbs have been exercised. Hence, for most people in sedentary occupations, little or no strength differences between the two sides would be expected, but for those engaged in physically active occupations or sporting pursuits where unequal demands on strength or endurance of the two sides are made, then significant right/left differences are much more probable, depending on the relative level of the demands being placed on each side (see e.g. Buskirk, Anderson & Brozek, 1956). This would certainly explain the somewhat mixed findings in the literature on strength testing described earlier.

It has also been shown that if the muscle activity is confined to a lightly graded isometric contraction of the (index) finger flexors, then there may be no significant difference in the accuracy of reproduction of the attempted movement between the preferred and non-preferred hands (Provins, 1956, 1958). The same series of experiments showed that in tests of the accuracy of reproduction of movements confined to an isometric contraction of the flexors of the big toe, again no significant differences between the right and left sides were recorded. But when the same subjects were tested for speed of oscillation of the index finger or big toe by rapid alternation of attempted flexion and extension movements, quite different results were obtained. For the finger, the speed of oscillation was significantly greater for the preferred compared with the non-preferred hand, but for the big toe there were no significant differences between sides. As the feet of these subjects would have been contained within the confines of closely fitting shoes and socks throughout the day and every day for most of their lives, it was concluded that the lack of any significant differences in the performances of the big toe on the two sides was due to the absence of any opportunities for developing differences in the degree of usage and hence skill. In contrast, the hands of the subjects would have had ample opportunity for differential experience and usage. But why should there be no difference between hands for discrete movements compared with significant differences between sides for rapid oscillatory movements of the index finger?

To determine to what extent these effects with the finger may be ascribed to differential experience or usage, the result of practicing both tasks daily for five days a week for four weeks was tested on the index finger of six subjects (Provins, 1958). It was found that whereas little improvement occurred in the grading of isolated contractions of the finger flexors, significant increases in speed of alternate flexion and extension movements with practice were recorded, and of a sufficient size to readily account for the previously observed differences between sides.

A similar improvement with practice for speed of finger tapping, has been reported by Peters (1976). In this experiment on one subject (himself), he carried out daily practice at the rate of 20 trials per day with the index finger of each hand for a total of 1,300 trials. He reported that at the beginning of the experiment, the mean tapping rates for the right and left hands were 56.8 and 49.9 taps per 10 second trial respectively, but that after about 1,200 trials both sides had improved markedly - especially the left which was now as fast as the right. But interestingly, the improvement recorded by the index finger appeared to be specific to that finger and had not generalized to the performances of the other fingers of the same hand. However, in a second similar experiment on the effect of daily practice on finger tapping rates of 14 subjects, Peters (1981) reported that there were wide individual differences, and that while a significant overall improvement during the training period occurred for both hands, there was no significant interaction between the hand used and practice.

In an experiment designed to examine the effects of practice on Annetts' pegboard task, Annett, Hudson and Turner (1974) studied the effects of five trials per hand per day for 26 consecutive weekdays on three subjects. These authors recorded a significant improvement for both hands of each subject over the 130 trials, and a significantly greater improvement in the performance of the non-preferred hand compared with the preferred hand for one subject. On a somewhat similar task, Perelle, Ehrman and Manowitz (1981) reported the results of an experiment in which 36 subjects (17 left-handers and 19 right-handers) were required to undertake six practice trials with each hand each day on the Crawford Small Parts Dexterity Test (Part 1) over a period of five consecutive days. These authors found that training significantly decreased their subjects' time to complete the task with each hand and also decreased the time for the non-preferred hand more than for the preferred hand.

In the more complex task of handwriting, Thorndike, Bregmen, Tilton and Woodyard (1928) gave 33 right-handed university students 15 hours of practice at copy-writing with the left hand spread over 90 periods of 10 minutes each. The published data reveal a significant improvement in writing speed with the left hand from a mean value of 38.1 letters/minute before practice to 63.1 letters/minute after practice. This compares with a mean writing speed for the right hand of 116.1 letters/minute for the same subjects. Thorndike et al. also reported that the average improvement from the beginning to the end of the experiment in both speed and quality of performance was equivalent to two years normal school handwriting progress - achieved between grade II and grade IV. No practice or before and after effects were reported for the preferred hand although it is unlikely that there would have been much (if any) change in such an overlearned activity with that hand.

However, the results of investigations of practice which continue for the very limited periods of time described in the tapping, pegboard, and other training experiments reviewed here, although suggestive, can give little realistic indication of the importance of usage. As Crossman (1959), Ericsson et al. (1993), and Klemmer (1962) have shown, improvements in occupational or professional motor skills continue for millions of trials, so that it is necessary to consider the individual performances of each hand on unimanual and bimanual tasks which have been subject to a lifetime of experience or similar extensive practice to obtain a more accurate and relevant estimate of the effects of training under normal conditions.

In this respect, handwriting and typewriting both present examples of well-practiced and overlearned highly skilled activities with one essential difference between them. Whereas only one (and the same) hand is always used for handwriting, typewriting requires both hands to be used to approximately the same extent. Consequently, Provins and Glencross (1968) designed an experiment to assess the performances of each hand separately on both tasks using the same subjects in each case. For the typewriting task, three different exercises were employed that provided matched tests for the right and left hands, and the achievements of two groups of trained and experienced touch typists were compared with a control group of non-typists. All subjects were unselected with regard to handedness, and two of each group of 20 trained typists said they wrote with the left hand while three of the 20 control group of subjects were also left-handed for writing. The results showed that there were few or no between-hand differences in typewriting performance of the typists, but highly significant differences in favour of the right hand for the non-typists. Where significant differences between the sides were recorded for the typists, they were in favour of the left hand and could be attributed to the slightly greater use (and hence prior practice by the trained typists) of the left hand in the layout of the standard typewriter keyboard. When the typists and non-typists were given a simple handwriting task to be completed by each hand in turn, they both recorded a highly significant difference in performance between the two sides in favour of the preferred hand.

Also relevant here is the evidence that while the speed of handwriting with the preferred hand for self-classified left-handers is usually no different from that of self-classified right-handers (Reed & Smith, 1962; Smith & Reed, 1959; Ziviani, 1984), left-handers tend to be significantly better at writing with the non-preferred hand (Provins & Magliaro, 1993). In this latter study, the left-handers were also significantly less consistent in their expressed hand preferences (on a 20 item questionnaire) than the right-handers, thus suggesting that their better handwriting performance with the non-preferred side was probably due to a more frequent general usage of this hand.

But to conclude from these results that the level of dexterity, typewriting, or handwriting proficiency attained with either hand is solely dependent on practice would be to ignore other likely and powerful influences such as motivation and ability. Most investigators soon discover that in testing non-preferred hand performances on almost any task, subjects usually display at least some diffidence or lack of confidence - especially if the task is one on which the preferred hand is highly skilled. For example, if, after obtaining a sample of an individual's normal handwriting, the subject is asked to provide a similar sample of writing with the other hand, the reply is more often than not, "Oh no! You must be joking - I'd be hopeless with that hand", and some considerable encouragement is frequently needed to persuade them to try.

It is of special interest then, to note the report of a case of spontaneous change of writing hand (Provins & Dalziel, 1969). This subject was a mature-age first year university student who developed severe cramps from writing in the absence of any neurological basis for the symptoms. He was a consistent right-hander who had always written with the right hand, but in desperation and in order to continue successfully with his university work, he spontaneously tried writing with his left hand. Fortunately, as the result of a referral from the Student Health Service, a sample of his handwriting with each hand had been obtained at interview prior to his changeover, and some ten months later when he called in to report his successful solution to the problem, further samples of his handwriting were obtained. Whereas his initial performance on the set handwriting task was 1 minute 34.9 seconds for the right hand and 3 mins. 17.4 secs. for the left hand, the corresponding results ten months later were 1 min. 29 secs. and 1 min. 33 secs. respectively, with a parallel improvement in the quality of his left-handed writing. Unfortunately, the change-over of writing hand did not provide a lasting solution as he later developed the same difficulties with cramps in his left hand. However, it appears that this is not an altogether uncommon problem, as Sheehy and Marsden (1982) have since reported on 29 similar cases, 14 of whom changed hands and learned to write with the unaffected limb although five of these later developed difficulties on that side as well. Further reports have been published by Sheehy, Rothwell and Marsden (1988) and Marsden and Sheehy (1990), while the successful rehabilitation of brain damage patients (Needles, 1942), peripheral accident victims (Schott, 1980) and war wounded soldiers (Hutt, 1917) suffering the loss of their right hand and hence forced to master writing with the left have also been reported. Undoubtedly, the importance of self-motivation in such situations is paramount and has been strongly emphasized in a description of the wide range of skills acquired by Count Zichy who lost his right arm accidentally at the age of 14 years (Anonymous, 1916).

In the case of individuals who are unimanually incapacitated from birth, there is little doubt that strong motivation is often needed to enable them to acquire many of the motor skills developed by physically normal children. And it may be assumed that this would be no different for those children who are manually incapacitated on the left as compared with those similarly afflicted on the right side, although the evidence is somewhat divided on this point. C.K. Hiscock et al. (1989) examined 30 right-hemiplegic and 27 left-hemiplegic patients for speed of finger-tapping, Purdue pegboard performance, and grip-strength, testing each subject on both the right and left sides. Apart from finding a significant difference between the left and right hands of the two groups of hemiplegics in the expected direction, a comparison between groups using the performances of the physically normal side in each case showed a significant tendency for the left hemiplegics to outperform the right hemiplegics on two of the three tests (grip strength and finger-tapping). In contrast, Weinstein, Sersen and Vetter (1964) reported that in the same two tests of motor performance (grip strength and finger-tapping) conducted on the physically intact hands of 44 patients with unilateral upper extremity aplasia, no significant differences on either task were found between their unimanual right- and their unimanual left-handed subjects. It is, perhaps, hazardous to suggest reasons for the difference in findings between these two investigations, but one obvious possibility is that whereas the hemiplegic patients were still able to use (to a varying extent) the limb on the affected side, the aplasic patients were not. Consequently, social conformity would have favoured the left hemiplegics in gaining a more extensive and consistent experience in manual activities with their unaffected right hand than the right hemiplegics would have attained with their left. No such considerations could apply to the aplasic patients who would have been unequivocally unimanual in their range of manual experiences, and hence not susceptible to the influence of social or cultural pressures.

It is also worth noting here the high degree of motor competence and dexterity often attained by those individuals who are born without arms at all. The present author was privileged to witness the accomplishments of one such person (Tommy Jacobsen) who, during the 1950's earned his living demonstrating his outstanding skills at English country fairs. He was so adept with his feet that, after sitting down on a stool, it almost passed without notice during his usual introductory commentary, that he had already taken a packet of cigarettes out of his jacket pocket, selected one and put it in his mouth. After selecting a match from a matchbox and lighting the cigarette, he then further displayed his versatility by, *inter alia,* successfully threading a needle and playing a variety of popular tunes on a piano. The degree of voluntary control over individual movements of his toes was indeed remarkable, and clearly demonstrated that, given the circumstances and the motivation, these movements are well within the normal limits of human achievement. A similar account of an Australian girl who was born without arms but who successfully mastered a wide range of skills, including typewriting and the operation of a telephone switchboard with her feet, has also been published (Special Correspondent, 1972), while reports of the remarkable feats of other physically handicapped people have been gathered together by Jay (1987).

Handedness and stuttering

However, not all individuals appear to be able to acquire the use of either hand equally well, and many case reports appeared in the early literature relating to the development of speech and other behavioural problems associated with an enforced change of preferred hand usage (see e.g. Claiborne, 1917; Fagan, 1931; Whipple, 1911). If hand preference in a given situation is simply a reflection of the degree of motor skill achieved - which is the result of prolonged training or experience, then why is it that some people find difficulty in adapting to the right-sided population bias in spite of the sometimes extreme social pressures exerted upon them during childhood (see e.g. Barsley, 1966, 1970; Blau, 1946)? And if motor skills are so specific, why has ambidexterity or a change in writing hand from left to right often been associated with speech defects such as stuttering (see e.g. Burt, 1937; Lauterbach, 1933b; Travis & Johnson, 1934)?

One of the earliest reports providing evidence of an association between handedness and stuttering was published by Ballard (1912). This author conducted a survey of over 13,000 schoolchildren (4 to 14 years of age) through the head teachers of a sample of London elementary schools, and described 545 as left-handers, of whom 73% were said to have been converted to writing with the right hand. Of the total 13,000 surveyed, 1.2% were stutterers and of these, 10.6% had been switched to writing with the right hand. Thus, while the proportion of left-handers converted to right-handed writing only formed about 3% of the total population sample, they amounted to nearly 11% of the stutterers. In a second and more detailed examination of nearly 12,000 schoolchildren (8 to 14 years of age), the same author reported that 322 were left-handers, of whom 84% had been converted to writing with the right hand. However, whereas none of the left-handers who still wrote with the left hand were found to be stutterers, nearly 26% of the switched-handers were reported to stutter or to have stuttered at some time during their schooling.

Many similar investigations followed, with some authors reporting very high rates of occurrence of speech problems associated with a change from left- to right-handedess. For example, as a result of two studies comparing a total of 152 stutterers with 152 non-stutterers (from 4 to 31 years of age) matched for sex, age, social status, and mental age or school achievement (see Bryngelson & Rutherford, 1937), Bryngelson (1939) reported that 61% of the stutterers had experienced a shift in handedness compared with only 5% of the control group, and in a much wider survey of 1,421 stutterers attending his speech clinic, he reported an even higher incidence (73%) of switched-handers.

A comprehensive review of other early studies associating stuttering with mixed-handedness or an enforced change of handedness has been published by Travis and Johnson (1934). On the basis of such evidence and their own laboratory and clinical studies, Orton (see e.g. Orton, 1928, 1930, 1937) and Travis (see e.g. Travis, 1931; Travis & Johnson, 1934) formulated the view that stuttering is a function of incomplete lateralization of language in the brain, or a delay in the development of cerebral dominance - of which mixed-handedness is a peripheral indicator. Thus, Travis (1931) suggested that stutterers differ from normal speakers due to a relative lack of laterally differentiated cerebral motor control, with the seriousness of the speech problem being inversely related to the degree of functional differentiation existing between the two cerebral hemispheres.

However, later research has revealed little or no association between stuttering and the incidence of mixed- or left-handedness. For example, Records, Heimbuch and Kidd (1977) compared the handedness questionnaire responses of a group of 446 stutterers with those of a group of 356 non-stutterers and found no significant differences in the handedness classification of the two groups. Similarly, in a comparison of the hand preferences of 134 stutterers and 257 non-stutterers, Webster and Poulos (1987) found no evidence to suggest that left-handedness was more common among stutterers or that stutterers were less consistent in their hand preferences than fluent speakers. And in Bishop's (1986) analysis of data from a survey of over 12,000 British schoolchildren, the results again provide no support for an association between the incidence of speech disorder and individual differences in handedness classification as determined by performance on each of two particular tests of manual proficiency. Other investigators have also reported a low or insignificant correspondence between handedness assessment and the incidence of stuttering (e.g. Andrews & Harris, 1964; Johnson & King, 1942; Porfert & Rosenfield, 1978).

Similarly, a number of investigators have found few or no ill-effects on speech of a change of hand usage - usually for writing. For example, Haefner (1929) in a school survey of over 1,100 children, identified 6.2% as left-handed, 63% of whom had been trained to use the right hand, especially for writing. However, in a comparison of the incidence of speech defects in the changed and unchanged left-handers and a group of matched right-handed control subjects, although the proportion of children displaying some impediment of speech was higher for the switched-handers than for the unchanged left-handers and right-handers, the differences did not reach statistical significance. Other earlier and much larger but less detailed studies have also reported few individuals suffering from adverse effects of a change of writing hand (see e.g. Parson, 1924; Wallin, quoted by Travis & Johnson,

1934). Also relevant here is the report by Scheidemann (1930) and Scheidemann and Colyer (1931a, 1931b) of a surprisingly high incidence of left-handed writing in a group of schoolchildren who were otherwise right-handed. They found that of 16 left-handed writers in a second grade class of 34, only 4 could be classified as left-handed on other criteria, and all but one of the remaining 12 had no left-handed relatives. None of these children were reported as having any psychopathic tendencies and no mention was made of any speech problems, although four were said to be somewhat nervous or emotionally unstable. However, as both the first and second grade teachers were left-handed, the authors concluded that they had probably influenced the children either directly or indirectly by displaying an overly sympathetic or permissive attitude towards them.

In an investigation that attempted to determine the cause of any association between handedness and stuttering, Lauterbach (1933b) assessed the direction and degree of the hand preferences of 1,061 individuals (half of whom said they were left-handed and the other half, right-handed) on a 50 item handedness inventory. Subjects were also classified with regard to their writing hand, i.e. whether they had (a) always used the right hand for writing, (b) always used the left, (c) successfully changed from left to right for writing, or (d) unsuccessfully attempted to change their writing hand. The author also reported (on a subgroup of 356 subjects) the type of training methods (and punishments) used to effect a transfer of writing hand, and the incidence of speech disorders associated with them. Two interesting findings emerged from this study. First, an association was found between the degree of left-handedness and success or failure to change the writing hand; the more strongly left-handed the subject, the less successful were any attempts at conversion to right-handed writing. Second, it was reported that in 50% of the speech disorder cases involving transfer, punishment was used as a corrective measure. The author concluded that it was not the transfer *per se* which caused the stuttering, but the methods used and the personality of the individuals concerned.

Shortly afterwards, as a result of his large survey of London schoolchildren, Burt (1937) concluded that the severity of school discipline (of which insistence on right-handed writing was simply an example), and the temperament of the child, were both highly influential in the development of stuttering. Meyer (1945) also stressed the close association between emotion and speech disorder, with enforced change of writing hand being but one possible precipitating agent. In a study of the case histories of 104 stutterers, he reported that only 25 showed evidence of ambilaterality or attempted handedness conversion, whereas in well over half the cases, he identified fear as the common factor primarily responsible for the speech disorder. And this, he suggested, should be regarded as simply one form of expression or manifestation of a more general neurotic problem. Similarly, Despert (1946) in an investigation of 50 stuttering children, reported that emotional factors rather than handedness *per se*, or change of hand usage, appeared to be associated with the observed speech defects, and Bryngelson (1939) noted that there were 50% more enuretics among his group of 152 stutterers than among his controls. Other more recent authors (e.g. Caruso et al., 1994; Cox, 1986; Craig, 1990) have also associated stuttering with anxiety and stress.

Although there are many different opinions on the causes of stuttering (see e.g. the review by Bloodstein, 1993), its past association with an enforced change of writing hand may, in some cases, have been due to a lack of ability to adapt. For example, the incidence of stuttering has been reported to be much higher in the mentally retarded (see e.g. the review by Homzie & Lindsay, 1984), and in Ballard's (1912) study of retarded children, the incidence of stuttering among left-handers who were converted to right-handed writing was much higher than for switched-handers in normal schools. This is certainly in keeping with what might be expected from less able subjects, since learning to write with the previously unpracticed hand in such cases would clearly be a much more difficult task than for children of normal intelligence. And on this basis, for people at the other end of the ability scale, it might be expected that those with an undisputed expertise at some manual task on one side, would be *more* able to develop a similar skill with the corresponding limb on the other side, unaccompanied by any ill effects. In this regard, there are certainly cases of well-known sportsmen who have demonstrated a capacity to develop their highly acclaimed skill equally well on each side without any apparent undesirable side-effects. For example, Jeremy Dale (a professional golfing coach), who originally played golf right-handed, not only taught himself to play equally well left-handed, but also developed a wide range of practice shots on both sides for the benefit of his (left-handed) students, as well as to demonstrate his golfing expertise (see Australian Golf, *121,* March, 1999). Similarly, in tennis, Luke Jensen (who played the international circuit in doubles during the 1980s with his brother Murphy), displayed his acquired ambidexterity by unpredictably changing the racquet from one hand to the other during a match, and often between serves. And in soccer, Tom Finney (Finney, 1955), having recognized that, as a right-footed winger, he had no hope of representing England as outside-right while Stanley Matthews occupied the position, consequently forced himself to practice kicking with his left foot until he was selected to play for England in the complementary position of outside-left. But to what extent is an ability to learn, or the opportunity to acquire skill, relevant to the development of the usual human characteristic of *right-* rather than left-handedness?

CHAPTER 8
Handedness and Disability

Chapter outline

Investigations of the hand preferences of mentally retarded groups of subjects reveals that there is a higher incidence of non-right-handedness amongst such individuals than in non-retarded populations, and that the incidence increases with severity of handicap. There is also evidence to indicate that this higher level of non-right-handedness in retarded subjects is associated with a lowered overall motor proficiency or reduced capacity to acquire motor skills with either hand. However, there are no grounds for suggesting that learning capacity or IQ is different in left- and right-handed normal subjects.

Elevated levels of non-right-handedness have been shown to be even more pronounced in groups of autistic children, presumably due to the associated lack of social awareness in such individuals as well as their usually low level of intellectual development. On the other hand, schizophrenic patients appear to be closer to the social norm with respect to the incidence of non-right-handedness, which may be accounted for by the later onset of this disorder. In children described as clumsy or lacking normal motor coordination, most investigators have also reported an increased incidence of nonright-handedness. However, in cases of developmental dyslexia, the evidence is inconsistent. In regard to twinning, it appears that the higher incidence of nominal left-handedness reported for both identical and fraternal twins cannot be ascribed to any genetic determination of manual asymmetry, but is likely to be the result of social and/or parenting circumstances. Cases of premature birth and/or extremely low birth weight have been shown to be associated with delays in development that adversely affect learning ability, the acquisition of motor skills, and hence, adoption of the usual right-sided bias for hand preference.

The incidence of left-handedness has been reported to be raised in severe and chronic cases of epilepsy, but not in the milder forms. With regard to immune disorders, there is now good evidence that where an increased incidence of non-right-handedness occurs, this is associated with neurodevelopmental disorders that involve generalized brain damage, rather than left hemisphere dysfunction.

In the normal population, there is a general bias towards the preferential use of the right hand, with the incidence of left- or mixed-handedness usually being assessed at about 11% (see e.g. Gilbert & Wysocki, 1992). But in a wide variety of clinical or pathological conditions, the incidence of nonright-handedness has often been reported as being higher than in the normal population. Clearly, in patients with anatomically verified early unilateral brain lesions that impair motor functions of the limbs on the opposite side, or in cases with some chronic unilateral peripheral limb injury or disease, it is to be expected that such conditions are likely to induce a preferential use of the corresponding member on the unaffected side. For example, the early development of a marked preferential use of one hand rather than the other during the first year of life has been reported to be a strong predictor of hemiplegic cerebral palsy, depending upon the severity of the condition (Cohen & Duffner, 1981; Uvebrant, 1988). However, it has also been shown that, whereas a relatively wide range of neurodevelopmental disorders may be associated with cerebral palsy, their incidence (in terms of both number and type) appears to be unrelated to the side of the hemiparesis (see e.g. Goodman & Graham, 1996; Goodman & Yude, 1997; Uvebrant, 1988). Assuming that in any group of patients such afflictions are equiprobable for the right and left sides, an enforced usage of the unaffected side would then be expected to result in a higher than normal incidence of left-handedness.

Yet there are many other clinical conditions that do not necessarily involve any readily identifiable lateralized organic basis for dysfunction of a limb, but which are nevertheless reported to be associated with an increased incidence of left-handedness. For example, mental retardation, clumsiness, developmental dyslexia, autism, schizophrenia, epilepsy, twinning, prematurity or extremely low birth weight, and various immune disorders have all been reported to be associated with elevated levels of left-handedness. Although a number of different theories have been advanced to account for this connection (for relevant reviews, see e.g. Bakan, 1990; Bishop, 1990a; Harris & Carlson, 1988), one of the commonest features in most of these explanations, is a postulated disturbance of the normal functional predominance of the left cerebral hemisphere. This is, perhaps, not surprising, since the bias towards right-handedness in the normal population has been widely assumed to be the result of some genetic or inborn predetermination of cerebral specialization of function (for relevant reviews, see e.g. Bishop, 1990a: McManus & Bryden, 1992).

However, if environmental pressures and learning provide the basis for the normal development of manual asymmetry, are these particular sources of influence also sufficient to explain the increased incidence of left-handedness in those clinical conditions where there is no readily identifiable organically based unilateral limb dysfunction? Certainly, in some clinical populations, the learning abilities of patients, or their opportunities for manual exercise and experience, are not at all normal. Indeed, in the following critical examination of the evidence, it will be seen that decreased learning capacity, and restricted experience, appear to be two key features limiting the motor skills that may be achieved in such populations, with the consequence that not only are recorded differences in preference or proficiency between hands reduced, but that the effects of a right-dominant cultural bias are also necessarily lessened.

Hand preference and intelligence

Early workers such as Gordon (1921) and Wilson and Dolan (1931) who recorded approximately double the incidence of left-handedness amongst retarded groups as compared with normal children, recognized two specific difficulties associated with reporting handedness. First, that a simple dichotomous classification did not adequately reflect the observed wide variation in hand preference from task to task; and second, that the social pressures present at that time frequently inhibited a free choice of hand use, and in many instances resulted in a change of hand (e.g. for writing). Gordon also noted (p.315) that in retarded children "of a very low-grade type" who appeared to use either hand indiscriminately, their lack of manual dexterity was marked for both hands. Burt (1937) similarly recorded a much higher incidence of left-handedness amongst backward and retarded children, and also remarked on the effect of the nature of the task on the expression of hand preferences. But of somewhat more significance, he likewise observed that "many of the dull and backward who pass for left-handed might truthfully be designated ambi-sinistral; they seem not so much dexterous with the left hand as gauche with the right" (p.287). In recognizing the relevance of level of achievement as well as performance differences between the two sides, Burt clearly implied that recording both the absolute and relative hand proficiencies rather than preferences was needed to provide a more complete assessment of an individual's handedness characteristics.

However, most investigators in this area have continued to use hand preference data with rather mixed results. Certainly, the finding of an increased incidence of left and/or mixed hand preference in low I.Q. populations is now well established (see e.g. Batheja & McManus, 1985; Bradshaw-McAnulty, Hicks & Kinsbourne, 1984; Hicks & Barton, 1975; Lucas, Rosenstein & Bigler, 1989; Morris & Romski, 1993; Porac, Coren & Duncan, 1980a; Soper et al., 1987; and for relevant reviews of the evidence, Pipe, 1988; 1990). Some authors have also reported that the incidence of left-handedness increases with the severity of retardation (see e.g. Hicks & Barton, 1975; McManus, 1983). But Gordon (1921, p.348) further reported that "very frequently the teachers in ordinary schools asserted that the children found to be left-handers were the most intelligent and the best at school work". Evidence has now been published which supports this view and suggests that the incidence of left- or non-right-handedness may be significantly higher than normal in intellectually more able schoolchildren (e.g. Annett & Manning, 1990a; Benbow, 1986; Hicks & Duseck, 1980), while others have reported that children developmentally advanced for their age may display a more definite or consistent hand preference irrespective of direction (Cohen, 1966; Gottfried & Bathurst, 1983; Kaufman, Zalma & Kaufman, 1978; Kee, Gottfried & Bathurst, 1991).

Certainly, many large-scale studies of the attributes of *normal* schoolchildren and adults have provided strong evidence against any distinction between left- and right-handers in overall intellectual ability (e.g. Douglas, Ross & Cooper, 1967; Fagan-Dubin, 1974; Hardyck, 1977; Hardyck, Petrinovich & Goldman, 1976; Newcombe et al., 1975; Satz & Fletcher, 1987). And there are many examples of high profile individuals of undoubted ability who also happen to be left-handers and pre-eminent in their particular field. For example, retired Australian cricket captains Alan Border and Mark Taylor both batted left-handed, as does Brian Lara who is also a holder of the highest number of runs scored in test cricket. In tennis, Wimbledon champions Martina Navratilova, Rod Laver, and John McEnroe all played left-handed, and in the field of entertainment, Paul McCartney and Ringo Starr of Beatles fame are also left-handed. Amongst those eminent in public life, there have been many left-handed presidents of the United States, viz. Harry Truman, Gerald Ford, Ronald Reagan, George Bush, and Bill Clinton.

However, a few investigators (e.g. Hicks & Beveridge, 1978; Levy, 1969; Miller, 1971; Nebes, 1971) have reported relatively subtle effects in their university subjects, suggesting that right-handers may do better on certain perceptuo-spatial aspects of cognitive tests, but not on the verbal components. In contrast, some authors (e.g. Gibson, 1973; Heim & Watts, 1976; Newcombe et al, 1975; Sheehan & Smith, 1986) have found no such distinction between handedness groups in normal population samples for either verbal or spatial tests, while yet others (e.g.

Burnett, Lane & Dratt, 1982; Harshman, Hampson & Berenbaum, 1983; Johnson & Harley, 1980; Mascie-Taylor, 1980; Sanders, Wilson & Vandenberg, 1982; Searleman, Herrmann & Coventry, 1984; Yen, 1975) have reported results which suggest that findings in this area may be subject to the influence of a wide variety of factors, including the particular population examined, the specific intellectual or cognitive tests used, the sex of the subjects, the method of classification, the strength of hand preference, familial sinistrality, and/or some combination of these variables.

Asymmetry of manual proficiency and intelligence

It has been shown in a number of studies of mentally retarded subjects, that the motor proficiency of such people on a variety of manual tasks is significantly lower than that recorded for comparable individuals in either clinical or non-clinical populations with normal I.Q.s (see e.g. Attenborough & Farber, 1934; Boldt, 1955; Chiarenza, 1993; Howe, 1959; Rapin, Tourke & Costa, 1966; Sloan, 1951; Weaver & Ravaris, 1972; see also reviews by Malpass, 1963; Reid, 1986). Furthermore, as the severity of intellectual deficit increases, the level of motor function decreases, and there is evidence to suggest that the difference in motor proficiency between retarded and nonretarded subjects becomes progressively larger with age (Bruininks, 1974).

It has also been shown that the normal developmental progress in ability to eliminate unintended muscle activity when attempting to initiate a preselected voluntary movement is markedly delayed in mentally retarded children (Cohen et al., 1967; Fog & Fog, 1963). For example, Fog & Fog found that in attempting to make a finger and thumb pinch with one hand, unintended associated movements of the opposite hand which were observable in the 265 normal children they tested, were not only more evident in the 184 retarded subjects of their study, but continued to be evident in the majority of cases at an age (14 to 16 year-olds) by which 90% of normal children had successfully eliminated the effect. And in a detailed study of 17 adolescent and young adult patients (between 13 and 24 years of age) who had been referred for neuropsychiatric examination due to pronounced and persisting associated movements in the absence of any gross neurological disorder, Rasmussen (1993) described nine of the cases as being left-handed or ambidextrous. Two others considered to be right-handed were said to be late in developing a hand preference, and were reported to be almost as competent with the left hand as with the right. Unfortunately, details relating to each subject's I.Q. level were not reported, although various neuropsychiatric deficits (including learning and speech problems, fine and gross motor dysfunction, or delayed psychomotor development) were recorded in about half the cases.

However, surprisingly little information is available relating to more general manual performance differences between the right and left sides in retarded cases compared with normals. In an investigation of the grip strength of 189 minimally brain injured (MBI) children and 226 educable mentally retarded (EMR) students ranging from 8 to 13 years of age, Broadhead (1975) published the means for the right and left hands separately for boys and girls. Overall, the data showed much the same trends with age as for normal children, with the right side some 0 to 6% stronger than the left. Although the author suggested that the results demonstrated a relatively low level of performance compared with normal children, no data from a control group were included in the report.

More recently, in an experimental investigation of finger tapping performances with the preferred (for writing) right hand and the (non-preferred) left hand of two groups of retarded adult subjects (one group with Down syndrome, the other comprising undifferentiated cases of retardation), Elliott (1985) found no significant differences between sides, although in a control group of non-retarded subjects, he recorded a significant superiority for the (preferred) right hand. In a subsequent study comparing the finger tapping performances of Down syndrome subjects (mean

CA = 18.5 yrs; mean MA = 5 yrs) with a non-retarded control group (mean CA = 20.1 yrs), Elliott, Weeks and Jones (1986) also reported significant group, hand, and group x hand interactions for tapping frequency. The retarded group were much slower overall and their differences between hands relatively small (7.8%) compared with the differences between sides for the control group (17.2%). The same trend with regard to differences between Down syndrome and normal subjects in overall speed of finger-tapping and in terms of the size of the performance differences between hands was later reported by Elliott et al. (1987) in a dual task experiment. Irrespective of whether or not subjects were required to undertake a secondary task (sound shadowing) concurrently with finger-tapping, the control subjects in both conditions outperformed the retarded group and showed a significantly greater difference between sides in favour of their preferred (right) hand.

Handedness in clumsy children

A lack of significant lateralization was also found by Beaumont (1976) in subjects diagnosed as suffering from minimal brain damage (i.e. MBD, although the concept is open to question - see e.g. Rutter, 1984a, 1984b). Beaumont compared his MBD children (mean age = 7.1 yrs; mean I.Q. = 87) with normal subjects (matched for age, sex and intelligence) on their hand preferences in executing twelve motor tasks on two occasions and the right and left hand performances on a pegboard and a formboard task. The MBD children were found to be significantly less right preferent than the control subjects and to be less stable in their preferences from one occasion of testing to another. In discussing these results, Beaumont suggested that the evidence supports the idea that MBD children are more clumsy in all their motor activities and that this deficiency can be related to a lesser degree of lateralization in the performance of motor tasks.

Other studies of children with some minor neurological deficit (including clumsiness or motor dysfunction) but with apparently normal intelligence (see e.g. Bullock & Watter, 1987; Dreifuss, 1963; Gillberg & Rasmussen, 1982; Gubbay, 1975; Walton, Ellis & Court, 1962; Watter & Bullock, 1989) have also indicated that many of these individuals showed some evidence of poorly developed or atypical hand preference. In discussing the evidence from his patients, Dreifuss contrasted the behavioural effects of bilateral and unilateral cerebral lesions of early onset, and concluded that whereas a severe unilateral brain disease of early onset does not necessarily affect the normal development of hemisphere dominance, even a mild bilateral hemisphere deficit could seriously jeopardise the acquisition of speech and handedness.

That this poorer development of right-handedness (or apparent increase in left-handedness) in clumsy children may be found in a more representative sample of the population became evident in the study by Calnan & Richardson (1976) of the relationship between a number of different variables and handedness in a British national survey of 11 year old schoolchildren. These authors found a slight but statistically significant association between non-right-handedness and deficits of performance on both cognitive tests and teacher's ratings of "poor control of hands". In a subsequent investigation of 170 unselected schoolchildren (8 to 9 years old), Bishop (1980b) recorded the errors made with each hand by subjects on a unimanual performance task (square-tracing), and found that when the children with the worst 20% of non-preferred hand scores (called the target group) were compared with the rest of the subjects (non-target group), there were significantly more left-handers in the target group. However, when the worst 20% of scores with the preferred hand were used to subdivide the subjects into target and non-target groups, there was no significant difference between groups in the incidence of left-handers. In a similar study using the same (square-tracing) task and methods in a comparison of groups of Swedish 9 to 10 year old schoolchildren,

Gillberg, Waldenstrom and Rasmussen (1984) reported much the same findings. Although the subjects of both studies came from normal schools, the authors found that the children with the worst non-preferred hand performances (i.e. the target group) on the tracing task also scored significantly lower on a variety of cognitive and/or achievement tests.

In a later investigation of a large sample of (12,000) 11 year old children using the same strategy, Bishop (1984) examined the incidence of left- and right-handedness in such target and non-target groups assessed on two different manual tasks (square-marking and match-sorting) and obtained somewhat mixed results. While a significant excess of left-handers was found amongst the worst performers on the match-sorting task, no significant effect was recorded using poor performance on the square-marking test to subdivide the subjects. However, it was also reported in this investigation, that a comparison between the target and non-target groups on three cognitive tests also showed no significant differences between groups except on the non-verbal part of one of them (a paper and pencil general ability test).

In each of the investigations by Bishop (1980, 1984) and Gillberg et al. (1984), the subjects were dichotomously classified as either left- or right-handed, so that the excess of left-handers in the group of poor performers on the manual tasks may more appropriately be described as non-right-handers since this group would include the majority of mixed-handers as well. That it may well be those with mixed or uncertain handedness rather than left-handers *per se* who are associated with clumsiness in these studies, gains support from the work of Tan (1985) and others in a somewhat different context. Combining results from the use of a wide range of motor ability tests on a group of 4 year olds attending pre-school, this author compared the performances of subjects classified as left-handers and right-handers with those identified as having no established hand preference. He found that whereas the left-handed subjects were not significantly different from the right-handers with regard to motor ability, those lacking a definite hand preference obtained significantly lower mean ability scores. A study by Tierney et al. (1984) using the McCarthy Scales of Children's Abilities to test 128 Scottish 5 year old children split into three groups of (i) left-handers, (ii) right-handers, and (iii) those with handedness not established (i.e. mixed or inconsistent), also showed that this latter group scored significantly lower on the general cognitive and perceptual performance tests than the right-handers. And in a study comparing the limb preferences of 40 clumsy children with an equal number of normal controls, Armitage and Larkin (1993) found an increased incidence of mixed laterality in their clumsy subjects.

At the other end of the scale, some evidence has been offered by Kilshaw and Annett (1983) and Annett (1985) regarding performance on a pegboard task to suggest that the left-handers in their population samples were superior to right-handers in respect of both the preferred and non-preferred hands, although the results were much clearer for the non-preferred side. The same tendency with regard to non-preferred hand performance was found by Demarest (cited by Annett, 1985) using the same pegboard task and a bean test on Guatemalan children. This investigator found little or no difference between right- and left-handed subjects with regard to their preferred hand performances, but a significantly better achievement by left-handers concerning non-preferred hand performances. Similar results have also been reported for finger-tapping (Peters & Durding, 1979), square-tracing (Peters & Servos, 1989) and handwriting (Provins & Magliaro, 1993).

Of particular interest here then, is the study by Doane and Todor (1977) who subdivided their university subjects into four groups: (i) right-handers, (ii) left-handers, (iii) high ability ambilaterals, and (iv) low ability ambilaterals according to their performances on the Fitts (1954) tapping task. They found that the performances of the high ability ambilateral group on this task with either hand were equal to the preferred hand performances of the right- and left-handers, whereas the low ability ambilaterals were inferior with both hands.

Thus, the evidence from both preference and performance testing of subjects is in general agreement that amongst groups with poor motor and/or cognitive ability, there is a disproportionately high incidence of individuals with mixed or inconsistent handedness. However, the weight of evidence suggests that normal left- and right-handed individuals perform tasks requiring motor skills equally well with their preferred hand but not necessarily with their non-preferred side. Where it has been reported that nominal left-handers achieve better non-preferred hand motor performances than right-handers, it seems likely that such an effect can be ascribed to a greater usage or experience with the non-preferred hand by left- and/or mixed-handers compared with right-handers due to the pressures of a right-handed world, rather than to any difference in their natural endowment.

However, It is also likely that not all the deficits in motor proficiency reported for mentally retarded subjects or clumsy children may be directly attributable to a lower native endowment in their ability to acquire skill. Indeed, there are good grounds for believing that the poorer performances recorded for either the preferred or non-preferred hand in such individuals may also be a result of their particular lifestyle and limited participation in physical exercise and sports or games (and hence restricted experience or training) compared with normal subjects (see e.g. Larkin & Hoare, 1991; Molnar & Alexander, 1983; Rarick, 1973).

Handedness in developmental dyslexia

In his examination of the problem of what he called "specific educational disability", Zangwill (1960b) suggested that a specific defect of reading is one of the most prevalent types of learning difficulty in children of normal intelligence. And on the basis of the evidence then available, he tentatively concluded that such a disability seems to be in some way associated with left- or mixed-handedness in at least a subgroup of cases. However, he also considered that it is unlikely that the disability is ever so circumscribed as the name suggests, and certainly, this is one of the many problems (including the definitions of handedness and dyslexia) highlighted by Bishop (1990a) in her more recent literature review.

Consequently, in critically evaluating the information available for an association between handedness and dyslexia, Bishop (1990a) adopted a number of stringent criteria for accepting evidence on both these aspects of behaviour, and reported on 25 studies that appeared to meet her requirements. She found that by one method of evaluating the data, a significant excess of left-handedness was evident amongst dyslexics, but that by treating the results differently, no such effect could be detected. A subsequent third method of analysis of the same data conducted by Eglinton and Annett (1994) suggested that although the effect of an increased incidence of non-right-handedness amongst the dyslexics may be small, it was nevertheless reliable.

However, in her own epidemiological survey of an unselected population of some 12,000 schoolchildren, Bishop (1984) found no increased incidence of left-handedness amongst retarded readers, in keeping with the negative findings remarked on by Rutter and Yule (1970) for other epidemiological studies. This suggests that the type of population examined may well have a major influence on the results reported in any particular investigation. The incidence of dyslexia recorded in cases referred for clinical investigation will almost certainly include more severe cases of reading disability than those detected in a survey of unselected subjects, even if the performance standard adopted for identification of the impairment is the same. Hence variations in the frequency of left- or mixed-handedness reported in studies of dyslexia appears likely to depend on the severity of the reading deficit.

In a more general discussion of the topic, Bishop (1990a) suggested that one difficulty in drawing any clear conclusions from the available data could be due to an inability to extract evidence relating to an unstable or delayed development of hand preference in dyslexics, as compared with an established and stable but nevertheless mixed or mostly non-right-handed usage of the hands. She believed that such a distinction could provide valuable information in helping to differentiate between the frequently postulated alternative explanations of dyslexia, i.e. whether it is due to an atypical cerebral lateralization or to a neuromaturational lag - for which an unstable or variable handedness would be a peripheral sign. Certainly, a maturational lag explanation finds some support in the evidence of Harris (1957) who compared 245 unselected schoolchildren with 316 clinically referred disabled readers, and presented data to show that whereas the proportion of those with mixed-handedness amongst the controls was 18.0% at 7 years of age and only 8.2% at 9 years of age, the corresponding figures for the reading disabled children were 40.0 % and 25.0% respectively.

However, it may also be suggested that the apparently inverse relationship between reading ability (Annett & Manning, 1990b), academic success (Annett, 1993a) or other measures of cognitive ability (see evidence cited by Crowe, Done & Sacker, 1996, p.184, of an unpublished study by Crowe, Crowe & Done) and failure to develop performance differences between hands for manual dexterity in a normal school population, could be a function of lifestyle differences for the children concerned. Thus, the more academically successful students and good readers may be less clearly dextrous with either the right or left hand because they spend more time studying or reading than indulging in physical activities, while the more dominantly-handed individuals prefer developing their manual skills - albeit at the expense of their academic work or reading competence (see e.g. Anderson, Wilson & Fielding, 1988). However, some investigators have found no such association between intellectual ability (McManus, Shergill & Bryden, 1993) or reading proficiency (Palmer & Corballis, 1996) and performance differences between hands, suggesting that the variation in findings from one investigation to another may be the result of population or other (e.g. environmental) differences. For example, it has often been reported that girls tend to be more proficient readers than boys (see review by Sheridan, 1976), and insofar as reading has been considered to be a stereotypically feminine pursuit (see e.g. Downing, May & Ollila, 1982), while boys generally participate more frequently in sport and physical activities than girls (see e.g. Branta, Painter & Kiger, 1987; Coakley, 1987; McManus & Armstrong, 1995), lifestyle differences rather than gender may again be a most likely source of influence, depending upon the cultural climate.

Autism and handedness

The results from studies of the cognitive development of autistic individuals are in general agreement that the majority of such cases fall into the mentally retarded range, with many being classified as severely and profoundly retarded. Volkmar, Burack and Cohen (1990) estimate that between 50% and 75% of autistic children have I.Q.'s below 50, while Gillberg and Coleman (1992) suggest that between 65% and 85% have I.Q.'s of less than 70. In addition, autistic individuals characteristically display a lack of social interaction and maturation (Bryson, 1990), so that language development is markedly affected, with as many as 50% remaining mute throughout their lives (Volkmar et al., 1990).

One of the first studies to examine the hand usage preferences of young autistic children (between 2 and 7 years of age), was carried out by Colby and Parkison (1977) who observed their 20 subjects in a free play situation. They scored the handedness of the children on at least 5 of 14 possible activities, and compared their hand preferences with a group of 25 normal children in the same age range. The authors reported a highly significant difference between the groups, with as many as 65% of the autistic subjects being classified as non-right-handers compared with 12% non-right-handers in the normal group. A similar investigation was conducted by Barry and James (1978) employing the same range of test items and means of evaluating hand preference scores, but using 20 autistic subjects ranging from

4 to 18 years of age. They also tested two control groups of 25 subjects each, matched by sex and age - one group drawn from the normal school population, and the other comprising retarded individuals additionally matched with the autistic children by I.Q. On the same criteria employed by Colby and Parkison, the results obtained by Barry and James showed that 41.4% of the autistic group were non-right-handers compared with 26.5% of the retarded subjects and 26.4% of the normal individuals.

Boucher (1977) also assessed 44 autistic subjects (between 6 and 17 years of age) on a 7 item test of hand preference, and compared the results with two control groups matched for verbal or non-verbal ability, sex and age. The author reported that there were no significant differences in hand preference patterns across groups, although the results show that the numbers (52%) of non-right-handers amongst the autistic subjects tended to be excessive relative to the data presented for a comparable group of normal boys (38%) and girls (24%). And in a comparison between 26 autistic children (ranging from 7 to 21 years of age) and a group of 52 age, sex and I.Q. matched controls, Gillberg (1983) reported that 62% of the autistic subjects and 37% of the controls were non-right-handed (on a 5 item test of handedness), with the corresponding figures for consistent left-handedness being 23% and 6% respectively.

McManus et al. (1992) similarly compared a group of 20 autistic children (mean age 11.1 years) with a sex and mental age matched group of 20 normal controls (mean age 4.9 years) and 12 mentally retarded children (mean age 12.2 years). Hand preferences were assessed on 13 unimanual test items and evaluated on a conventional laterality index for direction and degree of handedness. Since each test item was presented three times, consistency within tasks was also calculated. Significant differences were reported between groups for degree of handedness and for consistency, with the autistic children being less dominantly-handed than the normal controls and less consistent within tasks than either of the other groups. But perhaps even more explicitly informative in this respect are the findings of Soper et al. (1986) who examined the hand usage preferences of two autistic subject samples - one comprising 48 cases (mean age = 21.5 years) in the moderate to profound mental retardation category, and a second group of 31 cases (mean age = 11 years) classified as predominantly in the dull normal/borderline range of intelligence. Hand preference was assessed on an 8 item test administered three times in each of two sessions that were separated by at least a week. A laterality index was computed for the results of each subject to combine the hand preferences demonstrated both within and between tasks. A histogram of the laterality indices for the total 79 cases yielded a U shaped distribution in clear contrast to the form of distribution (i.e. J shaped) characteristic of a normal human population. Interestingly, a comparison of laterality indices between the two subject samples to determine if there might be an effect of severity of mental retardation, produced similar proportions of right-, mixed-, and left-handers in each. However, a data analysis which examined the within-task consistency for both autistic groups indicated a very high frequency of inconsistency (92% to 94%) associated with the classification by laterality index of mixed-handedness, suggesting an overall pattern of incomplete or unstable preference which the authors considered to be more appropriately described as ambiguous handedness.

Many studies have attempted to examine the handedness of family members of autistic children but have found no evidence to suggest any corresponding excess of mixed- or non-right-handedness amongst near relatives (Boucher, 1977; Boucher et al., 1990; Fein et al, 1985; Tsai, 1982) except for the report by Gillberg (1983). Indeed, some tendency for an increased incidence of *right-handedness* amongst the parents of autistic children has even been reported by Boucher and her colleagues.

Thus, the evidence suggests that autism is associated with an unstable or inconsistent hand usage which may be manifest in most test situations as mixed- or non-right-handedness, depending on the criteria used. Such findings appear to be in accord with expectation if it is acknowledged that learning plays a large part in the acquisition of motor skills and hand preference patterns. The low learning ability of mentally retarded subjects and autistic children would clearly hinder the handedness development of both groups, while those individuals who are additionally lacking in social awareness and social interaction (i.e. the autistic subjects), would be even less likely to conform to societal norms of motor behaviour (see e.g. Dawson & Adams, 1984: DeMyer et al., 1972). Thus, they would tend to maintain a relatively independent, random and/or unstable choice of hand usage in keeping with the severity of their disorder.

Schizophrenia and handedness

It has been pointed out that it is often difficult to distinguish diagnostically between autism and childhood schizophrenia (Fein et al., 1985), and since one of the most prominent features of schizophrenia is social withdrawal - a characteristic that is reported to be relatively unresponsive to medication or psychoeducational intervention (Walker, Davis & Baum, 1993), some lack of conformity with respect to population norms of handedness might also be predicted in this disorder. Although these social effects could be somewhat limited as the onset of schizophrenia often occurs during adolescence, there is evidence to suggest that such patients may nevertheless demonstrate some form of relevant impairment or retardation during their childhood. For example, Offord and Cross (1971) and Jones et al (1994) found that early onset schizophrenic adults also tend to have low childhood I.Q.'s and/or poor scholastic records, while Watt et al (1970), and Watt and Lubensky (1976) reported that such patients were often described by their school teachers as emotionally immature or behaviourally maladjusted. And more recently, Crow et al. (1996) have reported an analysis of data collected in the U.K. National Child Development Study which showed that children who later developed schizophrenia (by 28 years of age), tended to demonstrate retardation in establishing unambiguous hand preferences as seven-year-olds, or were less asymmetric in proficiency on a test of manual skill (compared with normal controls) as eleven-year-olds.

Of special interest then is a report by Walker and Birch (1970), who recorded the I.Q. levels, right/left awareness, and hand preference consistency (using a 4 item test) of a group of 80 schizophrenic children ranging from 8 to 11 years of age, although no child with an I.Q. below 70 was included in the subject sample (mean I.Q. approximated 85, with a range from 70 to 115). The authors reported that 32% of these children were right-handed, 12% left-handed, and 55% mixed-handed, which differed significantly from the corresponding values of 80%, 12%, and 8% respectively, reported by Belmont and Birch (1963) for normal children (with a mean I.Q. of approximately 120) in the same age range using the same criteria. Most subsequent investigators have not specified the I.Q. levels of patients or controls, but simply reported that any cases which could be classified as mentally retarded were excluded from their subject sample. And certainly, in the majority of studies that have been undertaken on schizophrenic adults, an excess of left- or mixed-handed individuals has been reported (e.g. Chaugule & Master, 1981; Clementz, Iacono & Beiser, 1994; Dvirskii, 1976; Green et al., 1989; Gur, 1977; Lishman & McMeekan, 1976; Manoach, Maher & Manschreck, 1988; Nasrallah et al., 1981, 1982a; Nelson et al., 1993; Piran, Bigler & Cohen, 1982; Yan et al., 1985).

On the other hand, some have failed to find any significant difference in the handedness distributions between schizophrenic and normal population samples (Guruje, 1988; Merrin, 1984; Oddy & Lobstein, 1972; Wahl, 1976). And in a number of studies, investigators have recorded a greater proportion of fully right-handed subjects in their psychiatric patients than in a group of apparently normal controls (Fleminger, Dalton & Standage, 1977a; McCreadie et al., 1982; Taylor, Brown & Gunn, 1983; Taylor, Dalton & Fleminger, 1980, 1982).

Various reasons have been offered to account for this wide discrepancy in findings. For example, Taylor et al. (1980) discuss the relevance of differences in handedness criteria used in different studies and the problems of interpretation associated with self-reporting by subjects. They also suggest that the most obvious potential source of variation between studies is in the classification of schizophrenia and the within-group variation in the type of schizophrenic illness. Certainly there is some evidence to support this latter view in relation to patients with and without formal thought disorder (see e.g. Taylor et al., 1982; Manoach et al., 1988) and with respect to paranoid compared with nonparanoid schizophrenic cases (Nasrallah et al., 1982a). However, careful examination of the control groups used for the comparisons made in each investigation reveals another interesting but curious feature common only to those studies which reported a significantly higher proportion of fully right-handed individuals in their schizophrenic patients. Each of these studies (i.e. Fleminger et al., 1977a; McCreadie et al., 1982; Taylor et al.,1980, 1982, 1983) employed the same control group of "normal" subjects (dental patients at Guys Hospital, London) in which the reported incidence of expressed preference for the right hand was only 46.2% for all males and 52.1% for all female subjects, and even less for those under 40 years of age, i.e. 40.2% and 47.0% respectively (Fleminger et al., 1977a). These are extraordinarily low right hand preference values for a normal population and are much lower than those given by Annett (1970a) using the same handedness criteria for university students (males 66.6% and females 65.11%) or for service recruits (70.95%). There can be little doubt that if the Annett data for normal subjects had been used instead of those collected by Fleminger et al., most of the studies concerned would have reported no effect or an excess of non-right-handedness rather than an unaccountably high level of right-handedness in their schizophrenic patients. Hence, it seems likely that the large variation in findings for schizophrenia may be due to a measurement anomaly, and that in comparison with a non-clinical population, a higher incidence of left- or mixed-handedness is more often to be found in association with this disorder.

In an attempt to extend the work on schizophrenia to an examination of the handedness of psychosis-prone people in the normal population, Chapman and Chapman (1987) collected data by questionnaire from a large sample of university students. Using scales developed to measure symptoms reported to be characteristic of pre-schizophrenic individuals, the authors found that their hypothetically psychosis-prone subjects tended to be significantly more mixed-handed than their controls. Similar findings were reported by Kelley and Coursey (1992) for their university undergraduates, and Kim et al. (1992) also found an association between certain schizotypal personality characteristics and mixed-handedness in a normal population sample, although Overby (1993) has since reported finding no such supportive evidence, and in two further studies, Poreh (1994) has recorded both positive and negative results.

In an examination of the handedness data reported in studies of identical and non-identical schizophrenic twins, Boklage (1977) reported that non-right-handedness was three times higher in identical twin pairs, and that the incidence of non-right-handedness was significantly higher in schizophrenic twins compared with normal twins. He concluded that the biological basis was associated with a dysfunction of the left cerebral hemisphere. However, in an attempt to replicate this study, Lewis, Chitkara and Reveley (1989) examined comparable data in a further series of psychotic twin pairs and found no such difference in the incidence of non-right-handedness between identical and non-identical twins or between schizophrenic twins compared with normal twin pairs.

Twinning and handedness

Apart from the relative rarity of their occurrence, there is general agreement that twins are a somewhat special population. For example, in a study of 358 twins amongst 4,754 babies delivered in a Birmingham (England) maternity hospital, Dunn (1965) found that the perinatal mortality for twins was more than three times the national average for all births. And despite the considerable advances that have been made in both obstetric and neonatal care since then, this increased risk of perinatal mortality for twins has not lessened (Bryan, 1992). Furthermore, approximately 50% of twin births have been recorded as weighing less than 2,500 gms. at parturition compared to 6% for singletons, with the average length of a twin pregnancy being some three weeks shorter than for a single birth.

Many investigators have also reported that various aspects of the postnatal development of twins may be adversely affected. For example, Mittler (1970) studied an unselected sample of 200 four-year-old twins and found an average retardation of six months in language development compared with a control group of 100 singletons of the same age. Furthermore, he reported that the effect was the same for both identical and fraternal twins, and for both right- and left-handed twins. Other research workers have reported similar delays in the development of speech and language (e.g. Hay et al., 1987), reading ability (e.g. Johnston, Prior & Hay, 1984) and other cognitive capacities (e.g. Record, McKeown & Edwards, 1970; Wilson, 1975).

It has also been reported that the incidence of left-handedness is increased amongst twins compared with the single born, although McManus (1980a) has questioned the validity of some of the earlier work by e.g. Rife (1940). Nevertheless, a more recent postal survey by Davis and Annett (1994) of 33,401 randomly selected people of whom 2.7% were born a twin, showed that of the twin-born, 11.7% wrote with the left hand compared with 7.1% of single birth. And Coren (1994) found that in a study of 224 monozygotic (MZ) and 74 dizygotic (DZ) twin pairs, 14.5% were left-handed compared with 9.9% in a control sample of 1,192 singletons. Furthermore, he found that the incidence of left-handedness was not significantly different between the MZ and DZ twins. Derom et al. (1996) have similarly reported a higher frequency of left-handedness in their study of 808 twin pairs irrespective of zygosity.

While the factors of prematurity and low birth weight may well contribute to some of the findings (see next section), the current consensus with respect to the retarded speech and cognitive development of twins favours a more significant role for postnatal environmental influences. For example, Luria and Yudovich (1959) in their case study of five-year-old twins concluded that an essential feature for speech development in infancy was the need for speech communication. In this respect, they found that their twins were a somewhat self-sufficient pair and interacted relatively infrequently with others. Lytton, Conway and Sauve (1977) and Conway, Lytton and Pysh (1980) came to much the same conclusion in their studies of infant socialization, and reported that parent-child verbal interactions were significantly less for twins than for singletons. However, Record, McKeown and Edwards (1970) went somewhat further, and in a very large survey that compared the verbal reasoning scores of over 2,000 twins with those of nearly 50,000 singletons eleven years of age, they also separated out the scores of single survivors of twin births where the twin-pair was stillborn or died within the first four weeks of birth. The authors found that although there was the usual adverse effect of twinning on ability scores relative to the single born, the effect was virtually eliminated in a comparison between the scores of single survivors of twin births and singletons. In a follow-up study by Mohay et al. (1986) on a much smaller sample of twins, singletons and single survivors of twin pairs, these authors also found that the achievements of the single survivors tended to follow those of the singletons more closely than those of the twins, who were significantly delayed (relative to the single born) on a number of developmental criteria at eight and twelve months of age, including motor ability.

Thus, as summarized by Hay et al. (1987), the two primary influences adversely affecting the general development of twins compared with the single born, appear to be (i) the reduced opportunity for interaction with parents due to the extra time demands on parents of twins with respect to the routine care involved; and (ii) the extent of activity confined to within twin-pairs, e.g. in their own study, these authors remarked that even at the age of 30 months, hardly any of their twins had done anything of significance when not in each others company. Hence, it is important to note that at the very stage when the usual elementary manual skills are being acquired, the twin situation tends to delay normal motor development and to impede social interaction. It is not surprising to find then, that less social conformity is evident in the handedness characteristics developed by twins, i.e. a greater incidence of left- or mixed-handedness, irrespective of zygosity.

Prematurity, extremely low birth weight (ELBW) and handedness

Consideration of the evidence concerning the effects of prematurity and/or low birth weight on neonatal head-turning has shown that there is a close association between them with respect to both spontaneous head movements and head responses to stimulation (Provins, 1992). But whereas most normal healthy newborns have been reported to develop a right-sided bias for head-turning by about the end of the first week of life and to relinquish it by about 3 months of age, it appears that young premature and poor condition babies not only demonstrate an initial absence of bias but often lack a more general responsiveness. However, with improving health, age and maturity, most of these neonates eventually progress from a stage of relative inactivity to inconsistent (i.e. unbiased) and finally, biased movements which may then persist for far longer than in full-term infants. Thus, it appears that poor physical condition in neonates not only affects general motor development but may also retard the acquisition and extinction of normal infant motor asymmetries.

It is of some significance then, that ELBW children (i.e. below 1,250 gms, but without any major neurological disorder, receiving mainstream education) have been found to display a wide range of minor impairments of motor and other functions when tested at six years of age (Marlow, Roberts & Cooke, 1989a). These authors found significantly more associated (i.e. unintended) movements amongst these children than amongst classroom control subjects, and reported that both teachers and parents noted a greater tendency for them to be clumsy and/or overactive. The same authors (Marlow et al., 1989b) questioned parents of 240 infants born prematurely (< 31 weeks gestation, but free of any major neurological impairment) about the handedness characteristics of their offspring when they were between 24 and 104 months of age (median = 56 months). Significantly more pre-term children were found to be left-handed (26.7%) compared with their mothers (7.8%), fathers (13.8%) or full-term siblings (14.4%). Furthermore, although cerebral ultrasound scans taken during their neonatal period revealed that 53 had bilateral and 42 unilateral abnormalities with 145 normal, the incidence of left-handedness associated with each of these groups was similar and unaffected by the side of the lesion.

Similar data have been reported by Ross, Lipper and Auld (1987, 1992). In their first report, 98 premature infants (mean birth weight = 1,175 gms) were assessed for handedness at four years of age, and out of a total of 63 who were reported to be free of any neurological abnormality, 17.5% were recorded as left-handed, 17.5% as mixed-handed and the other 65% as right-handed. Comparable percentages for their control group of 54 full-term children were 11%, 9%, and 80% respectively. In their later study, Ross et al. reported on 88 of the same premature subjects still available who had again been examined for handedness at seven to eight years of age. They found that of the 66 subjects diagnosed as neurologically normal at this stage, 15.2% were left-handed, 10.6% were mixed-handed and the other 74.2% were classified as right-handed. Comparable data for a control group of 80 full-term children yielded corresponding percentages of 12%, 8% and 80% respectively. The authors also reported that among the premature children there was an association between non-right-handedness and cognitive and behavioural deficits.

Similar findings have been reported by Saigal et al, (1992) and O'Callaghan et al. (1987, 1993a. 1993b) in their longitudinal studies of premature and/or ELBW cases. Saigal et al. recorded handedness at eight years of age for 114 low birth weight (< 1,000 gms) and 145 comparable full-term children, and found that non-right-handedness amongst the ELBW group (31%) was significantly higher than amongst the controls (19%). These authors also reported that the ELBW children had significantly lower cognitive, language achievement, and motor scores than the controls but that right-handed and non-right-handed children were no different from each other. In the first report by O'Callaghan et al. (1987) on 39 ELBW (< 1,000 gms) children assessed at four years of age, as many as 54% were considered to be left-handed, although in a subsequent study (1993b), on a total of 71 ELBW children tested at either four or six years of age, 34% were found to be left-handed as compared with 15% or 10% left-handedness recorded for two different but comparable control groups included in their 1993a report. In this latter investigation which focussed on the motor, intellectual and behavioural attributes of 115 children with premature, low birth weight or high risk neonatal records tested at four and six years of age, the authors reported that children with lower intellectual abilities were also more likely to be both poorly coordinated and left-handed.

Epilepsy and handedness

A review of the early literature by Bingley (1958) revealed that most investigators up to that time had recorded a higher than usual incidence of left-handedness amongst epileptic patients. However, in his own study of 90 cases of temporal lobe epilepsy, he reported that only six could be classified as left-handers. Since then, several other studies have also reported no raised incidence of left-handedness amongst their epileptic subject samples compared with appropriate control groups (e.g. Douglas, Ross & Cooper, 1967; McManus, 1980b; Rutter, Graham & Yule, 1970). Yet other investigators have continued to find unusually high levels of non-right-handedness in their particular studies (e.g. Dellatolas et al., 1993; Kurthen et al., 1994; Roberts, 1959; Satz, Baymur & Van der Vlugt, 1979). But the fact that the handedness criteria used have varied from one investigator to another, or have not been clarified in the reports, does not help in the interpretation of the evidence.

Nevertheless, there are grounds for believing that at least part of the conflicting findings may be due to the differences in type of population studied. For example, each of the studies by Douglas et al. (1967), Rutter et al. (1970) and McManus (1980b) were undertaken in relation to large scale epidemiological surveys of normal populations of children. From the evidence presented in these reports, it appears that relatively few of the cases classified as epileptic would have been chronic or severe. In contrast, Roberts's (1959) cases were patients at the Montreal Neurological Institute who had been referred for surgical treatment of chronic and intractable cerebral seizures. This author reported that of his 522 adult epileptic patients, he found that 17% were predominantly left-handed and the rest predominantly right-handed. Similarly, Satz et al. (1979) noted that in their sample of 151 adult epileptics who had been referred to an epilepsy clinic in The Netherlands, 17% could be classified as left-handers. And in the more recent study by Kurthen et al. (1994), out of 173 patients with long-standing and medically intractable complex-partial seizures, 17.9% were assessed as left-handed or ambidextrous.

But the situation becomes further clarified when available information relating to the age of occurrence of cerebral injury is also considered. For example, Roberts's (1959) data indicate that out of his 522 chronic epileptics, 136 suffered cerebral damage before the age of two, and of these, 41.2% were left-handed compared with 8.5% of the 386 whose injury was sustained after two years of age. And in an examination of data obtained from 76 outpatients from seizure clinics in Los Angeles (U.S.A.), the report of Orsini and Satz (1986) shows similar effects of the age of lesion onset. While 73 of the cases had medical histories of seizure disorders, two had head injuries incurring loss of consciousness and one suffered from a cerebro-vascular accident. However, out of the 50 patients whose lesion occurred on or before six years of age, 32% were found to be left-handed, compared with 19.2% of the 26 patients with lesion onset occurring after the age of six. And in their study of 405 epileptic patients without symptoms of hemiparesis, Dellatolas et al. (1993) reported that a significantly increased incidence of left-handedness was associated with (i) a high frequency of seizure occurrence, (ii) early onset of seizures (< 3 years of age), and (iii) moderate or severe cognitive deficit. Thus, the severity of the epileptic condition and the maturational age of the individual at the onset of brain damage both appear to be important influences on the handedness characteristics developed.

Handedness in immune disorders

As a result of two investigations in which they noted the relative rates of occurrence of certain immune disorders and developmental learning problems in left- and right-handed patients and normal individuals, Geschwind and Behan (1982) reported that both immune disorders and learning disabilities were recorded more often for left-handers. In a third study in which patients who had severe migraine or who suffered from a variety of immune disorders were assessed for handedness, the authors reported a higher percentage of left-handers amongst the migraine patients compared with controls, and a similar excess of left-handers amongst those suffering from myasthenia gravis. However, the numbers of left-handers found in certain other patient groups examined (i.e. those with rheumatoid arthritis, or mixed-collagen vascular diseases, and a group with multiple sclerosis) were not significantly different from normal population levels.

These findings were later elaborated upon by Geschwind (1983, 1984) and by Geschwind and Behan (1984) and Behan and Geschwind (1985) to form the basis for a more extensive review and theoretical treatment by Geschwind and Galaburda (1985a, 1985b, 1985c, 1987). In essence, these authors proposed that the development of non-right-handedness, immune disorders, and learning disabilities (including dyslexia) were causally inter-related, insofar as they resulted from a retarded *in utero* maturation of the left cerebral hemisphere and thymus, both due to the action of the hormone testosterone. Since the foetal testes secrete testosterone, the effect was postulated to be greater in boys, which would also account for the higher observed incidence of left-handedness and learning disorders in males. And as the thymus was believed to play a significant role in the development of the immune system, suppression of thymic growth was seen as favouring later immune disorder.

Since Geschwind and Behan's (1982) original paper. many other authors have provided evidence compatible with, or supporting their hypothesis (see e.g. Biary & Koller, 1985; Chengappa et al., 1991; Hassler & Birbaumer, 1988; Hassler & Gupta, 1993; Searleman & Fugagli, 1987, 1988; Smith, 1987), although many more appear to have found conflicting or confounding results (e.g. Betancur et al., 1990; Bishop, 1986; Bryden, McManus & Steenhuis, 1991; Chavance et al., 1990; Cosi, Citterio & Pasquino, 1988; Dellatolas et al., 1990; Flannery & Liederman, 1995; McKeever & Rich, 1990; McManus, Naylor & Booker, 1990; Pennington et al., 1987; Salcedo et al, 1985; Satz et al., 1991; Schur, 1986; Stanton et al., 1991; Steenhuis, Bryden & Schroeder, 1993; Temple, 1990). Several authors (e.g. Bishop, 1990a; Satz & Soper, 1986) have questioned the adequacy of the evidence offered by Geschwind and Behan (1982) and their supporters, and have also criticized the theoretical basis for the explanatory model postulated by Geschwind (1983), and Geschwind and Galaburda (1985a, 1985b, 1985c).

Certainly, in an extensive review of the literature, Bryden, McManus and Bulman-Fleming (1994) not only concluded that there was no convincing evidence to support the testosterone model, but that whatever reported associations do exist between handedness and immune disorders, they are not immunologically meaningful. And in a very large-scale test of the Geschwind hypothesis, Flannery and Liederman (1995) found that while non-right-handedness was not associated with neurodevelopmental disorders that could be considered to be secondary to left hemisphere dysfunction, it *was* associated with disorders that involve generalized brain damage.

CHAPTER 9
Cultural Values and Social Pressure

Chapter outline

It appears that from time immemorial, there has been a strong and widespread belief in the notion that right-handedness denotes all that is good, clean, and desirable, whereas left-handedness is symbolic of things that are evil, dirty, or unworthy. In Western societies, an increased incidence of left-handedness amongst criminals or delinquents has often been reported as lending some support to these views.

Anthropological studies reveal that a more universal antipathy towards left-handedness can be seen in the severe, and sometimes harsh measures employed in a wide range of cultures designed to discourage children from using the left hand. This is in keeping with the variation in child-rearing practices between cultures, especially with regard to the use of the right hand for such activities as eating and writing.

Accordingly, a relatively low incidence of left-handedness has been found in strict and conforming societies, while a high incidence is evident in populations associated with permissive community attitudes.

Such evidence and a considerable volume of social anthropological information is in keeping with Hertz's (1909) concept of a symbolic dualism, which he considers to be a feature of primitive thought and a powerful influence on the development of religious beliefs and practices. He suggests that the familiar oppositions of e.g. life and death, light and dark, strong and weak, right and left, are associated with ideas of the sacred and the profane, and have a culturally symbolic as well as an intrinsic meaning. He considers that systems of such oppositions are fundamental to primitive ways of life and account for the widespread acceptance of belief in the righteous right and sinister left still evident today in many religious and cultural contexts.

It has been said that "Since antiquity, left-handedness has been viewed as a sign of aberrancy or abnormality in the individual" (Satz & Fletcher, 1987), and certainly, the universal acknowledgement of right-handedness as the norm and left-handedness as deviant or socially undesirable, is clearly evident in the range of behaviours that are commonly associated with the words right (e.g. dexterous, righteous) and left (e.g. sinister, gauche). Indeed, Barsley (1970) collected together an extensive list of clearly derogatory slang terms such as "cack-handed" or "molly-duker" to further emphasize the stigmatizing label attached to left-handedness. And Blau (1946) even suggested that sinistrality is simply a symptom or manifestation of an attitude of opposition or negativism. But what evidence is there to substantiate the view that left-handedness and such negative characteristics are necessarily inter-related?

Handedness and delinquency

Two of the earliest studies which are frequently cited as providing relevant data on the question are those by Lombroso (1903) and Smith (1917). For example, Lombroso reported that amongst 1,029 normal subjects he studied, 4% of men and 5% to 8% of women were left-handed, but that these figures rose to 13% and 22% respectively in a group of criminals. Similarly, Smith surveyed over 2,000 normal schoolchildren and reported that 4.5% of girls and 5.5% of boys were left-handed, but that in 500 delinquent children, 6% of the girls and 11% of the boys were left-handed.

Many other studies since then have provided evidence in general agreement with these results. For instance, Pringle (1961) obtained data on left-handedness from a questionnaire survey of the headmasters of special schools for maladjusted children in England, and out of a total of over 2,000 boys surveyed, 14.3%

were said to be left-handed, and out of 537 girls, the corresponding figure was 13.6%, in comparison with prevailing estimates at that time of 8% and 5.9% (boys and girls respectively) for left-handedness in the normal school population. Fitzhugh (1973) also reported that in her small sample of 19 court-referred delinquent male adolescents, there was a significantly higher incidence of left-handedness than in non-court-referred emotionally disturbed controls. And Orme (1970) found that in a group of 300 adolescent girls passing through an approved school classifying centre, there was a significantly raised incidence of left-handers amongst those classified as emotionally unstable compared with a control group. Andrew (1978) also reported that amongst 70 adolescent and young adult male offenders placed on probation, a significantly higher (18.6%) proportion were left-handed than among military recruits, while Hanvik and Kaste (1973) found a non-significantly higher incidence of mixed-handedness among those children referred to a child guidance centre than amongst the normal school population. Similarly, Feehan et al. (1990) reported that for their sample of normal children, although there was no significant difference in the overall distribution of hand preferences between the delinquent and non-delinquent subgroups, a higher proportion of delinquents demonstrated a mixed hand preference. And in a study of incarcerated male adolescent delinquents, Grace (1987) reported that while the dichotomous measure of writing hand was not associated with any of the measures of delinquency used, left-handedness assessed by a continuous measure (multiple item questionnaire) was positively related to conduct disorder. Furthermore, in a study of three small groups of male adolescent psychiatric inpatients, Krynicki (1978) reported that while the number of left-handers in his sample was no different from normal, there was a significantly increased incidence of mixed-handedness amongst those who were considered to be the most unsocialized and aggressive individuals.

    In a prospective study of the distinguishing characteristics of individuals who later became delinquent, Gabrielli and Mednick

(1980) examined the relationship between measures of handedness obtained in 1972, and police records of offences committed in 1978. Both questionnaire and neurological assessments of the handedness of the subjects concerned showed that left-handers were significantly over-represented amongst those later registered as delinquent. And in an investigation of normal college students, Ellis and Ames (1989) found that for their male subjects, questionnaire assessments of left-handedness were significantly and positively associated with delinquency scores as determined by responses to a self-report measure of delinquent behaviour.

Nevertheless, not all studies have reported an excess of non-right-handedness in their criminal or delinquent populations. One of the first to provide contrary evidence was Goring (1913) who compared various physical characteristics (including the handedness) of prison inmates with those of other unconvicted individuals. He reported an incidence of 3.8% left-handedness in the 996 convicts he sampled, which appeared to be similar to other data available at the time on a comparable non-criminal population. However, although the author classified his subjects as left-handed if they always held a knife or pen in their left hand, he recognized that hand preferences varied greatly from one type of activity to another. More recently, West and Farrington (1977) stated that in their long term study of young males, left- or right-handedness was not associated with delinquency, although in subsequent correspondence with Ellis (1990), the authors reported that 16.8% of their delinquent sample were left-handed compared with 13.2% amongst their non-delinquent youths. They further stated that in their recidivist group, 21% were left-handed, which, as Ellis comments, is at least suggestive of a positive association with persistent delinquency. And in an investigation of 99 adolescent residents of a treatment centre for persistent delinquents, Yeudall, Fromm-Auch and Davies (1982) reported an incidence of 12.5% and 11.5% left-handedness for their male and female inmates respectively, as assessed by handedness questionnaire. In comparison with a non-delinquent group of high school students, the authors reported no significant differences in handedness although the incidence of left-handedness in the controls was 10% for males and a somewhat surprising 28% for the females. In commenting on their results, the authors suggested that the high school sample may have been subject to a volunteer bias since left-handedness was given some emphasis when the subjects were being recruited.

Hare and Forth (1985) similarly reported that in a sample of 258 prison inmates given a self-report questionnaire on lateral preferences, the handedness characteristics of the prisoners did not differ significantly from those of a large normative group of non-criminals. Further within-group analysis of the data from the prison inmates also revealed no evidence of differences in handedness between three groups of subjects classified for psychopathy as high, medium, or low, in spite of findings from subsamples of the same subject groups (reported by Porac & Coren, 1981) that suggested otherwise. And in a prospective study of a sample of 800 normal children (selected from 2,958 participants in a major hospital perinatal project) whose handedness was assessed at 7 years of age, Denno (1985) reported no evidence of an association between the distribution of right and left hand preferences and the numbers of police offences recorded by their subjects when they were between 10 and 17 years of age. Furthermore, a later analysis of the data for a much larger sample (1,066) of males taken from the same research project, reported that significantly more non-offenders (12.9%) than offenders (8.6%) were found to be left-handed! To what extent these negative and contrary findings may be due to changing social attitudes towards left-handedness over the years, or to the influence of certain other e.g. economic or educational factors, are topics that are addressed later (in Chapter 10).

Certainly, the common and persistently widespread belief in the righteous right and sinister left has often been responsible for a range of strong and sometimes severe measures taken to "correct" such left-handed and purportedly deviant tendencies. While much of the evidence on these corrective measures is anecdotal in nature, some investigators have provided more detailed data. Thus, the type and extent of disciplinary action which used to be taken against schoolchildren (e.g. in the U.S.A.) who attempted to use the left hand in certain situations, has been well documented by Haefner (1929) and Lauterbach (1933b). In the systematic study by Lauterbach which was designed to examine the desirability of requiring left-handers to use the right hand for eating and writing, this author found that amongst 356 cases of attempted reversal of hand use, whereas persuasion, explanation, encouragement, and methods of reward were used with 64%, outright punishment was resorted to in at least 27% of instances. The nature of the punishments included:- (i) tying a cloth or gourd over the left hand or compelling the wearing of a mitten; (ii) tying the left hand behind the back or tying the fingers together; (iii) cuffing, slapping, spanking, whipping, boxing ears, cracking the knuckles with a ruler; (iv) ridiculing, scolding, threatening; (v) leaving the table, sending to bed, confining in a closet; (vi) keeping after school, giving low grades for or refusing to accept work done with the left hand.

Symbolic significance of right and left

But what is the basis for these strong, underlying beliefs in the sinister nature of left-handedness? Unless it is suggested that left-handedness and wickedness have some common inborn determination, can they be attributed entirely to the effects of culture? Hertz (1909) clearly thought so, and presented strong sociological arguments based on extensive anthropological evidence to propose a universal conceptual association of positive attributes with the right hand or side, and negative attributes with the left. This was seen as part of a general acceptance of the fundamental distinction in human thought between the sacred and the profane associated with good and evil, originating in primitive societies and dominating spiritual beliefs and religious practices throughout history, including such current world religions as e.g. Judaism, Christianity, and Islam. Needham (1973) has assembled additional anthropological evidence from a wide range of cultures to further illustrate and support Hertz's ideas.

Certainly, there are a number of published accounts of strong sanctions being imposed against an improper use of the left hand in various societies. For example, Lttlejohn (1967, 1973) reports that the Temne people of Sierra Leone believe that the right hand should be the more actively used hand in everyday life and that the wrong use of the left hand by a child should be severely punished. This usually involves rapping the knuckles of the offending hand in the first instance, or in persisting cases, wrapping the hand tightly with rags so that it cannot be used. However, wrong use of the left hand can only be appreciated by reference to the Temne code of conduct with respect to the two hands. For them, the right hand serves the upper half of the body so that only the right hand may be used in handling food and eating, or washing the face, whereas the left hand must be used for the lower half of the body in such activities as handling the genitals or anus, as in excretion or sexual activity.

Such mores with regard to the respective roles of the right and left hands are not, of course, peculiar to the Temne people but are widely observed in many societies and religions. For instance, Westermarck (1926) in his extensive and detailed treatise on ritual and belief in Morocco writes "The disfavour with which a left-handed person is regarded is due to the notion that the left side is bad and the right side good, which is found among so many other peoples and also prevailed among the ancient Arabs. It is bad *fal* to use the left hand for good acts, which in accordance with custom are performed with the right, such as eating, giving alms, offering and receiving food and drink or other things, greeting a person, telling the beads of one's rosary; whereas the right hand should not be used for dirty acts, such as cleaning one's anus or genitals or blowing one's nose, and when you spit you should do it to the left. Whatever the left hand writes is bad *fal*, and even a left-handed man tries to use his right hand in writing words from the Koran." (Vol. 2, p.14). Clearly, the concepts of clean and dirty applied to the use of the right and left hands in regard to eating have both social and individual implications, as in many societies, taking food with the fingers from a common bowl is a communal activity.

Elsewhere in Africa, it has been reported (see Wieschoff, 1938) that sanctions against the use of the left hand are such that if a child persists in using that hand, boiling water is poured into a hole in the ground and the offending hand placed therein with the earth rammed down around it so that the hand is so scalded the child is forced to use the other (right) hand. It has even been reported that in some societies, the symbolic association of right and left with good and evil may lead to self-mutilation. For example, Evans-Pritchard (1953) writes of the Nuer people of southern Sudan that Nuer youths emphasize the contrast between the two sides through pressing a series of metal rings into the flesh of the left arm thereby inducing sores and great pain as well as incapacitating the arm completely for several months or even a year or more.

However, while many other societies observe the important contrasting roles of the right and left hands, they do not resort to physical punishment. For example, Beidelman (1973) reports that in the Kaguru, the right hand is used for eating and greeting etc., and is associated with cleanliness, strength, and auspiciousness, but if a child grows up to be left-handed he is not physically admonished but may be subject to verbal abuse. Similarly, In Tanzania, Rigby (1966) points out that in the Gogo culture, although many actions which are considered polite should be performed with the right hand (including eating), left-handed children are not physically coerced into being right-handed. And in Timor, on the island of Roti, Fox (1973) reports that the right hand is favored as the "knowing hand" and the left referred to as the foolish or ignorant hand, while left-handed individuals are apparently tolerated but mockingly called "monkey left" i.e. ignorant of proper human conventions - like monkeys.

But why should there be this universal association between what is good, desirable and right, and what is undesirable, dirty and left with apparently no culture adopting the reverse association? Although the origin of this general association is unknown, there is evidence to suggest that the use of symbolism in hominid evolution emerged as a primitive attribute that may well have predated the development of tool-making (see e.g. Oakley, 1981). And clearly, once made, there are many ways by which such symbolic associations may be transmitted down through time to ensure their survival, e.g. through religious beliefs, rites, traditions, customs, legends, myths or superstitions (Tylor, 1903).

Certainly, Hertz (1909) believed that his concept of dual symbolic classification is a fundamental and all-pervasive feature of human thought and ideas, and made it clear that the contrast between right and left must be seen in the context of a wide range of similar oppositions which occur in nature such as light and dark, day and night, high and low, sky and earth, east and west, strong and weak, male and female. He suggested that in primitive thought these oppositions represent the two contrary supernatural powers - the sacred, which supports and increases light, life, and all that is good and desirable, and the profane which is a weakening influence associated with darkness, evil, and death. That such associations have been perpetuated throughout history and into modern times is evident in the frequently occurring references to relevant customs, prejudices, and prohibitions found in so many cultures and religions. Reference to any number of passages in the New Testament of the Bible (as well as the Old Testament) for example, indicates the close relationship in Christianity between the right hand, goodness, light, and God, in contrast to the association between the left, damnation, darkness, and the devil (for a review of these biblical references, see e.g. Fabbro, 1994). Similar references appear in other religious contexts (see e.g. Chelhod, 1964; Wile, 1934).

The literature on the subject is extensive and a brief review here cannot do justice to the wealth of material available. Furthermore, as Middleton (1968) has pointed out, societal classifications and concepts must be considered and analysed in terms of their functions in the particular culture of which they form a part and should not be taken out of context. While Hertz's complementary dualism appears to be universally applicable as a broad generalization, all of the dichotomous pairs of opposites are not always associated in the same way in different cultures. Neither are they necessarily consistent within a particular culture, since it is not uncommonly reported that the associations are reversed in conditions such as sleep, or more particularly, death. For example, Littlejohn (1973) states that in the Temne culture, after an individual dies, his left hand becomes stronger than his right and it is through the left hand (which has been a silent witness during life) that the individual is then able to communicate with God. Neither the mouth nor the right hand could give a true account to God of the individual's life since they were effectively accomplices. Needham (1967) also provides examples from several different cultures of reversals of association after death e.g. with hand use, so that the left then becomes the more actively used hand instead of the right (Toraja people of Celebes), or e.g. with relationships in the language spoken in the afterlife such that right now becomes left and up becomes down, etc (Ngaju of southern Borneo).

Such apparent contradictions serve to emphasize that in the world of religious beliefs, the oppositions of right and left, up and down, strong and weak, or life and death, etc. have both a culturally symbolic as well as intrinsic meaning. As Evans-Pritchard (1953) points out in discussing evidence from the Nuer people, the ideas of weakness, femininity, and evil are not determined by or derived from a possible organic inferiority of the left hand, but rather the reverse. It is as a symbol and not a functional entity that the left hand attains significance. He further illustrates this by reference to the fact that in the Nuer culture, the left half of a sacrificial animal is not in itself evil - it is always eaten by someone - it is only bad symbolically.

With regard to the implications of these two levels of meaning for left and right, Hertz (1909) drew attention to the fact that in most Indo-European languages, while a single term for right is quite extensively employed and shows considerable stability, the concept of left is often expressed in many less widely adopted forms which, over time, are frequently replaced by new ones. He suggested that the number and variety of words for left and their instability was a reflection of the inevitable fate of any such terms ultimately becoming contaminated with the other (religious) associations, and hence changed to eliminate or avoid these unwanted connotations. Wile (1934) has also reviewed and extensively discussed the linguistic origins and associations of left and left-handedness in comparison with right and right-handedness, and provided considerable additional evidence on the topic. He confirmed Hertz's wide range of dichotomous associations, remarking that right and left, light and dark, male and female, health and disease, good and evil etc. were not only closely correlated with the sacred and profane, south and north, east and west, but were also significantly associated with the role of the sun and sun worship in the beliefs and religions of many cultures.

Handedness and social conformity

It is a matter of common observation that there is a considerable variation from one society to another in both the strength and extent of religious and other cultural pressures influencing the lives and lifestyle of people within each community. Consequently, on Hertz's (1909) hypothesis, it may be expected that such differences in social pressure would produce variations in the observed relative incidence of left- and right-handedness, not only between cultures but also between identifiable groups subject to socialization differences within cultures.

For example, the child-rearing attitudes of parents may vary along a continuum from one of authoritarian control to the other extreme of laissez-faire permissiveness, and a number of measures have been devized to assess such social attitudes (see e.g. Barry, Child & Bacon, 1959; Berry, 1967; Schaefer & Bell, 1958). Some investigators (e.g. Silverman, Adevai & McGough, 1966; Pizzamiglio, 1974) have also shown that right-handers tend to be more field independent than left- or mixed-handers, but little or no systematic use has been made of these ideas in evaluating inter-cultural differences in the incidence of left- or right-handedness, except for the work of Dawson (1972, 1977). This author compared three groups of societies which he classified as (1) permissive and Independent (Alaskan Eskimo, Australian Arunta, and Hong Kong boat people); (2) Permissive (the Scots, English, and North Americans); and (3) Strict and Conforming (the Sierra Leone Temne, Congolese Katanga, and Hong Kong Hakka villagers), according to results he obtained from employing recognized measures of field dependency, and a parental strictness rating scale. Handedness was determined by the hand predominantly used on three criteria - writing, receiving an object, and cutting with scissors. His results showed that the recorded frequencies of left-handedness were greatest in the most permissive societies (Group 1 = 9.4% to 11.3%), less amongst the communities in Group 2 (5.1% to 7.0%), and least in the strictest societies (Group 3 = 0.6% to 3.4%). An even lower incidence of only 0.5% left-handedness was reported by Verhaegen and Ntumba (1964) in their investigation of 1,047 Katanganese schoolchildren using three similar criteria, while other authors (e.g. Berry, 1967; Littlejohn, 1973; Wieschoff, 1938) have presented evidence which supports the relative emphasis given to strictness in child-rearing practices or conformity in the African societies Dawson studied.

Several studies of other African societies have reported data in keeping with these findings. For example, Bakare (1974) investigated 360 Nigerian primary schoolchildren using a 10 item hand preference inventory and reported an overall preference for the left hand of only 0.28%. In discussing his results, the author pointed out that in the traditional Nigerian family, children are consciously trained to use the right hand, leftness is associated with evil, the left hand with deviance, and the left leg with ill-luck. A later and much larger survey of 56,779 Nigerian schoolchildren conducted by Payne (1981a) employing the writing hand as the sole criterion of handedness, found that a decidedly higher proportion (4.51%) of those questioned used the left hand for this purpose, which the author suggested may be a result of an increasing tolerance of left-handedness and/or a lack of formal writing instruction for many primary age children in the schools visited. Between a quarter and a third of all teachers interviewed said that they took action against writing with the left hand although this appears to be surprisingly lenient since the region from which the data were compiled is predominantly Muslim (Payne, 1987). It is interesting therefore that in her survey, Payne (1981a, p.237) noted that in some cases, students were allowed to use the left hand for writing in their state school but were forced to write Arabic with their right hand at Quranic school.

In a subsequent study of Nigerian university students (employing a 60 item hand preference inventory), Payne (1987) reported that of the subjects she identified as overall left-handers, approximately half wrote with the left hand compared with three-quarters who said they would use the left hand to paint or draw. A detailed and useful discussion of the attitudes of Muslim and non-Muslim university students towards the use of the left hand for writing in Nigeria is presented in an earlier paper by Payne (1981b). In another study by Brain (1977) who also obtained data on the hand used for writing by Tanzanian schoolchildren, it was reported that of 2,124 students only 0.8% used the left hand and that this was again in accordance with strong cultural pressures against the use of the left hand for any but "unclean" purposes. And in his investigation of 1,057 university students in Turkey, Tan (1988) noted an overall incidence of 3.4% left-handedness in this predominantly Muslim country, although no separate information was reported relating to the writing hand.

Evidence is also available from other parts of the world where strict child-rearing practices are embraced - especially China and Japan. For example, in the case of Chinese populations, Guo (1984) reported the incidence of left-handedness in a large sample (10,314) of the inhabitants of mainland China to be 0.24% although this author gave no details of the handedness criteria used. Other studies by Teng, Lee, Yang and Chang (1976, 1979) similarly recorded a very low incidence of left-handedness among 4,143 Taiwan Chinese, particulary for writing (0.7%) and for eating (1.5%). And in an even larger survey of 15,865 Taiwan Chinese primary schoolchildren, Hung, Tu, Chen and Chen (1985) reported an overall incidence of 3.55% left-handedness with only 0.34% using the left hand for writing. This trend is also supported by the evidence of Dean, Ratten and Hua (1987) who found a significant group difference in a comparison between the handedness characteristics of a Taiwan Chinese sample of university undergraduates and a similar group of Anglo-American students. And in a further study on Hong Kong Chinese university students, Hoosain (1990) recorded an overall incidence of 4.8% left-handedness with a much smaller proportion (1.6%) using the left hand for writing and drawing.

It is clear that traditionally, Chinese parents have been strongly influenced by Confucian values in relation to child-rearing practices and the need for children to respect their elders and to obey their parents, with children being indoctrinated with the ideas of filial piety both at home and at school (Ho, 1994; Hsu, 1955; Wong, 1970). Certainly, Chiu (1987) has shown that in comparison with Anglo-American and Chinese-American customs, Taiwan Chinese mothers are the most restrictive and Anglo-American mothers the least restrictive in attitudes to their children, with Chinese-American mothers somewhere in between. Hence, as Teng et al. (1976, 1979) suggested, the low incidence of left-handedness - particularly for eating and writing in their study - no doubt reflects the strong social pressure to conform to the use of the right hand for these activities in the Chinese culture. And an even more revealing comment may be found in the discussion of results by Hung et al. (1985, p.155) who referred to the strong disapproval of any such left-handed tendencies by Chinese people as the correction of *bad* handedness - i.e. equating left with bad.

A similarly low incidence of left-handedness has been reported for samples of Japanese populations by a number of authors. For example, in their study of 1,199 people living in Osaka, Hatta and Nakatsuka (1976) reported an overall incidence of 3.09% left-handedness, derived from a 10 item handedness inventory. And in a survey of 4,282 high school students in the Toyama district of Honshu, Shimizu and Endo (1983) classified

3.2% as left-handed on the basis of a 13 item questionnaire. However, these latter authors also found that for writing and eating, the percentages using the left hand were 0.68% and 1.7% respectively, which are only slightly greater than the corresponding figures reported some 50 years earlier by Komai and Fukuoka (1934) for Grade 8 students. And again, Rymar et al. (1984) in a study of 725 Tokyo elementary and junior high school students reported an incidence of 3.7% left-handedness based on a range of seven hand preference items, while Kameyama, Niwa, Hiramatsu and Saitoh (1983) in an investigation of 688 Tokyo adult subjects using a 14 item handedness questionnaire in which four questions related to writing, drawing and eating, reported that whereas 1% could be classified as left-handed on these particular tasks, 4.8% may be considered left-handed using the other 10 criteria. Similarly, Ida and Bryden (1996) in a 47 item questionnaire study comparing the hand preferences of Japanese and Canadian university students, found that the significantly lower incidence of left-handedness amongst their Japanese subjects was particularly evident with respect to writing (1.4%).

In Japan, traditional social values owe much to the influence of Confucian ideas and the central role of filial piety. Hence, children have been taught to worship their ancestors, to respect their elders, and to revere and obey their parents absolutely (Yokoe, 1970). Considerable importance has been attached to the development of good manners and the avoidance of any behavior which may bring shame or disgrace on the family. Hence it is not surprising that the very strong cultural pressures against the use of the left hand (Hendry, 1966, p.91) especially for eating and writing (Komai & Fukuoka, 1934, p.38) have been so effective.

Further cross-cultural evidence confirming the influence of social pressure also comes from a few studies of relatively isolated communities. For example, in an investigation of the 264 inhabitants of the islands of Tristan da Cunha, Provins (1990) reported that in a sample of some 28% of the total population, only 2.6% could be classified as left-handed on the basis of eight hand preference criteria, while no subjects interviewed wrote with the left hand (they had been taught to write with the right hand at school). Reference to the prevailing social attitudes of this community suggest that these handedness characteristics may be attributed to two strong community pressures, viz deference to authority, and group conformity (see e.g. Munch, 1964; Rawnsley & Loudon, 1964). And in studies of Tocano adolescents in Colombia by Ardila et al. (1989) and Bryden, Ardila and Ardila (1993) who assessed the hand preferences of 116 subjects using a range of criteria, not one individual could be classified as left-handed. Although Bryden et al. report that the cultural attitudes of this community are said to be generally permissive, the authors nevertheless state that left-handedness is considered to be "pathological" and that there is a strong bias towards the use of the right hand and to become skilled with it. Connolly and Bishop (1992) have also presented some handedness data gathered on children and adults from nine villages in the highlands of Papua New Guinea, only 45% of whom had received any schooling. These authors found that on the basis of nine hand preference criteria, these subjects were significantly more right-handed than a comparison group of English children. Unfortunately, the report provides no information on the cultural traditions or religious beliefs of these Melanesian villagers.

The effect of social pressure can also be seen in a comparison of children's handedness at different ages in different cultures. For example, Komai and Fukuoka (1934) found that in a survey of 17,000 children in 20 primary schools in Kyota (Japan), the incidence of left-handedness (assessed on the basis of using the left hand for any one of seven different tasks) gradually decreased from an average of 17.1% for boys and 15.6% for girls in Grade 1 to 9.4% and 5.7% respectively by Grade 8. But of more particular interest, the incidence of left-handedness for writing decreased from 5.1% for boys and 2.9% for girls to 0.2% and 0.0% respectively from Grade 1 to Grade 8. In contrast, Enstrom (1962) reported that in his survey of over 92,000 children in the U.S.A. where a permissive attitude towards the left hand for writing had been in force for many years, he found a mean incidence for left-handed writing of 12.5% for boys and 9.7% for girls from Grade 1 to Grade 6 with only small and unsystematic variations from one grade to another. And in an investigation by Marrion (1986) also relating to the hand used for writing, comparisons were reported between two Canadian subgroups differing in their ethnic origin but matched for age, sex, and geographic location. These authors found that their North American Indian sample of 180 Kwakiutl were significantly more left-handed than their 180 Caucasian subjects at each of three age levels (4-6 years, 10-12 years and adult). In particular, they found that whereas the Caucasian group showed no significant change in the incidence of left-handedness (7%) with age, the Kwakiutl group showed a marked effect - from 21% left-handed and 17% ambi-lateral for writing in the 4-6 year-olds to 15% left- and 7% ambilateral at the adult level - reflecting the social pressure on the Kwakiutl in the Caucasian education system to employ the right hand. Evidence that the results for the Kwakiutl may be a reflection of inherent cultural bias in hand preference comes from a further study by Marrion and Rosenblood (1986) on the relative frequency of right and left hand use depicted on Kwakiutl totem and house poles. These authors examined 110 examples of poles carved in the late 19th and early 20th centuries and reported that in 56% of the cases, both hands were employed simultaneously, but in the remainder, 20% depicted right hand use and 24% left hand activity.

Other studies have also attempted to consider the effect of racial origin on the relative incidence of right- and left-handedness within the broader North American (U.S.) society. For example, Hardyck, Goldman and Petrinovich (1975) collected hand preference data (using three criteria) from 7,688 primary school children in California where the stated education policy was not to interfere in any way with the individual's handedness tendencies. The authors subdivided the results from their subjects into groups identified as White, Black, Oriental or Mexican-American with a recorded incidence of left-handedness of 10.2%, 9.5%, 6.5% and 8.8% respectively. Although these results were found to be significantly different, the authors suggested that the effect was weak and due almost entirely to the small number of orientals in the sample. In another investigation, Thompson and Marsh (1976) compared the hand preferences (derived from writing hand, self-classification, and any non-preferred hand usage) of 722 adult subjects identified as White and 575 as Black residents in the Los Angeles area of California. These authors reported no significant differences between the two subject groups, but a particularly low incidence of left-handedness (Black, 4.9% and White, 4.0%) due to their method of classification, which produced correspondingly high levels of mixed-handedness - i.e. 31.3% and 33.8% respectively. A further study of Black and White subjects carried out on over 7,000 seven-year-old children in Philadelphia by Nachson, Denno and Aurand (1983) assessed handedness in terms of hand consistency in the use of a pencil, but again, no statistically significant racial differences were recorded.

Thus, where significant differences in hand preference have been found between different ethnic groups, the evidence suggests there are clear cultural differences in attitude towards social conformity. Where pressures to conform are high, the incidence of left-handedness is low and vice-versa. Conversely, where no significant handedness differences have been found between subject groups of widely different ethnic origin within a heterogeneous population, there are good grounds for proposing an explanation in terms of a common cultural or religious upbringing.

CHAPTER 10
Intracultural Differences

Chapter outline

Cross-sectional data from people of different age groups show that the incidence of left-handedness decreases systematically with age and that this trend is most pronounced with respect to the hand used for writing. Comparisons with other hand preference criteria suggest that this effect is the result of an increasing social tolerance of left-handedness with time during the 20th century, a conclusion which is supported by the absence of any such age effects for between-hand differences in manual proficiency. Similarly, where differences between the sexes (demonstrating a higher incidence of left-handedness for males) have been found, these can also be attributed to a difference in social attitudes in child-rearing practices, corresponding to a pressure on females to fulfill their traditional role of nurturance, obediance and responsibility in caring for others, and for males to develop self-reliance and demonstrate achievement.

There is also evidence to suggest that parental levels of education and socio-economic status may contribute to the formation of social attitudes towards the acceptance of left-handedness. It appears that a higher parental education and either a low or high socio-economic status favors a permissive attitude towards handedness development (right or left), with middle-class attitudes being associated with conformity and uniform right-handedness. Hand preference comparisons between professional groups yield little or no consistency of findings, and in the sporting arena it seems likely that the apparent increase in successful left-handers in e.g. tennis and baseball is most simply explained in terms of a practical playing advantage for the unorthodox. It is suggested that the difficulty of interpreting the available evidence relating to a possible association between left-handedness and creativity or talent is due to the absence of relevant information concerning the social environment in which the respective subject populations have been raised.

Generational differences in handedness

Differences in the recorded incidence of left- and right-handedness between generations may be a function of aging, or they may be an effect of secular trends due to changes over time in social attitudes, and the question naturally arises as to which interpretation may be placed on the evidence available. In this regard, it is helpful to first examine the findings from investigations of hand preference and then to compare these with the results obtained from experiments on manual asymmetry using proficiency tests of performance.

(a) Hand preference and age

In a typical investigation of hand preferences, Brackenridge (1981) surveyed the parents of seven Melbourne private schools and besides obtaining information on their writing hand, he also collected data on the writing hand of each of their children, parents and grandparents. Although there was considerable ethnic variation between the subjects of his study, the author confined his anaysis of results to those born in Australia or New Zealand. He found that the incidence of left-handed writers increased over the generations from about 2% for those born around 1890 to approximately 13% for those born in the late 1960's. Similarly, Smart, Jeffery and Richards (1980) obtained data from 1,094 mothers of six to seven-year-old children born in Manchester (England) in 1971 concerning the hand used for writing by the child, its parents and its grandparents. They found that the incidence of left-handed writers increased significantly from one generation to the next, with 6.2% being recorded amongst the grandparents, 10.6% for the parents, and 17.5% amongst the children. Also, Tambs, Magnus and Berg (1987) recorded the writing hand of three generations of Norwegian subjects and reported that the incidence of left-handed writers had increased systematically from approximately 1% to 10% corresponding with year of birth from about 1900 to 1980.

Confirmatory data for these trends may be found in a number of other studies. For example, in England, Fleminger, Dalton and Standage (1977b) found that in their survey of 800 adult subjects, the incidence of left-handed writers dropped from 10.8% for those aged between 15 and 24, to 2.9% for the 55 to 64 year-olds, while Davis and Annett (1994) reported a similar decrease in a sample of 33,401 British subjects, from 11.2% to 2.0% between their 18 to 30 and 81+ year-old age group. And in a very large survey of over one million American (U.S) men and women between 10 and 86 years old, Gilbert and Wysocki (1992) reported much the same effects of age. These authors investigated age trends for both writing and throwing and reported that irrespective of which hand they used for throwing, those who wrote with the left hand declined in numbers with age, whereas those who wrote with the right hand but threw with the left (presumably "shifted" left-handers) increased in numbers with age. And Hugdahl et al. (1993) who presented data on 2,787 subjects for both left-handed writing and the self-reported incidence of forced use of the right hand for writing have recorded confirmatory results. They found a decreasing prevalence of left-handedness from 15.2% in their 20 to 30 year-olds to 1.7% in subjects older than 80 years, with a corresponding increase in switched handedness (to the right hand) for writing from 2.7% in the youngest group to 6.8% for the 80 year-olds or older.

That the age effect is more clearly evident with respect to writing hand than it is for other handedness criteria has been shown many times in those studies where other systems of classification have been employed. For example, in a study by Beukelaar & Kroonenberg (1986), these authors reported on the incidence of left-handed writing but confined their investigation to

331 Dutch volunteers who classified themselves as left-handed. An analysis of the data showed that of the 73 left-handers born prior to 1940, not one wrote with the left hand, whereas for the 94 left-handers born after 1960, all but one wrote with the left hand. And in a broad survey of adults in Spain, Greece, Italy, and France, Dellatolas et al. (1991) classified 365 subjects as left-handers on 10 items (excluding writing and drawing) of a 12 item questionnaire and subdivided them into seven different age groups. The authors found that amongst these otherwise identified left-handers, there was a steady decrease in the incidence of left-handed writers with increasing age, from a high of 81.8% in the 15 to 19 year-old group to a low of 17.9% in the over 50 year-olds.

Further, in studies where trends on individual hand preference criteria have been presented, the age effect with regard to writing hand is clearly greater than for any other single item. For example, Plato, Fox and Garruto (1984) obtained data from 461 male and 244 female American (U.S.) adults using a 10 item hand preference questionnaire and subdivided their subjects into three age groups. They found a general decrease in the reported left hand preferences with increasing age for both sexes, but the effect was significant and most marked for the writing hand which decreased from 11.8% for males under 40 years old to 3.5% for those aged over 60 years, with figures of 9.2% and 1.0% for females in the corresponding age groups. Similarly, in a 14 item questionnaire survey of 1,425 Australian tertiary students and their families, Tan (1983) subdivided his subjects into two age groups (above and below 35 years of age). While he found a significantly lower overall incidence of left-handedness (5.9%) in the older compared with the younger (11.9%) subjects, he also reported that for every item, the older subjects gave significantly fewer left-handed responses than did the younger ones, with the incidence of left-handed writing (3.7%) for the older subjects being the lowest of all.

Other investigators using multiple item handedness questionnaires with respect to English (Ellis, Ellis & Marshall, 1988), Italian (Salmaso & Longoni, 1985) and North American

(Coren, 1995; Lansky, Feinstein & Peterson, 1988; Porac, 1993; Porac, Coren & Duncan, 1080b) population samples, have reported similar overall significant decreases in the incidence of left-handedness with increasing age of their subjects. And Brito, Brito and Paumgarten (1985) have also reported a corresponding age effect for men in their survey of Brazilian adults, but not for women.

The most frequently offered reason for these trends is in terms of an increased social tolerance with regard to left-handedness and particularly in respect of left-handed writing in schools since the early part of the 20th century. Certainly, this interpretation would account well for most of the data presented from the above Australian, North American and European sources of evidence, and would share a common basis of explanation for the cross-cultural differences discussed earlier. Support for this proposal may also be found in the data presented by Levy (1974) relating to the changing percentage of left-handed writers recorded in the literature over the years. She found that the reported incidence increased from about 2% for studies published around 1930 to over 11% for work reported in 1970. And in a similar survey of reports relating to overall hand preference published between 1913 and 1976, Porac, Coren and Duncan (1980b) found a corresponding but smaller reduction in the reported incidence of right-handedness from the earlier to the later studies.

However, in a series of papers by Halpern and Coren (1988, 1990, 1991, 1993) and Coren and Halpern (1991), an explanation of the decreasing incidence of left-handedness with age has been attributed to a reduced longevity in left-handers. It has been suggested that a variety of factors may contribute to such a decreased survival fitness, both environmental and neuropathological, and as reviewed in Chapter 8, there are many debilitating clinical and non-clinical conditions which could have some relevant association with non-right-handedness. However, in a lengthy critical examination of the concept, Harris (1993a, 1993b) has provided good grounds for doubting that a longevity hypothesis can be sustained by the evidence.

(b) Hand proficiency and age

Also inconsistent with a longevity explanation is the evidence from performance studies of manual asymmetry which show that the incidence of left-handedness measured by proficency tests changes little, if at all, with age. One of the earliest such investigations was undertaken by Miles (1931), who recorded the speed of rotary movements of each hand separately in operating a small hand drill on 863 subjects ranging from 6 to 95 years of age. He found that the proficiency of each hand on this task improved rapidly up until about 25 years of age, and then declined gradually thereafter. But whereas the reported improvement was markedly greater for the preferred hand until peak performance was reached in the mid-twenties, the steady decline from the 25 year-olds to the 80+ age group occurred at about the same rate for both sides. Jebson et al. (1969) also assessed the performance of 300 subjects (between 20 and 94 years of age) on seven different manual activities using each hand in turn, and reported that there were no significant differences in performance between age groups less than 60 years of age, but that subjects older than this were significantly slower in six of the seven functions irrespective of the side tested. And a comparison of the mean times taken by the preferred and non-preferred hands to undertake each task, shows that the relative differences between sides remained remarkably stable. For example, the times taken for each age group to complete a writing task with the preferred side expressed as a percentage of the time taken by the non-preferred side were 37.8% and 38.7% for the younger males and females respectively, compared with 40.5% and 40.4% for the older subjects. Comparable figures for the other six tasks showed a similar stability (an average of 90.4% and 90.0% respectively for the lower age group, and of 86.7% and 91.1% for the older subjects) although the differences between sides on these other tasks are clearly much less.

Similar declines in speed of overall manual performance in older people have been demonstrated by many other investigators (see e.g. Bornstein, 1985; Era, Jokela & Heikkinen, 1986; Joseph,

1988) with few reporting changes in the asymmetry of manual proficiency associated with the aging process. For example, Bornstein (1986) tested 365 subjects ranging from 18 to 69 years of age on grip strength, finger tapping, and pegboard tasks, and found that there was no significant correlation with age for performance differences between the preferred and non-preferred hands on any of the three tasks. Furthermore, he noted that approximately 30% of the males, and 20% of the females in his subject sample obtained scores on each test indicating a better performance with the nominally non-preferred hand. Kilshaw and Annett (1983) similarly reported no trends with age in performance differences between hands on a pegboard task, and Mitrushina et al. (1995) also found no correlation with age (albeit in a much narrower age range, i.e. from 60 to 88 years of age) on a finger tapping or a pegboard task. However, these latter authors reported a low but significant correlation with age for performance on a third, hole-piercing task called a pin test.

Meudell and Greenhalgh (1987) compared the performances of 21 school- children (mean age 15 years) with 21 elderly subjects (mean age 72 years) on three manual tasks (a pegboard, "fairground", and a square-tracing test), and only found a significant effect of age on differences between sides for the pegboard task. On this task, the older group of subjects took relatively longer to perform the test with their non-preferred hand than did the younger subjects. Weller and Latimer-Sayer (1985) also found the same differential effect with age for the two hands on a similar pegboard task (using 119 subjects, 16 to 87 years of age), although these authors reported that the effect was nullified by practice on the task; the older subjects improving more with the non-preferred hand than the younger ones.

Handedness and gender

Many investigators who have looked for sex differences in their hand preference data have found no effect. This includes studies using reliable handedness criteria, carried out on populations of different nationalities, and employing substantial numbers of subjects (e.g. Ellis, Ellis & Marshall, 1988; Porac et al., 1980b; Silverberg et al., 1979). Yet other equally large surveys using similar handedness measures to gather data from a variety of ethnic groups have reported significant differences in the hand preference patterns of males and females (e.g. Bryden, 1977; Maehara et al., 1988; Oldfield, 1971; Teng et al., 1979). Why should this be so and is there a consistency in the type or pattern of hand preference differences between the sexes in those studies which have found such an effect?

Certainly, there is overwhelming evidence that where gender differences in hand preferences occur, the incidence of left-handedness is higher in males than in females (see e.g. Brito et al., 1985; Hatta & Nakatsuka, 1976; Oldfield, 1971; Shimizu & Endo, 1983; Teng et al., 1976, 1979). Some authors have reported that the sex difference is primarily the result of an excess of mildly left-handed or mixed-handed males rather than an effect of extreme left-handedness (Oldfield, 1971; Brito et al., 1989), a tendency which is also reflected in the results of Shimizu and Endo (1983) and Teng et al. (1979). In noting a similar effect for his overall handedness scores (obtained from a five point scale) on both the Crovitz & Zener (1962) and Oldfield (1971) handedness inventories applied to 1,107 Canadian university students, Bryden (1977) also examined the data separately for each item and found that the mean response for males was significantly more left-handed than for females on 18 of the 24 comparisons. Nevertheless, in two further analyses, he obtained evidence to indicate that his male subjects were significantly less extreme in their stated hand preferences than the females irrespective of direction (i.e. left or right). An interesting and notable exception for both questionnaires was the item for throwing, on which men were clearly more extreme in their preferences than the women. And in a subsequent (13 item) questionnaire survey on large numbers of American (U.S.) university students, Chapman and Chapman (1987) reported that whereas significantly more of their female subjects than males were consistently right-handed (i.e. over the 13 items), significantly more males than females claimed to be consistently left-handed.

Since it has already been shown that cultural differences in hand preference are strongly influenced by differences in social tolerance of left-handedness, can the same be said of child-rearing attitudes towards boys compared with girls? There is certainly very strong and consistent evidence for sex differences in childhood socialization from a number of sources (Fagot & Leinbach, 1993). Probably the most comprehensively relevant study for the present discussion is the cross-cultural survey of socialization practices in 110 (mostly non-literate) cultures carried out by Barry, Bacon and Child (1957). These authors used two judges to assess ethnographic reports relating to five areas of cultural pressures on children, viz. nurturance, obedience, responsibility, achievement, and self-reliance. They found that there was considerable agreement across cultures such that pressure towards nurturance, obedience and responsibility is most often stronger for girls, whereas pressure towards achievement and self-reliance is most often greater for boys. These observed differences were noted as being consistent with well-recognized universal tendencies in the adult roles allotted to each sex. For example, men are more frequently required to participate in activities such as hunting or warfare which take them away from home and put a premium on self-reliance and achievement. In contrast, women are usually entrusted with tasks in and around the home, ministering to the needs of others and conforming to an established pattern of family maintenance in respect of the preparation of food, cooking, water-carrying and the like. As Barry et al. point out, most of these allotted adult sex roles are not inevitable, although the biological differences relating to child-bearing ensure that where such distinctions are made, they are consistently in the same direction.

These findings have been largely confirmed in a detailed investigation by Whiting and Edwards (1973) which involved field studies of children 3 to 11 years of age belonging to particular communities located in the six different cultures of Kenya,

Okinawa, India, Philippines, Mexico, and the U.S.A. (New England). These authors found that whereas boys interacted less frequently with adults and were assigned tasks such as feeding and herding animals more frequently than girls, domestic activities which involved cleaning, food preparation, cooking, grinding, and caring for younger siblings were significantly more frequently undertaken by girls. In discussing their results, Whiting and Edwards suggested that the nature of the assigned tasks were clearly predictive of the observed gender behaviour stereotypes - with girl's jobs demanding compliancy and a willingness to service the needs of others, and boy's activities fostering self-reliance and a need to achieve. Other observational evidence supporting this differential treatment of boys and girls in Western societies through parents, teachers or peers encouraging certain types of activity or behaviour and discouraging others has been provided by Fagot (1977) and Tauber (1979).

It is not surprising then, to find that handedness differences between the sexes depend on whether the criteria being employed take account of what is relevant to males and females in the society being examined. For example, de Chateau and Anderson (1976) investigated the preferred methods of holding a doll by children of various ages, and found that whereas girls soon displayed a proficiency and preferred side for cradling the imaginary baby in their arms, boys were not only less consistent and less comfortable with the procedure but that some were thoroughly uncooperative. Similarly, in a stereotypically male activity such as throwing a ball, for which gender differences in hand preference are strong (Bryden, 1977; Plato et al., 1984) throwing proficiency develops more rapidly with age in boys than in girls (Espenschade, 1960; Halverson, Robertson & Langendorfer, 1982), consistent with their more frequent involvement in sport and motor skills (Branta, Painter & Kiger, 1987; Coakley, 1987).

It also appears true that in those cultures displaying the strongest tendency to a right-sided population bias (e.g. Chinese and Japanese), significant gender differences in favour of a greater female conformity to the norm is usually evident (e.g. Hatta &

Nakatsuka, 1976; Maehara et al., 1988; Rymar et al., 1984; Shimizu & Endo, 1983; Teng et al., 1979), whereas in more permissive (e.g. European and North American) cultures, gender differences tend to be either small or absent (Ellis et al., 1988; Fleminger et al., 1977b; Hardyck, Petrinovich & Goldman, 1976; Kilshaw & Annett, 1983; Levander & Scalling, 1988).

If it is assumed from the evidence so far that the degree of social conformity or permissiveness is the basis for differences in the population incidence of left-handedness, then it may be expected that the effect of variations in social pressure would be detected not only in comparisons between cultures, over time, or according to gender, but also between certain population subgroups which may be distinguished by other relevant differences in background or experience.

Socio-economic status and education

If part of the purpose of education in an advanced society is the cultivation of independent thought and original ideas, as well as the acquisition of a body of knowledge and the development of each individual's natural talents, then it might be expected that those with a good educational background and wide experience would be less susceptible to social pressures to conform in respect of their own behaviour, and would display a greater understanding or tolerance of behavioural non-conformity in others, than those with a more restricted upbringing. Certainly, Kahl and Davis (1955) have shown that level of educational achievement is highly correlated with other indicators of socioeconomic status, and in this regard, parental education is one of the three criteria (together with occupation, and labour force status) usually employed in the U.S.A. to assess the socioeconomic level of a household (Entwistle & Astone, 1994; Hauser, 1994). It is noteworthy then, that Zuckerman et al. (1958) in their study of 222 mothers using the Parental Attitude Research Instrument (PARI) found that the level of parental education was a highly significant variable, and that the less educated mothers tended to possess more authoritarian, suppressive, and hostile social attitudes than their more educated contemporaries. Similar findings have been reported by Schaefer and Bell (1958).

However, studies of child-rearing practices in relation to socioeconomic indicators of the home environment have often shown that, contrary to expectation on an educational hypothesis, lower class parents may be more permissive and middle class parents more demanding of their children. For example, Ericson (1946) interviewed 48 middle class and 52 lower class white American mothers of 107 and 167 children respectively who lived in Chicago. She found that the middle class children were expected to assume responsibilities in the home at an earlier age and were more carefully and closely supervised in their movements (e.g. going to the movies) than their lower class contemporaries. And in an associated but more comprehensive report which included the results of interviews with both black and white American mothers from middle and lower class Chicago families, Davis and Havighurst (1946) obtained essentially the same results. These authors found that the social class differences in child-rearing practices were considerable and appreciably greater than differences between black and white families of the same social class. Again, middle class families placed greater emphasis on the early assumption of responsibility for the self and on individual achievement of their children, and were less permissive than the lower class families.

How then may this apparently contradictory information be reconciled with the evidence available relating socioeconomic status to the incidence of left- and right-handedness? Clearly, if degree of permissiveness is the relevant variable and not socioeconomic status *per se*, it may be expected that (i) families with the advantage of a high level of education, and (ii) families of low socioeconomic status, are both more permissive than those in the middle income bracket. In this regard, both Thompson and Marsh (1976) and Lansky, Feinstein and Peterson (1988) assessed the influence of socioeconomic status on the hand preferences of their subjects, and found that the incidence of left- or left/mixed-handedness was significantly associated with a higher socioeconomic standing, and right- or right/mixed-handedness with lower status levels. And in a third survey by Laponce (1976) who obtained handedness data on over 40,000 children from teachers at various schools in the Vancouver area of Canada, the incidence of left-handedness was again found to be higher amongst students living in high socioeconomic parts of the city. Highly relevant here then, is the report by Thompson and Marsh that their subject sample was over-representative of people with high educational attainment, and in the study by Lansky et al., there is also evidence to suggest that it was the higher levels of both their educational and occupational subdivisions that contributed most to their findings.

At the other extreme, Douglas, Ross and Cooper (1967) in their national survey of 11 year-old British schoolchildren found that a significantly higher proportion of manual working class boys had inconsistent hand preferences compared with middle class boys, although there were no significant differences between the two social classes for the girls. Furthermore, there is evidence that the population they investigated was under-representative of children from the private school sector (Douglas, 1964, p.10) so that their comparison between social classes was relatively depressed towards the lower end of the socioeconomic range. But a similar survey of 11 year-old British schoolchildren by Calnan and Richardson (1976) reported no social class differences in the proportion of left and mixed hand preferences in the data they obtained. However, the absence of a social class effect in this later study could well be the result of a continuing secular trend towards increasing social tolerance towards left-handedness, since, as these authors observed, the incidences of left-handedness reported in their study (11.7% for boys and 8.5% for girls) were appreciably higher than those given in the Douglas et al survey (7.2% for boys and 4.7% for girls).

Of further relevance is the evidence presented by Williams and Scott (1953) who investigated the child-rearing practices of parents in two contrasting socioeconomic groups from the black American population of Washington, D.C. These authors examined the gross motor development of 104 infants from the two groups and found that the low socioeconomic group infants showed a significant acceleration in motor development during their first 18 months compared with their peers in the higher social class. But they also found that the home atmosphere (assessed as permissive/accepting v rigid/rejecting) was more directly and significantly related to motor development than socioeconomic status, with children from permissive families being the more developmentally advanced. These latter findings are of particular importance here since there is evidence that the early development of hand preference (for either side) is significantly related to advanced mental or motor development in infancy (Cohen, 1966; Kaufman, Zalma & Kaufman, 1978).

Thus, the variables of parental income or education appear to be relevant to the development of hand preference but only indirectly, inasmuch as they relate to the degree of permissiveness that characterizes the child-rearing attitudes of the home and school environment.

Profession or occupation

Another social variable which has often been considered to be relevant to the recorded incidence of left-handedness, is the particular area of professional competence or vocational expertise of the subjects surveyed. For example, although a number of authors have suggested that a high incidence of left-handedness is significantly associated with success in certain games such as tennis (Annett, 1985) or baseball (McLean & Ciurzak, 1982), it has been suggested that this may be due to the fact that in face to face situations, it is the unorthodox player who has the advantage (Hemenway, 1983). Certainly, in a large-scale survey of participants in a wide range of sports, Porac and Coren (1981) reported that right- or left-handedness alone did not appear to be associated with level of achievement. A subsequent detailed review of the available evidence by Wood and Aggleton (1989) that included an examination of the world sporting records for tennis and cricket, and a questionnaire survey of English and Scottish league goalkeepers in soccer, came to much the same conclusion. These latter authors found that although there was evidence of an over-representation of left-handers in certain games (such as tennis, or for bowling in cricket) where they could have a strategic playing advantage over an opponent, in situations (e.g. goal-keeping in soccer) where no such advantage could be attached to being left-handed, no excess of left-handers was found. A further comparison of the incidence of left-handed spin and seam bowlers in cricket by Edwards and Beaton (1996) has also provided evidence to support the view that the over-representation of left-handed bowlers has a strategic rather than a neuropsychological basis.

Other workers have been more interested in the incidence of left-handers in the creative arts and associated professions. For instance, Oldfield (1969) was intrigued by the fact that in playing many musical instruments, both hands are required to exercise considerable motor skill, but that in most cases, one hand plays the major role and the other has a relatively minor function. In order to examine if this posed any special problems for players with particular handedness characteristics, Oldfield surveyed 129 students and staff in two university schools of music and found that the distribution of their hand preferences on his questionnaire did not differ significantly from that derived from the corresponding responses of 1,128 psychology undergraduates. Furthermore, he found that amongst the musicians, left-handedness did not appear to have posed any special problems for them in learning to master their particular instruments, although conducting was said to present a difficulty for some.

However, most investigators who have undertaken research in this field have been inspired to do so by the speculations made concerning the possible specialization of the two cerebral hemispheres and the frequently made distinction between verbal/linguistic behaviour and visual/spatial functions.

Accordingly, Byrne (1974) followed up Oldfield's (1969) work by conducting a similar survey of the hand preferences of university music staff and students, but in this study the author divided his subjects into vocalist (n = 134) and instrumentalist (n = 108) subgroups and compared them with a sample of 864 unselected university students. He found no significant differences in hand preference between the vocalists and the unselected students, but a significant difference in handedness for the instrumentalists compared with the control group, primarily due to an excess of mixed-handers amongst the instrumentalists. Pursuing the possibility of handedness differences between subgroups of musicians, Christman (1993) compared those instrumentalists who played instruments requiring integrated and temporally coordinated bimanual activity (e.g. strings, woodwinds) with performers of instruments requiring relatively independent bimanual activity (e.g. keyboards). Using the Edinburgh Handedness Inventory (Oldfield, 1971), he reported no significant differences between groups in respect of left- and right-handedness, but found that musicians who played instruments requiring a greater bimanual coordination (i.e. strings and woodwinds) displayed an overall weaker degree of handedness irrespective of direction. In a study of professional classical musicians who described themselves as strong right-handers, Jancke, Schlaug and Steinmetz (1997) compared the manual dexterity and finger-tapping performances of these subjects with three groups of nonmusicians. They reported that, although the musicians recorded a performance asymmetry favouring their right hand, the difference in skill between hands was less than for a control group of consistent right-handed nonmusicians, and was due to the musicians' greater skill with the left hand compared with that of the nonmusicians. The authors also reported that although keyboard musicians were better than string players on the tapping task, they did not differ with respect to performance asymmetry, and that the results for a smaller tapping asymmetry in their musicians overall appeared to be related to the age at which they began musical training.

Similar studies have been undertaken on other specialist groups. For instance, Peterson and Lansky (1974) reported that a survey of staff and students in a university department of architecture revealed that a high 29.4% of the staff and 16.3% of the students were left-handed. And in a follow-up study, the same authors (Peterson & Lansky, 1977) reported that proportionally more of the left-handed than right-handed students who had enrolled in the architecture program successfully graduated. Extending his enquiries further, Peterson (1979) reported that in a survey of 1,045 undergraduates in psychology, when hand preference was related to the area in which students were majoring, he found that there were significant differences between disciplines. Whereas the incidence of left-handedness was only 4.35% in the sciences, it was 14.89% for those majoring in music and 12.24% for those majoring in the visual arts. Gotestam (1990) also reported a higher incidence of left hand preferences amongst students of architecture and music than amongst a general student group. And in a questionnaire survey by Mebert and Michel (1980), a comparison of hand preferences was made between liberal arts undergraduates and students from two schools of art. These authors found that the distribution of left and right hand preferences differed significantly between the artists and general studies students, with more of the artists being left-handed.

In contrast, there have also been studies reporting no significant differences between artists or architects and professionals in other fields. For example, Shettel-Neuber and O'Reilly (1983) questioned 109 faculty members in four university departments (architecture, art, law, and psychology) concerning their hand preferences, and found no significant differences between disciplines in the incidence of left-handedness. And in a somewhat different type of survey, Wood and Aggleton (1991) reported the hand preferences of 236 fully qualified male architects employed professionally by architecture firms in Britain, together with similar information obtained from 78 male architecture students. Since these authors used the Oldfield (1971) questionnaire to obtain their handedness data, they were able to compare their results with those obtained by Bryden (1977) and Oldfield (1971) on large Canadian and British (respectively) samples of general university populations of students. However, Wood and Aggleton found no evidence of a significantly higher incidence of left-handedness amongst their qualified architects or architecture students than had previously been reported amongst these other general university groups.

Other studies of handedness differences between professional groups have also produced inconsistent findings. For example, in his early investigation of the incidence of right- and left-handedness in various professions or vocations, Quinan (1930) used throwing a ball as his sole criterion. From the data he presented for 100 professional musicians, 92% were clearly right-handed compared with 96% for 100 journeymen mechanics, while comparable figures for 70 students of fine arts and 41 students of religion were 82.7% and 95% respectively. More recently, Jones and Bell (1980) used the Oldfield (1971) questionnaire to examine the handedness of 137 psychology students who were considered to be representative of the general student body, and 162 more highly selected engineering students enrolled in a program with a strong mathematics requirement. These authors found a significant difference between the handedness distributions of the two groups, with the engineers more right-handed than the psychologists. In contrast, Annett and Kilshaw (1982) examined the hand preference data they had collected from three sources of male subjects, and in a comparison of the incidence of left-handedness amongst 86 maths students with 475 non-maths controls, they reported a significantly higher proportion of left-handers (assessed on two separate criteria) amongst the mathematicians than amongst the control group. But a similar comparison of female mathematicians with a non-maths group revealed no significant differences. Furthermore, in a study of the handedness characteristics of staff members in university departments of law and modern languages (classified as verbal academics), compared with mathematical academics (in the departments of mathematics, theoretical physics, theoretical chemistry, engineering, and computing), Temple (1990) found no statistically significant difference between the two groups, with a 12% incidence of left-handedness in each. In fact, when the strict mathematicians were separated from those working in the associated mathematical sciences, the author discovered that a significantly lower proportion of the mathematicians (5%) were left-handed in comparison with the mathematical scientists (16%). And in a questionnaire study by Schacter & Ransil (1996) of the handedness preferences of 1196 professionally qualified individuals comprising approximately equal numbers of mathematicians, architects, dentists, lawyers, librarians, orthodontists, orthopaedic surgeons, and psychiatrists, the reported mean laterality scores for the different professional groups were remarkably similar, although the incidence of left-handedness was highest for the architects and lawyers, and lowest amongst the orthopaedic surgeons and mathematicians.

In a very large survey of 16,590 applicants for places in a Brazilian university, Cosenza & Mingotti (1993) examined their data on handedness distributions obtained by using Oldfield's (1971) handedness questionnaire in three broad areas of study, viz. the biological sciences, mathematical sciences, and the humanities. They reported that significantly more left-handers applied for studies in the mathematical sciences than elsewhere, but found that this was mainly due to the relative numbers of males and females applying in the different areas, since there were no significant differences regarding career choice between left- and right-handers when male and female students were studied separately. However, in a comparison between the hand preferences of 225 science and graphic arts students, and 262 language and literature undergraduates confined to four handedness criteria, Coren and Porac (1982) reported a significantly greater incidence of consistent right-handers amongst the language and literature majors than amongst the science and graphic arts group. And finally, in a study using the Briggs and Nebes (1975) questionnaire, Schlichting (1982) compared the hand preferences of three different groups of naval personnel comprising advanced sonar trainees, paramedics and submariners (totalling 640 men), with those obtained by Briggs and Nebes on a university population. Although the author found that there were significantly more mixed-handers amongst the navy men, she concluded that the results were unlikely to be due to differences in intellectual ability.

Clearly, the absence of an adequate explanation for these widely varying results and inconsistent associations would suggest that the most important variables relevant to an understanding of the above evidence have not yet been identified or taken into account. If it is accepted that the origin and development of handedness is determined by a combination of (i) each subject's inborn capacity to acquire skill (irrespective of side), and (ii) the nature or pressures of his/her environment (Provins, 1997a), could this adequately explain the above evidence? Information on the learning capacity of subjects can be inferred from the fact that in most cases they are university students. Unfortunately, the circumstances under which individuals in the above experiments have been brought up is lacking, although some clues as to their social conformity may be derived from the reputations of different professional groups. For example, artists, whether they be painters, sculptors, musicians, architects or writers, besides possessing recognized skills, are renowned for their non-conformity to existing cultural norms, and their work tends to be judged *inter alia*, on the basis of their creativity, divergent thinking, freedom of expression, or novelty of technique. The traditional portrayal of an impoverished painter or poet living in a garret with like-thinking companions conveys very clearly the stereotypical view of an artist's characteristic scale of values or lack of concern for conventional ideas and practices. In contrast, bankers and accountants or practitioners of the legal and medical professions have usually been seen as highly conservative. But while only artistically inclined individuals may have enjoyed a reputation for non-conformity in the past, in more recent times, independent thinking and creativity are increasingly being seen as not only relevant, but highly valued in many other spheres, and may well determine whether an individual is, or is not, a successful achiever irrespective of the particular discipline.

Certainly, during the last thirty or forty years, there has been a considerable amount of rethinking of our concepts of intelligence, originality, and talent (see e.g. Collier, 1994; Ericsson, Krampe & Heizmann, 1993; Gardner, 1983; Sternberg, 1985, 1988), with an accumulation of evidence suggesting that the development of creativity is strongly influenced by a family environment which fosters a relatively permissive but stimulating parent-child relationship that encourages self-reliance, independence, and diverse interests in the child (Dacey, 1989; Dewing, 1970; Dewing & Taft, 1973; Nichols, 1964; Weissler & Landau, 1993).

Hence, it may be suggested that the difficulties presently encountered in interpreting the often confusing, if not conflicting evidence from studies of hand preference and creativity or unusual talent (see e.g. Aliotti, 1981; Coren, 1995b; Hicks & Dusek, 1980; Lewandowski & Kohlbrenner, 1985; McNamara et al. 1994; O'Boyle & Benbow, 1990; O'Boyle, Gill, Benbow & Alexander, 1994; Szesko, Madden & Piro, 1997; Wiley & Goldstein, 1991), are more likely to be resolved if investigators are encouraged to collect information on the home and school environments of subjects at the same time as obtaining data on their handedness and intellectual characteristics. Clearly, both creativity and an increased incidence of left hand preference appear to have a significant, but independent connection with a relatively permissive or tolerant social environment during infancy and childhood. It seems reasonable to suggest therefore, that the rather enigmatic association between the directional component of handedness and divergent thinking or talent, may well be due to this so far uncontrolled variable, i.e. a common underlying influence of type of home and school background.

CHAPTER 11
Inferences and Implications

Compatibility of the Social and Biological Evidence

Evidence from the archaeological material reviewed here suggests that a human bias favouring the use of the right hand first appeared during hominid evolution at or about the same time as tool-making skills developed. Comparative studies of tool-using and tool-making in other extant species of the higher primates reveal that these animals also employ some limited capacity for such manual activities, but there appears to be no consistent evidence for a general population bias favouring one particular hand more than the other. And from breeding and other studies of paw preference in nonhuman mammals, the results provide little or no grounds to support an explanation in terms of a genetic pre-determination for the direction of any lateral asymmetries they may display.

Thus, the peculiar characteristic of an apparently universal right-sided population bias, together with other particularly human attributes such as speech and language have, for a long time, given rise to considerable speculation and controversy over their likely origins. But the demonstration more than 100 years ago of an organic association (however imperfect), between speech and right-handedness through a common functional role of the left cerebral hemisphere, led to a widespread acceptance of the primacy of an innate predetermination of these asymmetries in human populations. And as a result of family studies of handedness, a wide range of theoretical explanations proposing a genetically determined lateral bias have been advanced over the years, although a detailed consideration of the evidence has shown that the results do not conform to the usual Mendelian laws of inheritance. Furthermore, the notion of a human inborn predetermination of lateral asymmetry favouring one particular side does not accord with what would be expected from the normal evolutionary processes of natural selection in providing a better adaptation to the environment than the inheritance of an equal potential for the two sides. In fact, it is suggested that such a predetermined asymmetry would be a considerable disadvantage in the struggle for survival - especially in early human societies.

Hence, the main purpose of the present review has been to examine at greater length than provided elsewhere (Provins, 1997a, 1997b), the suggestion that human handedness is no different in kind from the lateral asymmetries observed in subhuman mammalian species - just more sophisticated due to the greater learning abilities possessed by Homo sapiens. Further consideration has also been given to understanding the basis for a predominance of the right side in terms of Hertz's (1909) explanation of the universal symbolic significance of the right (as sacred) in contrast to that of the left (as profane). Considerable social anthropological evidence has accrued on this topic since Hertz's time and the origin of its symbolic significance coinciding with the development of tool-making appears to correspond with the evolutionary emergence of intelligent behaviour as the ultimate adaptation to the environment through a genetically determined inheritance of brain plasticity and learning ability.

Symbolism and Speech

Many investigators have stressed the correspondence between the conceptual demands made by learning to make tools and learning to speak, and both Oakley (1972) and Montagu (1976) for example, have suggested that a creature with the foresight and abilities demonstrated by the tool-making achievements of our early ancestors would have had the capacity to acquire at least the rudiments of speech.

But, how or when speech developed in relation to tool-using or tool-making is largely a matter of speculation in our present state of knowledge, although there are good grounds for believing that they may well have developed in parallel as a function of natural selection for learning ability. According to Jerison (1985, 1986, 1991), the essential feature of human intelligence is to be found not so much in the development of learning ability *per se*, but rather in *what* is learned. He has suggested that selection pressures for an ability to learn may lead to differently organized brains, but not necessarily to large ones. Rather, he proposed that the type of intelligence that evolved with higher grades of encephalization was related to the development of a means of creating the sort of mental representation of the external world described by Craik (1943). Thus, Jerison equated intelligence with cognition in the sense that it involves what an individual knows as well as the ability to acquire knowledge. Similarly with language, Jerison suggested that as a means of communication, von Frisch's (1971) discovery of the "language" of bees provides convincing evidence that effective communication between individuals of a species does not necessarily require a large brain. Rather, it is *what* is communicated that is relevant to a solution of the riddle of encephalization. And in this respect, he postulated that the cognitive characteristics of human language provide a means by which different individuals can not only make sense of their own external worlds, but are also able to share their mental constructs of reality. In this way, mental images or symbolic representations of an individual's world can be transmitted to others who can thereby experience the same experiences and come to know the same thoughts and ideas.

In such mental model-building, the transformation into symbolic form of some objects and events would be relatively simple and direct, while others would have complex associations. But the great advantage of this capacity for mental model-building is not only in the transmissibility of information between individuals, but in the potential for models to grow, knowledge to accumulate, and continuity to be facilitated between one generation and the next. This is certainly in keeping with Dobzhansky's (1963) view that the critical feature in human evolution was the genetic selection for a capacity to develop and maintain culture, or as Julian Huxley (1951) described it, selection for a mechanism of social heredity and the transmission of tradition. According to Huxley, whereas gestures, call-notes and facial expressions may be an early or primitive means of communication, human inter-individual transmission of information uses symbolic systems that include art and language. And in relation to what is transmitted, he postulated that both material and mental or psychological factors are always present. In some cases, the mental may be subordinate to the material, or in some instances, neither aspect may be given priority. Yet in other cases, ideas or beliefs, however generated, may have primacy and have a profound influence on what is conveyed.

But Huxley (1951) also emphasized that the initial function or significance of an object or event may change considerably, become distorted, or even be lost over time. Could this perhaps apply to the unknown origins of the significance of right-handedness? Certainly, as Hertz (1909) has pointed out, the universal association of right with sacredness, goodness, light, and life, in contrast to the association of left with profanity, evil, darkness, and death, appears at the very genesis of religious belief. And in the minds of our early hominid ancestors, mental models could only have had a somewhat superficial representation of their encounters with the world about them, since their understanding of reality (as we know it today) must have been extremely limited. Although their use and making of tools suggests that they possessed some appreciation of the principle of cause and effect, much of their experience of events at that time would undoubtedly have been in the form of contingent rather than dependent relationships. The coincidence of some right-sided activity with success, or a series of tragic outcomes associated with the left would have produced a lasting impression on those concerned. The likelihood of such chance associations developing into superstitions would have been quite high, with subsequent behaviour of the people involved modified accordingly, and the experience shared with others in their community. Certainly, the association of left with evil, darkness and death would have been a powerful influence to conform and a strong deterrent to any deviant behaviour. Hence, the social significance of right and left and their wider symbolic associations would then have become self-perpetuating. Thus if, as appears likely, the time and place of origin of these associations preceded the emerging growth of human populations (see e.g. Deevey, 1960), then it is not difficult to see how or why the symbolic importance of right- and left-handedness became universal with the subsequent dispersal of later generations through migration from their African birthplace.

Facilitation of Brain Lateralization of Function

Although social inheritance as an evolutionary vehicle enables knowledge and beliefs to be passed on from one generation to the next, it also requires that all individuals in each succeeding generation must, through personal experience and learning from others, build up their own mental model of reality. By the nature of things, each individual's model is unique, since personal experiences differ from one person to another and the range and depth of contacts between individuals also vary. But superimposed on these individual experiences is the wider, cohesive influence of community cultural values manifest in various traditions or conventions - including right- and left-handedness. Consequently, while individuals usually, but not necessarily, grow up developing many of the handedness characteristics of their close contacts such as parents, teachers, or peers, the overall strength and direction of their ultimate handedness tendencies will also reflect the effects of other more general and often quite subtle environmental influences.

Although Jerison's (1991) concept of mental modelling was primarily oriented towards explaining the integration and coordination of incoming sense data in some hierarchical system of organization, Sperry (1952) has pointed out that the primary business of the brain is the governing, directly or indirectly, of overt behaviour. In this respect, he stressed that "the entire output of our thinking machine consists of nothing but patterns of motor coordination". And in an attempt to provide a comprehensive explanation of the neurological basis for the development of adaptive behaviour during ontogeny, Edelman (1987) applied Darwin's concept of natural selection to the populations of neurones and synapses in the nervous system. He suggested that with the repeated sensory-motor activity of post-natal behaviour, there is a selective advantage for the grouping of some neurones and their connections compared with others, ultimately leading to the development of higher-order systems of neurophysiological organization. With respect to human motor behaviour, this is seen by Sporns and Edelman (1993) as underlying the progressive change from the highly variable and unpredictable limb movements of the newborn, to the more clearly controlled and well-directed actions of the adult (see e.g. Fetters & Todd, 1987).

Certainly, the acquisition of voluntary control over spontaneous motor activity in infancy and the subsequent development of motor competence and ultimately skill is a lengthy and arduous process that appears to be organized hierarchically (Elliott & Connolly, 1974). Hence, it may be suggested that the programs of motor skills controlling purposive bodily activity - especially of the limbs - also stem from mental models built up from the experiences of previous successes and failures to make particular intended movements. At the expert levels of motor proficiency, the output programs are highly sophisticated, and in each instance the skills are relatively specific to their particular designated neuromuscular components. As some sub-routines involving the same effector elements may be common to several other purposive movements, a certain degree of transfer of skill may be possible between them. However, unilateral motor skills developed by extensive experience or training of neuromuscular components on one side of the body, are unlikely to transfer to the other side insofar as the effectors in each instance are different - apart from those elements that may offer synergistic postural support. And the increasingly predictable outcome of movements made in acquiring a particular skill with one hand is likely to engender greater confidence in other movements made by that hand, so that the same side becomes more frequently used and hence, more generally proficient as well. Since the neural mechanisms of the left cerebral hemisphere are primarily responsible for the control and organization of movements of the right hand, then for most people, the left hemisphere would tend to become the dominant or more sophisticated side of the brain for most unilateral motor activity.

The directional bias favouring the right hand produced by cultural and environmental influences may be indirect, subtle, and/or circumstantial, or it may be through explicit instruction or training. But the bias is rarely 100% effective, and most people are to some extent mixed-handed, so that the population profile of relative frequency of use or proficiency of the right and left hands for most activities (considered separately), takes the form of a normal distribution with the mean shifted to the right. However, in highly specialized unimanual skills such as handwriting that are acquired by all members of a population, the distribution of relative use or proficiency becomes clearly bimodal, with most individuals clustering to the right and left extremes.

But how would differences in usage between the two sides in different activities affect the various motor control systems in the respective cerebral hemispheres? Training or experience on each *specific* task undertaken consistently by the same hand would produce a more refined motor program for each skill involving appropriate changes confined to the relevant neural substrate in the contralateral cerebral hemisphere. And an enhanced *generalized* usage of the same hand would have a similar but more general facilitatory effect on the neural circuitry of the contralateral hemisphere. Such a facilitatory effect would be a function of the relative frequency and duration of the general motor activity undertaken by one hand compared with the other, leading to a variation in the performance levels of the two sides as a result of the asymmetrical practice or training. Small achievement differences between sides would then lead to larger differences as the individual gained greater confidence in the more practiced side and the effects would become cumulative. Thus, from a fluctuating and purely chance level of imbalance of motor activity between the two sides evident at or soon after birth, minor differences in experience during infancy would lead to increasingly greater differences with age and maturation depending upon the motor activity history of the individual. For most infants, this means that the more frequent use of the right hand would provide a consistent facilitatory influence on the development of more competent motor control mechanisms in the left cerebral hemisphere. But some infants - for a variety of reasons including chance situational factors and/or social circumstances - would be less consistent in their manual preferences, or would even favour the more frequent use of the left hand, so that the facilitatory influences on the cerebral motor control systems would be more evenly distributed between hemispheres or might even favour the right.

However, it may be postulated that such a generalized cerebral facilitation stemming from the differential usage of the two hands, would affect *all* the motor control systems in the respective hemispheres, including those involved in speech. But this depends on the extent to which development of motor proficiency for the limbs, and particularly the hands, precedes or parallels the development of motor proficiency for speech in any given individual. Certainly, the longitudinal study by Molfese and Betz (1986) has provided evidence to suggest that there is an increasingly significant correspondence between measures of language competence and various aspects of motor development with age during infancy. And from an overall review of the available evidence, Molfese and Betz (1987) concluded that an explanation most compatible with the general trend of experimental findings would regard language and motor development during approximately the first two or three years of infancy as largely dependent on maturational processes in the nervous system. They suggested that there is little or no reason to believe that the lateralization of hemispheric function evident in children and adults is present from birth. Rather, they felt that it was only after a degree of consistency could be demonstrated in the overall relationships between hand preferences, relative manual proficiencies, and language production and comprehension, that lateralized behaviours could be used as markers for differences in hemispheric function.

But the notion of a behavioural marker for hemispheric differences requires some discussion. Since manual asymmetries tend to be specific to particular situations or tasks, any overall assessment of a person's handedness tendencies must be recognized as simply some sort of averaging of these particular biases or accomplishments, whether it is obtained by means of a self-estimation or by an objective measurement. Hence, differences between hemispheres in the degree of proficiency of control of motor activities at any given moment of time during the normal course of growth and maturation, are also best seen as primarily an historical summary of the differences between sides in the range and depth of the individual's manual learning experiences since birth. Insofar as the functional cerebral control of speech may be facilitated by the activity of cerebral control mechanisms for the limbs, then an increasingly consistent lateralization of manual activity would be expected to enhance or facilitate the lateralization of speech in the same hemisphere. And in this regard, speech may be seen as another specific skill to be acquired, i.e. one that is concerned with the organization and programming of the specialized neuromotor elements controlling language production.

Of particular importance here then, is the evidence reported by Johnson and Newport (1989, 1991) and Newport (1990, 1991) which indicates that the most successful period for language acquisition extends from infancy to about seven years of age, after which, learning effectiveness gradually declines until physical maturity is reached. Hence, on the basis of the biosocial explanation offered here, the cerebral control of speech laid down during the major phase of gaining mastery of speech production would be located according to the degree and direction of asymmetries in neural facilitation offered by the motor activities undertaken by the right and left sides of each individual child prior to, and during, this same period of time. However, as speech develops, it may be hypothesized that language production itself would become the more important agent in consolidating the cerebral (i.e. hemispheric) location of its neural control and hence, to become relatively less affected by facilitatory influences from other (e.g. manual) sources of motor activity.

Thus, early development of manual asymmetry would facilitate a greater development of the neural control of speech in the same hemisphere, whereas mixed or inconsistent hand usage in individuals of normal intelligence would cause speech control to be established to approximately the same extent in both hemispheres. Any interference with the development of manual skill at this time would tend to have a corresponding effect on the development of speech. Hence, a reduced mental capacity to acquire skill, or a chronic condition of ill-health limiting the opportunities for physical exercise and social interaction would also retard the development of manual asymmetries and speech. Similarly, generalized (i.e. bilateral) brain damage would be expected to have an adverse effect on the development of speech and handedness depending on the severity of the injury. Trauma sufficient to affect cognitive capacity would certainly qualify in this regard.

The concept of handedness and speech as simply acquired skills also provides a ready explanation for the evidence on clinical cases suffering from the effects of cerebral lesions confined to one particular hemisphere. When such a unilateral brain injury is sustained early in life, the course of motor development will clearly be affected in those instances where hemiplegia limits the use of the paretic limbs. In these circumstances, an asymmetrical development of manual skills will occur in accordance with an increased usage of the unimpaired side, and speech control would become established in the same (undamaged) hemisphere. But if the unilateral brain damage is sustained later in life, then speech would be affected to whatever extent the asymmetry of cerebral speech control had been established before the injury occurred. For those individuals who started life using the right hand more than the left - a tendency usually reinforced by cultural norms - both manual and speech skills are likely to be affected by a lesion in the left hemisphere but not the right. However, for those infants who initially started to use the left hand more than the right but later became subject to a cultural bias to change hands e.g. for writing and eating, the effects of a left-sided cerebral lesion would depend on the degree to which a switch in hand usage had occurred by the time the injury took place. But clearly, such effects are much more difficult to predict, as the degree and extent of asymmetrical usage of the two hands is known to vary considerably for nominal left-handers.

Bibliography

Abercrombie, M.L.J., Lindon, R.L. & Tyson, M.C. (1964). Associated movements in normal and physically handicapped children. *Developmental Medicine & Child Neurology, 6,* 573-580.

Adelson, E. & Fraiberg, S. (1974). Gross motor development in infants blind from birth. *Child Development, 45,* 114-126.

Ajuriaguerra, J. de & Tissot, R. (1969). The apraxias. In P. Vincken and G, Bruyn (Eds). *Handbook of clinical neurology, Vol. 4.* Amsterdam: North Holland. p.48-66.

Alekoumbides, A. (1978). Hemispheric dominance for language: quantitative aspects. *Acta Neurologica Scandinavica, 57,* 97-140.

Aliotti, N.C. (1981). Intelligence, handedness and cerebral hemisphere preference in gifted adolescents. *The Gifted Child Quarterly, 25,* 36-41.

Andersen, P. & Andersson, S.A. (1968). *Physiological basis of the alpha rhythm.* New York: Appleton-Century-Crofts.

Anderson, R.C., Wilson, P.T. & Fielding, L.G. (1988). Growth in reading and how children spend their time outside of school. *Reading Research Quarterly, 23,* 285-303.

Andrew, J. (1978). Laterality on the tapping test among legal offenders. *Journal of Clinical Child Psychology, 7,* 149-150.

Andrews, G. & Harris, M. (1964). The syndrome of stuttering. *Clinics in Developmental Medicine No 17.* London: The Spastics Society & William Heinemann.

Andy, O.J. & Bhatnager, S. (1984). Right hemispheric language evidence from cortical stimulation. *Brain & Language, 23,* 159-166.

Annett, J., Annett, M., Hudson, P.T. & Turner, A. (1979). The control of movement in the preferred and nonpreferred hands. *Quarterly Journal of Experimental Psychology, 31,* 641-652.

Annett, M. (1970a). A classification of hand preference by association analysis. *British Journal of Psychology, 61,* 303-321.

Annett, M. (1970b). The growth of manual preference and speed. *British Journal of Psychology, 61,* 545-558.

Annett, M. (1972). The distribution of manual asymmetry. *British Journal of Psychology, 63,* 343-358.

Annett, M. (1973a) Laterality of childhood hemiplegia and the growth of speech and intelligence. *Cortex, 9,* 4-33.

Annett, M. (1973b). Handedness in families. *Annals of Human Genetics, 37,* 93-105.

Annett, M. (1985). *Left, right, hand and brain: the right shift theory.* London: Lawrence Erlbaum Associates.

Annett, M. (1992a). Parallels between asymmetries of planum temporale and of hand skill. *Neuropsychologia, 30,* 951-962.

Annett, M. (1992b). Five tests of hand skill. *Cortex, 28,* 583-600.

Annett, M. (1993a). Handedness and educational success: the hypothesis of a genetic balanced polymorphism with heterozygote advantage for laterality and ability. *British Journal of Developmental Psychology, 11,* 359-370.

Annett, M. & Annett, J. (1979). Individual differences in right and left reaction time. *British Journal of Psychology, 70,* 393-404.

Annett, M. & Annett, J (1991). Handedness for eating in gorillas. *Cortex, 27,* 269-275.

Annett, M., Hudson, P.T.W. & Turner, (1974). The reliability of differences between the hands in motor skill. *Neuropsychologia, 12,* 527-531.

Annett, M. & Kilshaw, D. (1982). Mathematical ability and lateral asymmetry. *Cortex, 18,* 547-568.

Annett, M. & Manning, M. (1990a). Arithmetic and laterality. *Neiropsychologia, 28,* 61-69.

Annett, M. & Manning, M. (1990b). Reading and a balanced polymorphism for laterality and ability. *Journal of Child Psychology & Psychiatry, 31,* 511-529.

Anonymous. (1916). The acquirement of single-handed dexterity. *The Lancet, (ii),* 241.

Aram, D.M. & Eisele, J.A. (1992). Plasticity and recovery of higher cognitive functions following early brain injury. In I.Rapin and S.J.Segalowitz (Eds). *Handbook of neuropsychology, Vol. 6, Child neuropsychology.* Amsterdam: Elsevier, p.73-92.

Archer, L., Campbell, D. & Segalowitz, S. (1988). A prospective study of hand preference and language development in 18 to 30 month-olds. I. Hand preference. *Developmental Neuropsychology, 4,* 85-92.

Ardila, A., Ardila, O., Bryden, M.P., Ostrosky, F., Rosselli, M. & Steenhuis, R. (1989). Effects of cultural background and education on handedness. *Neuropsychologia, 27,* 893-897.

Armatas, C.A., Summers, J.J. & Bradshaw, J.L. (1994). Mirror movements in normal adult subjects. *Journal of Clinical & Experimental Neuropsychology, 16,* 405-413.

Armitage, M. & Larkin, D. (1993). Laterality, motor asymmetry and clumsiness in children. *Human Movement Science, 12,* 155-177.

Asanuma, C. (1991). Mapping movements within a moving motor map. *Trends in Neurosciences, 14,* 217-218.

Ashton, R. & Beasley, M. (1982). Cerebral laterality in deaf and hearing children. *Developmental Psychology, 18,* 294-300.

Attenborough, J. & Farber, M. (1934). The relation between intelligence, mechanical ability and manual dexterity in special school children. *British Journal of Educational Psychology, 4,* 140-161.

Autret, A., Auvert, L., Laffont, F. & Larmande, P. (1985). Electroencephalographic spectral power and lateralized motor activities. *Electroencephalography & Clinical Neurophysiology, 60,* 228-236.

Bair, J.H. (1901) Development of voluntary control. *Psychological Review, 8,* 474-510.

Bakan, P. (1990). Nonright-handedness and the continuum of reproductive casualty. In S.Coren (Ed). *Left-handedness: behavioral implications and anomalies.* Amsterdam: North-Holland. p.33-74.

Bakare, C.G.M. (1974). The development of laterality and right-left discrimination in Nigerian children. In J.L.M.Dawson & W.J.Lenner (Eds). *Readings in cross-cultural psychology.* Hong Kong: Hong Kong University Press. p.150-166.

Ballard, P.B. (1912). Sinistrality and speech. *Journal of Experimental Pedagogy & Training College Record, 1,* 298-310.

Barry, H., Bacon, M.K. & Child, I.L. (1957). A cross-cultural survey of some sex differences in socialization. *Journal of Abnormal & Social Psychology, 55,* 327-332.

Barry, H., Child, I.L. & Bacon, M.K. (1959). Relation of child training to subsistence economy. *American Anthropologist, 61,* 51-63.

Barry, R.J. & James, A.L. (1978). Handedness in autistics, retardates, and normals of a wide age range. *Journal of Autism & Childhood Schizophrenia, 8,* 315-323.

Barsley, M. (1966). *The left-handed book.* London Souvenir Press.

Barsley, M. (1970). *Left-handed man in a right-handed world.* London: Pitman.

Bashore, T.R., McCarthy, G. Haffly, E.F., Clapman, R.M. & Donchin, E. (1982). Is handwriting posture associated with differences in motor control? An analysis of asymmetries in the readiness potential. *Neuropsychologia, 20,* 327-346.

Basmajian, J.V. (1963). Control and training of motor units. *Science, 141,* 440-441.

Basmajian, J.V. (1973) Control of individual motor units. *American Journal of Physical Medicine, 52,* 257-260.

Basmajian, J.V. & Samson, J. (1973). Standardization of methods in single motor unit training. *American Journal of Physical Medicine, 52,* 250-256.

Basser, L.S. (1962). Hemiplegia of early onset and the faculty of speech with special reference to the effects of hemispherectomy. *Brain, 85,* 427-460.

Batheja, M. & McManus, I.C. (1985). Handedness in the mentally handicapped. *Developmental Medicine & Child Neurology, 27,* 63-68.

Bauer, R.W. & Wepman, J.M. (1955). Lateralization of cerebral functions. *Journal of Speech & Hearing Disorders, 20,* 171-177.

Baxter, B. (1942). A study of reaction time using factorial design. *Journal of Experimental Psychology, 31,* 430-437.

Bear, D., Schiff, D., Saver, J., Greenberg, M. & Freeman, R. (1986). Quantitative analysis of cerebral asymmetries: fronto-occipital correlation, sexual dimorphism, and association with handedness. *Archives of Neurology, 43,* 598-603.

Beaton, A.A. & Moseley, L.G. (1984). Anxiety and the measurement of handedness. *British Journal of Psychology, 75,* 275-278.

Beaton, A.A. & Moseley, L.G. (1991). Hand preference scores and completion of questionnaires: another look. *British Journal of Psychology, 82,* 521-525.

Beaumont, J.G. (1976). The cerebral laterality of "minimal brain damage" children. *Cortex, 12,* 373-382.

Beaumont, J.G. (1983). The EEG and task performance: a tutorial review. In A.W.K.Gaillard & R.Ritter (Eds). *Tutorials in ERP research: endogenous components.* Amsterdam: Elsevier. p.385-447.

Beck, C.H.M. & Barton, R.L. (1972). Deviation and laterality of hand preference in monkeys. *Cortex, 8,* 339-363.

Behan, P.O. & Geschwind, N. (1985). Hemispheric laterality and immunity. In R.Guillemin, M.Cohn & M.Melnechuk (Eds). *Neural modulation of immunity.* New York: Raven Press. p.73-80.

Beidelman, T.O. (1973). Kaguru symbolic classification. In R.Needham (Ed). *Right and left: essays on dual symbolic classification.* Chicago: Chicago University Press. p.128-166.

Bell, G.A. (1972). The origin of the alpha rhythm: a review of current theories. *Journal of Behavioral Science, 1*, 151-159.

Belmont, L. & Birch, H.G. (1963). Lateral dominance and right-left awareness in normal children. *Child Development, 34*, 257-270.

Benbadis, S.R., Dinner, D.S., Clune, G.J., Piedmonte, M. & Luders, H.O. (1995). Objective criteria for reporting language dominance by intracarotid amobarbital procedure. *Journal of Clinical & Experimental Neuropsychology, 17*, 682-690

Benbow, C.P. (1986). Physiological correlates of extreme intellectual precocity. *Neuropsychologia, 24*, 719-725.

Benecke, R., Dick, J.P.R., Rothwell, J.C., Day, B.L. & Marsden, C.D. (1985). Increase of the Bereitschaftspotential in simultaneous and sequential movements. *Neuroscience Letters, 62*, 347-352.

Benninger, C., Matthis, P. & Scheffner, D. (1984). EEG development of healthy boys and girls: results of a longitudinal study. *Electroencephalography & Clinical Neurophysiology, 57*, 1-12.

Benson, D.F. (1967). Fluency in aphasia: correlation with radioactive scan localization. *Cortex, 3*, 373-394.

Benson, D.F. (1988). Classical syndromes of aphasia. In: F.Boller, J.Grafman, G.Rizzoletti & H.Goodglass (Eds). *Handbook of Neuropsychology. Vol. 1.* Amsterdam: Elsevier, p.267-280.

Benton, A.L., Meyers, R. & Polder, G.J. (1962). Some aspects of handedness. *Psychiatria et Neurologia (Basel), 144*, 321-337.

Berker, E.A., Berker, A.H. & Smith, A. (1986). Translation of Broca's 1865 report. Localization of speech in the third left frontal convolution. *Archives of Neurology, 43*, 1065-1072.

Berry, J.W. (1967). Independence and conformity in subsistence-level societies. *Journal of Personality & Social Psychology, 7*, 415-418.

Betancur, C., Neven, P.J. & Le Moal, M. (1991). Strain and sex differences in the degree of paw preference in mice. *Behavioural Brain Research, 45*, 97-101.

Betancur, C., Velez, A., Cabanieu, G., Le Moal, M. & Neveu, P.J. (1990). Association between left-handedness and allergy: a reappraisal. *Neuropsychologia, 28*, 223-227.

Beukelaar, L.J. & Kroonenberg, P.M. (1983). Towards a conceptualization of hand preference. *British Journal of Psychology, 74*, 33-45.

Beukelaar, L.J. & Kroonenberg, P.M. (1986). Changes over time in the relationship between hand preference and writing hand among left-handers. *Neuropsychologia, 24*, 301-303.

Bianki, V.L. (1981). Lateralization of functions in the animal brain. *International Journal of Neuroscience, 15,* 37-47.

Biary, N. & Koller, W. (1985). Handedness and essential tremor. *Archives of Neurology, 42,* 1082-1083.

Bingley, T. (1958). Handedness and brainedness. *Acta Psychiatrica et Neurologica (Suppl 120), 33,* 32-151.

Bisazza, A., Rogers, L.J. & Vallortigara, G. (1998). The origins of cerebral asymmetry: A review of evidence of behavioural and brain lateralization in fishes, reptiles and amphibians. *Neuroscience & Biobehavioral Reviews, 22,* 411-426.

Bishop, D.V.M. (1980a). Measuring familial sinistrality. *Cortex, 16,* 311-313.

Bishop, D.V.M. (1980b). Handedness, clumsiness, and cognitive ability. *Developmental Medicine & Child Neurology, 22,* 569-579.

Bishop, D.V.M. (1981). Plasticity and specificity of language localization in the developing brain. *Developmental Medicine & Child Neurology, 23,* 251-255.

Bishop, D.V.M. (1983). Linguistic impairment after left hemidecortication for infantile hemiplegia? A reappraisal. *Quarterly Journal of Experimental Psychology, 35a,* 199-207.

Bishop, D.V.M. (1984). Using non-preferred hand skill to investigate pathological left-handedness in an unselected population. *Developmental Medicine & Child Neurology, 26,* 214-226.

Bishop, D.V.M. (1986). Is there a link between handedness and hypersensitivity? *Cortex, 22,* 289-296.

Bishop, D.V.M. (1988a). Language development after focal brain damage. In D.V.M.Bishop and K.Mogford (Eds). *Language development in exceptional circumstances.* Edinburgh: Churchill-Livingstone. p. 203-219.

Bishop, D.V.M. (1988b). Can the right hemisphere mediate language as well as the left? A critical review of recent research. *Cognitive Neuropsychology, 5,* 353-367.

Bishop, D.V.M. (1989). Does hand proficiency determine hand preference? *British Journal of Psychology, 80,* 191-199.

Bishop, D.V.M. (1990a). *Handedness and developmental disorder.* Oxford: Mac Keith Press.

Bishop, D.V.M (1990b). On the futility of using familial sinistrality to subclassify handedness groups. *Cortex, 26,* 153-155.

Blau, A. (1946). *The master hand.* New York: American Orthopsychiatric Association.

Bloodstein, O. (1993) *Stuttering: the search for a cause and cure.* Boston: Allyn & Bacon.

Boklage, C.E. (1977). Schizophrenia, brain asymmetry development and twinning: cellular relationship with etiological and possibly prognostic implications. *Biological Psychiatry, 12,* 19-35,

Boldt, R.F. (1955). Motor learning in college students and mental defectives. *Proceedings of the Iowa Academy of Science, 60,* 500-505.

Bonin, G. von (1962). Anatomical asymmetries of the cerebral hemispheres. In V.B.Mountcastle (Ed). *Interhemispheric relations and cerebral dominance.* Baltimore: Johns Hopkins Press. p.1-6.

Bonin, G. von (1963). *The evolution of the human brain.* Chicago, Ill.: University of Chicago Press.

Boring, E.G. (1929). *A history of experimental psychology.* New York: Appleton-Century.

Bornstein, R.A. (1985). Normative data on selected neuropsychological measures from a nonclinical sample. *Journal of Clinical Psychology, 41,* 651-659.

Bornstein, R.A. (1986). Normative data on intermanual differences on three tests of motor performance. *Journal of Clinical & Experimental Psychology, 8,* 12-20.

Borod, J.C., Caron, H.S. & Koff, E. (1984). Left-handers and right-handers compared on performance and preference measures of lateral dominance. *British Journal of Psychology, 75,* 177-186.

Boucher, J. (1977). Hand preference in autistic children and their parents. *Journal of Autism & Child Schizophrenia, 7,* 177-187.

Boucher, J., Lewis, V. & Collis, G. (1990). Hand dominance of parents and other relatives of autistic children. *Developmental Medicine & Child Neurology, 32,* 304-313.

Brackenridge, C.J. (1981). Secular variation in handedness over ninety years. *Neuropsychologia, 19,* 459-462.

Bradshaw, J.L. (1991). Animal asymmetry and human heredity: dextrality, tool use and language in evolution - 10 years after Walker (1980). *British Journal of Psychology, 82,* 39-59.

Bradshaw, J.L. & Rogers, L.J. (1993). *The evolution of lateral asymmetries, language, tool use, and intellect..* San Diego: Academic Press.

Bradshaw-McAnulty, C., Hicks, R.E. & Kinsbourne, M. (1984). Pathological left-handedness and familial sinistrality in relation to degree of mental retardation. *Brain & Cognition, 3,* 349-356.

Brain, J.L. (1977). Handedness in Tanzania: the physiological aspect. *Anthropos, 72,* 180-192.

Brain, W.R. (1945). Speech and handedness. *Lancet, 249 (ii),* 837-841.

Brain, W.R. (1965). *Speech disorders: aphasia, apraxia and agnosia.* London: Butterworth.

Branch, C., Milner, B. & Rasmussen, T. (1964). Intracarotid sodium amytal for the lateralization of cerebral speech dominance. *Journal of Neurosurgery, 21,* 399-405.

Branta, C., Haubenstricker, J. & Seefeldt, V. (1984). Age changes in motor skills during childhood and adolescence. *Exercise & Sport Sciences Reviews. 12,* 467-520.

Branta, C.F., Painter, M. & Kiger, J.E. (1987). Gender differences in play patterns and sport participation of North American youth. In D.Gould & M.R.Weiss (Eds). *Advances in Pediatric Sport Sciences. 2, Behavioral issues.* Champaign, Ill.: Human Kinetics. p.25-42.

Briggs, G.G. & Nebes, R.D. (1975). Patterns of hand preference in a student population. *Cortex, 11,* 230-238.

Briggs, P.F. & Tellegen, A. (1971). Development of the manual accuracy and speed test (MAST). *Perceptual & Motor Skills, 32,* 923-943.

Brinkman, C. (1984). Determinants of hand preference in Macaca fascicularis. *International Journal of Primatology, 5,* 325.

Brito, G.N.O., Brito, L.S.O. & Paumgarten, F.J.R. (1985). Effects of age on handedness in Brazilian adults is sex dependent. *Perceptual & Motor Skills, 61,* 829-830.

Brito, G.N.O., Brito, L.S.O., Paumgarten, F.J.R. & Lins, M.F.C. (1989). Lateral preferences in Brazilian adults: an analysis with the Edinburgh Inventory. *Cortex, 25,* 403-415.

Brito, G.N.O., Lins, M.F.C., Paumgarten, F.J.R. & Brito, L.S.O. (1992). Hand preference in 4 to 7 year-old children: an analysis with the Edinburgh Inventory in Brazil. *Developmental Neuropsychology, 8,* 59-68.

Broadbent, D.E. (1957). A mechanical model for human attention and immediate memory. *Psychological Review, 64,* 205-215.

Broadbent, D.E. (1958). *Perception and Communication.* London: Pergamon.

Broadhead, G.D. (1975). Dynamometric grip strength in mildly handicapped children. *Rehabilitation Literature, 36,* 279-283.

Brooker, B.H. & Donald, M.W. (1980). Contribution of the speech musculature to apparent human EEG asymmetries prior to vocalization. *Brain & Language, 9,* 226-245.

Brooker, R.J., Lehman, R.A.W., Heimbuch, R.C. & Kidd, K.K. (1981). Hand usage in a colony of Bonnett monkeys, Macaca radiata. *Behavior Genetics, 11,* 49-56.

Brooks, V.B. (1981). Task related cell assemblies. In O.Pompeiano and C.Ajmone-Marsan (Eds). *Brain mechanisms and perceptual awareness.* New York: Raven Press. p.295-309.

Brooks, V.B. (1983). Motor control: How posture and movement are governed. *Physical Therapy, 63,* 664-673.

Brooks, V.B., Kennedy, P.R. & Ross, H. (1983). Movement programming depends on understanding of behavioral requirements. *Physiology & Behavior, 31,* 561-563.

Brookshire, K.H. & Warren, J.M. (1962). The generality and consistency of handedness in monkeys. *Animal Behaviour, 10,* 222-227.

Brown, J.W. (1988). *Agnosia and apraxia: selected papers of Liepmann, Lange, and Potzl.* Hillsdale, NJ: Erlbaum. p.3-39.

Brown, J.W. & Hecaen, H. (1976). Lateralization and language representation. *Neurology, 26,* 183-189.

Bruininks, R.H. (1974). Physical and motor development of retarded persons. *International Review of Research in Mental Retardation, 7,* 209-261.

Bruml, H. (1972). Age changes in preference and skill measures of handedness. *Perceptual & Motor Skills, 34,* 3-14.

Bryan, E.M. (1992). *Twins and higher multiple births: a guide to their nature and nurture.* London: Edward Arnold.

Bryden, M.P. (1977). Measuring handedness with questionnaires. *Neuropsychologia, 15,* 617-624.

Bryden, M.P. (1982). *Laterality. Functional asymmetry in the intact brain.* New York: Academic Press.

Bryden, M.P. (1986). The nature of complementary specialization. In F.Lapore, M.Ptito & H.H.Jasper (Eds). *Two hemispheres - one brain: Functions of the corpus callosum. New York: Alan R.Liss. p.463-469.*

Bryden, M.P. (1987). Handedness and cerebral organization: data from clinical and normal populations. In D.Ottoson (Ed). *Duality and unity of the brain.* London: Macmillan. p.55-70.

Bryden, M.P. (1990) Choosing sides: The left and right of the normal brain. *Canadian Psychology, 31,* 297-309.

Bryden, M.P., Ardila, A. & Ardila, O. (1993). Handedness in native Amazonians. *Neuropsychologia, 31,* 301-308.

Bryden, M.P., McManus, I.C. & Bulman-Fleming, M.B. (1994). Evaluating the empirical support for the Geschwind-Behan-Galaburda model of cerebral lateralization. *Brain & Cognition, 26,* 103-167.

Bryden, M.P., McManus, I.C. & Steenhuis, R.E. (1991). Handedness is not related to self-reported disease incidence. *Cortex, 27*, 605-611.

Bryden, M.P., McRae, L. & Steenhuis, (1991). Hand preference in school children. *Developmental Neuropsychology, 7*, 477-486.

Bryngelson, B. (1939). A study of laterality of stutterers and normal speakers. *Journal of Speech Disorders, 4*, 231-234.

Bryngelson, B. & Rutherford, B. (1937). A comparative study of laterality of stutterers and non-stutterers. *Journal of Speech Disorders, 2*, 15-16.

Bryson, S.E. (1990). Autism and anomalous handedness. In S.Coren (Ed). *Left-handedness: behavioral implications and anomalies*. Amsterdam: Elsevier. p.441-456.

Bullock, M.I. & Watter, P. (1987). A review of the histories of children with minimal cerebral dysfunction. *Australian Journal of Physiotherapy, 33*, 145-149.

Burge, I.C. (1952). Some aspects of handedness in primary school children. *British Journal of Educational Psychology, 22*, 45-51.

Burgess, J.W. & Villablanca, J.R. (1986). Recovery of function after neonatal or adult hemispherectomy in cats. II, Limb bias and development, paw usage, locomotion and rehabilitative effects of exercise. *Behavioral Brain Research, 20*, 1-18.

Burgess, J.W., Villablanca, J.R. & Levine, M.S. (1986). Recovery of functions after neonatal or adult hemispherectomy in cats. III. Complex functions: open field exploration, social interactions, maze and holeboard performances. *Behavioral Brain Research, 20*, 217-230.

Burke, H.L., Yeo, R.A., Delaney, H.D. & Conner, L. (1993). CT scan cerebral hemisphere asymmetries: predictors of recovery from aphasia. *Journal of Clinical & Experimental Neuropsychology, 15*, 191-204.

Burnett, S.A., Lane, D.M. & Dratt, L.M. (1982). Spatial ability and handedness. *Intelligence, 6*, 57-68.

Burt, C. (1937). *The backward child*. London: University of London Press.

Buskirk, E.R., Anderson, K.L. & Brozek, J. (1956). Unilateral activity and bone and muscle development in the forearm. *Research Quarterly, 27*, 127-131.

Butler, S.R. & Glass, A. (1976). EEG correlates of cerebral dominance. In A.H.Riesen & R.F.Thompson (Eds). *Advances in Psychobiology, Vol.3*. New York: Wliey. p.219-272.

Butler, S.R. & Glass, A. (1985). The validity of EEG alpha asymmetry as an index of the lateralization of human cerebral functions. In D.Papakostopoulos, S.Butler & I.Martin (Eds). *Clinical and experimental neuropsychophysiology*. London: Croom Helm. p.370-394.

Butler, S.R. & Glass, A. (1987). Individual differences in the asymmetry of alpha activation. In A.Glass (Ed). *Individual differences in hemisphere specialization.* New York: Plenum Press. p.103-120.

Buxton, C.E. (1937). A comparison of preference and motor-learning measures of handedness. *Journal of Experimental Psychology, 21,* 464-469.

Buxton, C.E. & Humphreys, L.G. (1935). The effect of practice upon intercorrelations in motor skills. *Science, 81,* 441-442.

Byrne, B. (1974). Handedness and musical ability. *British Journal of Psychology, 65,* 279-281.

Byrne, R.W. (1995). *The thinking ape: evolutionary origins of intelligence.* Oxford: Oxford University Press.

Byrne, R.W. & Byrne, J.M. (1991) Hand preferences in the skilled gathering tasks of mountain gorillas (Gorilla g. berengei). *Cortex, 27,* 521-546.

Calnan, M. & Richardson, K. (1976). Developmental correlates of handedness in a national sample of 11 year olds. *Annals of Human Biology, 3,* 329-342.

Carlier, M., Duyme, M. Capron, C., Dumont, A.M. & Perez-Diaz, F. (1993). Is a dot-filling group test a good tool for assessing manual performance in children? *Neuropsychologia, 31,* 233-240.

Carlson, D.F. & Harris, L.J. (1985). Development of the infant's hand preference for visually directed reaching: preliminary report of a longitudinal study. *Infant Mental Health Journal, 6,* 158-174.

Carmichael, E.A. (1966). The current status of hemispherectomy for infantile hemiplegia. *Clinical proceedings Children's Hospital Washington, 22,* 285-293.

Carmon, A., Harishanu, Y., Lowinger, E. & Lavy, S.A. (1972). Asymmetries in hemispheric blood volume and cerebral dominance. *Behavioral Biology, 7,* 853-859.

Carter, R.L., Hohenegger, M.K. & Satz, P. (1982). Aphasia and speech organization in children. *Science, 218,* 797-799.

Carter-Saltzman, L. (1980). Biological and sociocultural effects on handedness; comparison between biological and adoptive families. *Science, 209,* 1263-1265.

Carter-Saltzman, L., Scarr-salapatek, S., Barker, W.B. & Katz, S. (1976). Left-handedness in twins: incidence and patterns of performance in an adolescent sample. *Behavior Genetics, 6,* 189-203.

Caruso, A.J., Chodzko-Zajko, W.J., Bidinger, D.A. & Sommers, R.K. (1994). Adults who stutter: responses to cognitive stress. *Journal of Speech & Hearing Research, 37,* 746-754.

Castro, A.J. (1977). Limb preference after lesions of the cerebral hemisphere in adult and neonatal rats. *Physiology & Behavior, 18,* 605-608.

Chamberlain, H.D. (1928). The inheritance of left-handedness. *Journal of Heredity,* *19*, 557-559.

Chapman, J.P. & Chapman, L.J. (1987). Handedness of hypothetically psychosis-prone subjects. *Journal of Abnormal Psychology, 96*, 89-93.

Chapman, J.P., Chapman, L.J. & Allen, J.J. (1987). The measurement of foot preference. *Neuropsychologia, 25*, 579-584.

Chapman, L.J. & Chapman, J.P. (1987). The measurement of handedness. *Brain & Cognition, 6*, 175-183.

Chateau, P. de & Anderson, Y. (1976). Left-side preference for holding and carrying newborn infants: II. Doll-holding and carrying from 2 to 16 years. *Developmental Medicine & Child Neurology, 18*, 738-744.

Chaugule, V.B. & Master, R.S. (1981). Impaired cerebral dominance and schizophrenia. *British Journal of Psychiatry, 139*, 23-24.

Chavance, M., Dellatolas, G., Bousser, M.G., Amor, B., Grardel, B., Kahan, M.F., Le Floch, J.P. & Tchobroutsky, G. (1990). Handedness, immune disorders and information bias. *Neuropsychologia, 28*, 429-441.

Chelhod, J. (1964). A contribution to the problem of the pre-eminence of the right, based upon Arabic evidence. *Anthropos, 59*, 529-545. Reprinted in R.Needham (Ed). *Right and left: essays on dual symbolic classification.* Chicago: Chicago University Press. p.239-262.

Chengappa, K.N., Cochran, J., Rabin, B.S. & Ganguli, R. (1991). Handedness and autoantibodies. *The Lancet, 338*, 694.

Chesher, E.G. (1936). Some observations concerning the relation of handedness to the language mechanism. *Neurological Institute of New York Bulletin, 4*, 556-562.

Chi, J.G., Dooling, E.C. & Gilles, F.H. (1977). Left-right asymmetries of the temporal speech areas of the human fetus. *Archives of Neurology, 34*, 346-348.

Chiarello, C. (1980). A house divided? Cognitive functioning with acallosal agenesis. *Brain & Language, 11*, 128-158.

Chiarenza, G.A. (1993). Movement-related brain macropotentials of persons with Down syndrome during skilled performance. *American Journal of Mental Retardation, 97*, 449-467.

Chiarenza, G.A., Iacono, W.G. & Beiser, M. (1994). Handedness in first-episode psychotic patients and their first-degree relatives. *Journal of Abnormal Psychology, 103*, 400-403.

Chisholm, R.C. & Karrer, R. (1983). Movement-related brain potentials during hand squeezing in children and adults. *International Journal of Neuroscience, 19*, 243-258.

Chisholm, R.C. & Karrer, R. (1986). Associated movement, motor control and the readiness potential. *In* W.C.McCallum, R.Zapploi & F.Denoth (Eds). Cerebral psychophysiology: studies in event-related potentials. *Electroencephalography & Clinical Neurophysiology, Supp.38.* Amsterdam: Elsevier. p.242-244.

Chisholm, R.C. & Karrer, R. (1988). Movement-related potentials and control of associated movements. *International Journal of Neuroscience, 42*, 131-148.

Chisholm, R., Karrer, R. & Cone, R. (1984). Movement -related ERP's during right vs. left hand squeeze. In R.Karrer, J.Cohen & P.Tueting (Eds). *Brain and Information: event-related potentials. Vol.425.* New York Academy of Sciences. p.445-449.

Chiu, L-H. (1987) Child-rearing attitudes of Chinese, Chinese-American and Anglo-American mothers. *International Journal of Psychology, 22*, 409-419.

Christman, S. (1993). Handedness in musicians: bimanual constraints on performance. *Brain & Cognition, 22*, 266-272.

Chugani, H.T. & Phelps, M.E. (1991). Imaging human brain development with positron emission tomography. *Journal of Nuclear Medicine, 32*, 23-26.

Chugani, H.T., Phelps, M.E. & Mazziotta, J.C. (1987). Positron emission tomography study of human brain functional development. *Annals of Neurology, 22*, 487-497.

Chui, H.C. & Damasio, A.R. (1980). Human cerebral asymmetries evaluated by computed tomography. *Journal of Neurology, Neurosurgery & Psychiatry, 43*, 873-878.

Claiborne, J.H. (1917). Stuttering relieved by reversal of manual dexterity. *New York Medical Journal, 105*, 577-581; 619-621.

Clark, J.E. (1988). Development of voluntary motor skill. In E.Meisami & P.S.Timiras (Eds). *Handbook of human growth and developmental biology, Vol 1. Pt B.* 237-250. Florida: C.R.C. Press.

Clark, J.E. (1995). On becoming skillful: Patterns and constraints. *Research Quarterly for Exercise & Sport, 66*, 173-183.

Clarke, S., Kraftsik, R., van der Loos, H. & Innocenti, G.M. (1989). Forms and measures of adult and developing human corpus callosum: is there sexual dimorphism? *Journal of Comparative Neurology, 280*, 213-230.

Coakley, J.J. (1987). Children and the sport socialization process. In D. Gould & M.R.Weiss (Eds). *Advances in Pediatric Sport Sciences. 2, Behavioral issues.* Champaign, Ill.: Human Kinetics. p. 43-60.

Cohen, A.I. (1966). Hand preference and developmental status of infants. *Journal of Genetic Psychology, 108*, 337-345.

Cohen, H.J.S., Taft, L.T., Mahadeviah, M.S. & Birch, H.G. (1967). Developmental changes in overflow in normal and aberrantly functioning children. *Journal of Pediatrics, 71*, 39-47.

Cohen, M.E. & Duffner, P.K. (1981). Prognostic indicators in hemiplegic cerebral palsy. *Annals of Neurology, 9*, 3530357.

Colby, K.M. & Parkison, C. (1977) Handedness in autistic children. *Journal of Autism & Childhood Schizophrenia, 7*, 3-9.

Cole, J. (1955). Paw preference in cats related to hand preference in animals and man. *Journal of Comparative & Physiological Psychology, 48*, 137-140.

Collier, G. (1994). *Social origins of mental ability.* New York: Wiley.

Collins, R.A. & Collins, R.L. (1971). Independence of eye-hand preference in mentally retarded: evidence of spurious associations in heterogeneous populations. *Journal of Optometry & Archives of American Academy of Optometry, 48*, 1031-1033.

Collins, R.L. (1968) On the inheritance of handedness. I. Laterality in inbred mice. *Journal of Heredity, 59*, 9-12.

Collins, R.L. (1969). On the inheritance of handedness. II. Selection for sinistrality in mice. *Journal of Heredity, 60*, 117-119.

Collins, R.L. (1970). The sound of one paw clapping: an enquiry into the origin of left-handedness. In G.Lindzey & D.D.Thiessen (Eds). *Contributions to behavior-genetic analysis: the mouse as a prototype.* New York: Appleton-Century-Crofts, p.115-136.

Collins, R.L. (1975). When left-handed mice live in right-handed worlds. *Science, 187*, 181-184.

Collins, R.L. (1977a). Origins of the sense of asymmetry: Mendelian and non-Mendelian models of inheritance. *Annals of the New York Academy of Sciences, 299*, 283-305.

Collins, R.L. (1977b). Toward an admissible genetic model for the inheritance of the degree and direction of asymmetry. In S.Harnad, R.W.Doty, L.Goldstein, J.Jaynes & G.Krauthamer (Eds). *Lateralization in the nervous system.* New York: Academic Press, p.137-150.

Collins, R.L. (1985). On the inheritance of direction and degree of asymmetry. In S.D.Glick (Ed). *Cerebral ateralization in nonhuman species.* New York: Academic Press, p.41-71.

Collins, R.L. (1991). Re-impressed selective breeding for lateralization of handedness in mice. *Brain Research, 564*, 194-202.

Collins, R.L., Sargent, E.E. & Neumann, P.E. (1993). Genetic and behavioral tests of the McManus hypothesis relating response to selection for lateralization of handedness in mice to degree of heterozygosity. *Behavior Genetics, 23*, 413-421.

Colon, E.J., De Weerd, J.P.C., Notermans, S.L.H. & De Graaf, R. (1979). EEG spectra in children aged 8, 9 and 10 years. *Journal of Neurology, 221*, 263-268.

Connolly, K. & Bishop, D.V.M. (1992). The measurement of handedness: a cross-cultural comparison of amples from England and Papua-New Guinea. *Neuropsychologia, 30*, 13-26.

Connolly, K. & Dalgleish, M. (1989). The emergence of a tool-using skill in infancy. *Developmental Psychology, 25*, 894-912.

Connolly, K. & Elliott, J. (1972). The evolution and ontogeny of hand function. In N.B.Jones (Ed). *Ethological tudies of child behaviour.* New York: Cambridge University Press. p.329-383.

Connolly, K. & Stratton, P. (1968). Developmental changes in associated movements. *Developmental Medicine & Child Neurology, 10*, 49-56.

Conway, D., Lytton, H. & Pysh, F. (1980). Twin-singleton language differences. *Canadian Journal of Behavioral Science, 12*, 264-271.

Conrad, K. (1954). New problems of aphasia. *Brain, 77*, 491-509.

Corballis, M.C. (1991). *The lopsided ape: evolution and the generative mind.* New York: Oxford University Press.

Coren, S. (1989). Left-handedness and accident-related injury risk. *American Journal of Public Health, 79*, 1040-1041.

Coren, S. (1992) The left-hander syndrome: *The causes and consequences of left-handedness.* New York: The Free Press.

Coren, S. (1994). Twinning is associated with an increased risk of left-handedness and inverted writing hand posture. *Early Human Development, 40*, 23-27.

Coren, S. (1995a). Age and handedness. Patterns of change in the population and sex differences become isible with increased statistical power. *Canadian Journal of Experimental Psychology, 49*, 376-386.

Coren, S. (1995b). Differences in divergent thinking as a function of handedness and sex. *American Journal of Psychology, 108*, 311-325.

Coren, S. & Halpern, D.F. (1991). Left-handedness: a model for decreased survival fitness. *Psychological Bulletin, 109*, 90-106.

Coren, S. & Porac, C. (1977). Fifty centuries of right-handedness: the historical record. *Science, 198*, 631-632.

Coren, S. & Porac, C. (1978). The validity and reliability of self-report items for the measurement of lateral preference. *British Journal of Psychology, 69*, 207-211.

Coren, S. & Porac, C. (1980). Family patterns in four dimensions of lateral preference. *Behavior Genetics, 10,* 333-348.

Coren, S. & Porac, C. (1982). Lateral preference and cognitive skills: an indirect test. *Perceptual & Motor Skills, 54,* 787-792.

Coren, S., Porac, C. & Duncan, P. (1979). A behaviorally validated self-report inventory to assess four types of lateral preference. *Journal of Clinical Neuropsychology, 1,* 55-64.

Coren, S., Porac, C. & Duncan, P. (1981). Lateral preference behaviors in preschool children and young adults. *Child Development, 52,* 443-450.

Cornil, L. & Gastaut, H. (1947). Etude electroencephalographique de la dominance sensorielle d'un hemisphere cerebral. *Presse Medicale, 37,* 421-422.

Cornwell, K.S., Harris, L.J. & Fitzgerald, H.E. (1991). Task effects in the development of hand preference in 9, 13, and 20 month old infant girls. *Developmental Neuropsychology, 7,* 19-34.

Cosenza, R.M. & Mingotti, S.A. (1993). Career choice and handedness: a survey among university applicants. *Neuropsychologia, 31,* 487-497.

Cosi, V., Citterio, A. & Pasquino, C.A. (1988). A study of hand preference in myasthenia gravis. *Cortex, 24, 573-577.*

Cowan, W.M., Fawcett, J.W., O'Leary, D.D.M & Stanfield, B.B. (1984) Regressive events in neurogenesis. *Science, 225,* 1258-1265.

Cox, M.D. (1986). The psychologically maladjusted stutterer. In K.O. Louis (Ed). *The atypical stutterer: principles and practice of rehabilitation.* Orlando, FL. p.93-122.

Craig, A. (1990). An investigation into the relationship between anxiety and stuttering. *Journal of Speech & Hearing Disorders, 55,* 290-294.

Craik, K.J.W. (1943). *The nature of explanation.* London: Cambridge University Press.

Craik, K.J.W. (1947). Theory of the human operator in control systems. 1. The operator as an engineering system. *British Journal of Psychology, 38,* 56-61.

Craik, K.J.W. (1948) Theory of the human operator in control systems. 2. Man as an element in a control system. *British Journal of Psychology, 38,* 142-148.

Critchley, M. (1958). A critical survey of our conceptions as to the origins of language. In: M.Perrin (Chairman, Wellcome Foundation). *The history and philosophy of knowledge of the brain and its functions.* Oxford: Blackwell. p.45-72.

Critchley, M. (1970). *Aphasiology and other aspects of language.* London: Edward Arnold.

Cronholm, J.N., Grodsky, M. & Behar, I. (1963). Situational factors in the lateral preferences of rhesus monkeys. *Journal of Genetic Psychology, 103*, 167-174.

Crossman, E.R.F.W. (1959). A theory of the acquisition of speed-skill. *Ergonomics, 2*, 153-166.

Crossman, E.R.F.W. (1964). Information processes in human skill. *British Medical Bulletin, 20*, 32-37.

Crovitz, H.F. (1962). On direction in drawing a person. *Journal of Consulting Psychology, 26*, 196.

Crovitz, H.F. & Zener, K. (1962). A group test for assessing hand and eye dominance. *American Journal of Psychology, 75*, 271-276.

Crow, T.J., Done, D.J. & Sacker, A. (1996). Cerebral lateralization is delayed in children who later develop schizophrenia. *Schizophrenia Research, 22*, 181-185.

Cunningham, D.J. (1902). Right-handedness and left-brainedness. *Journal of the Anthropological Institute, 32*, 273-296.

Curt, F., De Agostini, M., Maccario, J. & Dellatolas, G. (1995). Parental hand preference and manual functional asymmetry in preschool children. *Behavior Genetics, 25*, 525-536.

Curt, F., Maccario, J. & Dellatolas, G. (1992). Distributions of hand preference and hand skill asymmetry in preschool children: theoretical implications. *Neuropsychologia, 30*, 27-34.

Dacey, J.S. (1989). Discriminating characteristics of the families of highly creative adolescents. *Journal of Creative Behavior, 23*, 263-271.

Dargent-Pare, C., De Agostini, M., Mesbah, M. & Dellatolas, G. (1992) Foot and eye preferences in adults: relationship with handedness, sex and age. *Cortex, 23*, 343-351.

Davidson, R.J., Chapman, J.P., Chapman, L.J. & Henriques, J.B. (1990). Asymmetrical brain activity discriminates between psychometrically-matched verbal and spatial cognitive tasks. *Psychophysiology, 27*, 528-543.

Davis, A. & Annett, M. (1994). Handedness as a function of twinning, age, and sex. *Cortex, 30*, 105-111.

Davis, A. & Havighurst, R.J. (1946). Social class and color differences in child rearing. *American Sociological Review, 11*, 698-710.

Davis, R.C. (1942). The pattern of muscular action in simple voluntary movement. *Journal of Experimental Psychology, 31*, 347-366.

Davis, R.C. (1943). The genetic development of patterns of voluntary activity. *Journal of Experimental Psychology, 33*, 471-486.

Dawson, G. & Adams, A. (1984). Imitation and social responsiveness in autistic children. *Journal of Abnormal Child Psychology, 12,* 209-225.

Dawson, J.L.M.B. (1972). Temne Arunta hand-eye dominance and cognitive style. *International Journal of Psychology, 7,* 219-233.

Dawson, J.L.M.B. (1977). Alaskan Eskimo hand eye auditory dominance and cognitive style. *Psychologia, 20,* 121-135.

De Agostini, M., Pare, C., Goudot, D. & Dellatolas, G. (1992). Manual preference and skill development in preschool children. *Developmental Neuropsychology, 8,* 41-57.

Dean, R.S. (1982). Assessing patterns of lateral preference. *Clinical Neuropsychology, 4,ˆ 124-128.*

Dean, R.S., Rattan, G. & Hua, M. (1987). Patterns of lateral preference: an American-Chinese comparison. *Neuropsychologia, 25,* 585-588.

Dean, R.S. & Smith, L.S. (1982). Personality and lateral preference patterns in children. *Clinical Neuropsychology, 3,* 22-28.

Deecke, L. (1987). Bereitschaftspotential as an indicator of movement preparation in supplementary motor area and motor cortex. In G.Bock, M.O'Connor & J.Marsh (Eds). *Motor areas of the cerebral cortex. Ciba Foundation Symposium 132.* New York: Wiley. p.231-250.

Deecke, L., Engel, M., Lang, W. & Kornhuber, H.H. (1986). Bereitschaftspotential preceding speech after holding breath. *Experimental Brain Research, 65,* 219-223.

Deecke, L. & Kornhuber, H.H. (1977). Cerebral potentials and the initiation of voluntary movement. In J.E.Desmedt (Ed). *Attention, voluntary contraction and event-related cerebral potentials. Progress in Clinical Neurophysiology, Vol.1.* Basel: Karger. p.132-150.

Deecke, L., Scheid, P. & Kornhuber, H.H. (1969). Distribution of readiness potential, pre-motor positivity and motor potential of the human cerebral cortex preceding voluntary finger movements. *Experimental Brain Research, 7,* 158-168.

Deevey, E.S. (1960). The human population. *Scientific American, 203,* 195-203.

Dellatolas, G., Annesi, I., Jallon, P., Chavance, M. & Lellouch, J. (1990). An epidemiological reconsideration of the Geschwind-Galaburda theory of cerebral lateralization. *Archives of Neurology, 47,* 778-782.

Dellatolas, G., Luciani, S., Castresana, A., Remy, C., Jallon, P., Laplane, D. & Bancaud, J. (1993). Pathological left-handedness: left-handedness correlatives in adult epileptics. *Brain, 116,* 1565-1574.

Dellatolas, G., Yubert, P., Castresana, A., Mesbah, M., Giallonardo, T., Lazartou, H. & Lellouch, J. (1991). Age and cohort effects in adult handedness. *Neuropsychologia, 29*, 255-261.

Dellatolas, G., Tubert-Bitter, P., Curt, F. & De Agostini, M. (1997). Evolution of degree and direction of hand preference in children: methodological and theoretical issues. *Neuropsychological Rehabilitation, 7*, 387-399.

DeMyer, M.K., Alpern, G.D., Barton, S., Demyer, W.E., Churchill, D.W. Hingten, J.N., Bryson, C.Q., Pontius, W. & Kimberlin, C. (1972). Imitation in autistic, early schizophrenic, and non-psychotic subnormal children. *Journal of Autism &*

Childhood Schizophrenia, 2, 264-287.

Denckla, M.B. (1973). Development of speed in repetitive and sequential finger-movements in normal children. *Developmental Medicine & Child Neurology, 15*, 635-645.

Denckla, M.B. (1974). Development of motor coordination in normal children. *Developmental Medicine & Child Neurology, 16*, 729-741.

Dennis, M. & Whitaker, H.A. (1976). Language acquisition following hemidecortication: Linguistic superiority of the left over the right hemisphere. *Brain & Language, 3*, 404-433.

Dennis, W. (1958). Early graphic evidence of dextrality in man. *Perceptual & Motor Skills, 8*, 147-149.

Denno, D.W. (1985). Sociological and human developmental explanations of crime: conflict or consensus? *Criminology, 23*, 711-741.

De Renzi, E. (1989). Apraxia. In F.Boller and J.Grafman (Eds). *Handbook of neuropsychology, Vol. 2*. Amsterdam: Elsevier. p.245-263.

De Renzi, E., Motti, F, & Nichelli, P. (1980). Imitating gestures, A quantitative approach to ideomotor apraxia. *Archives of Neurology, 37*, 217-221.

Derom, C., Thiery, E., Vlietinck, R., Loos, R. & Derom, R. (1996). Handedness in twins according to zygosity and chorion type: a preliminary report. *Behavior Genetics, 26*, 407-408.

Despert, J.L. (1946). Psychosomatic study of fifty stuttering children. *American Journal of Orthopsychiatry, 16*, 100-113.

De Toffol, B., Autret, A., Markabi, S. & Roux, S. (1990). Influence of lateralized sensorimotor and neuropsychological activities on electroencephalographic spectral power. *Electroencephalography & Clinical Neurophysiology, 75*, 200-206.

Deuel, R.K. & Dunlop, N.L. (1980). Hand preferences in the rhesus monkey. Implications for the study of cerebral dominance. *Archives of Neurology, 37*, 217-221.

Dewing, K. (1970). Family influences on creativity: a review and discussion. *Journal of Special Education, 4,* 399-404.

Dewing, K. & Taft, R. (1973). Some characteristics of the parents of creative twelve-year-olds. *Journal of Personality, 41,* 71-85.

Doane, T. & Todor, J.I. (1977). Motor activity as a function of handedness. In D.M.Landers & R.W.Christine (Eds). *Psychology of motor behavior and sport.* Champaign, Ill.: Human Kinetics. p.264-271.

Dobbing, J. & Sands, J. (1973). Quantitative growth and development of the human brain. *Archives of Disease in Childhood. 48,* 757-767.

Dobzhansky, T. (1962). *Mankind evolving: the evolution of the human species.* New Haven: Yale University Press.

Dobzhansky, T. (1963). Cultural direction of human evolution - a summation. *Human Biology, 35,* 311-316.

Dobzhansky, T. & Montagu, M.F.Ashley (1947). Natural selection and the mental capacity of mankind. *Science, 105,* 587-590.

Donchin, E., Kutas, M. & McCarthy, G. (1977). Electrocortical indices of hemispheric utilization. In S.Harnad, R.W.Doty, L.Goldstein, J.Jaynes & G.Krauthamer (Eds). *Lateralization in the nervous system.* New York: Academic Press. p.339-384.

Donchin, E., McCarthy, G. & Kutas, M. (1977). Electroencephalographic investigations of hemispheric specialization. In J.E.Desmedt (Ed). *Language and hemispheric specialization in man: cerebral ERP's. Progress in Clinical Neurophysiology, Vol.3.* Basel: Karger. p.212-242.

Douglas, J.W.B. (1964). *The home and the school.* London: MacGibbon & Kee.

Douglas, J.W.B., Ross, J.M. & Cooper, J.E. (1967). The relationship between handedness, attainment, and adjustment in a national sample of school children. *Educational Research, 9,* 223-232.

Downing, J., May, R. & Ollila, L. (1982). Sex differences and cultural expectations in reading. In E.M.Sheridan (Ed). *Sex stereotypes and reading: research and strategies.* Newark, Del.: International Reading Association. p. 17-34.

Dreifuss, F.E. (1963). Delayed development of hemispheric dominance. *Archives of Neurology, 8,* 510-514.

Driessen, N.R. & Raz, N. (1995). The influence of sex, age, and handedness on corpus callosum morphology: a meta-analysis. *Psychobiology, 23,* 240-247.

Dunn, P.M. (1965). Some perinatal observations on twins. *Developmental Medicine & Child Neurology, 7,* 121-134.

Durost, W.N. (1934). The development of a battery of objective group tests of manual laterality with the results of their application to 1300 children. *Genetic Psychology Monographs, 16,* 225-335.

Dvirskii, A.E. (1976). Functional asymmetry of the cerebral hemispheres in clinical types of schizophrenia. *Neuroscience & Behavioral Physiology (Washington), 7,* 236-239.

Easter, S.S., Purves, D., Rakic, D. & Spitzer, N.C. (1985). The changing view of neural specificity. *Science, 230,* 507-511.

Edelman, G.M. (1987). *Neural Darwinism: The theory of neuronal group selection.* New York: Basic Books.

Edwards, A.S. (1948). Handedness and involuntary movement. *Journal of General Psychology, 39,* 293-295.

Edwards, J.M. & Elliott, D. (1987). Effect of unilateral training on contralateral motor overflow in children and adults. *Developmental Neuropsychology, 3,* 299-309.

Edwards, S. & Beaton, A. (1996). Howzat? Why is there an over-representation of left-handed bowlers in professional cricket in the U.K.? *Laterality, 1,* 45-50.

Efron, R. (1990). *The decline and fall of hemispheric specialization.* Hillsdale, N.J.: Lawrence Erlbaume.

Eglinton, E. & Annett, M. (1994). Handedness and dyslexia: a meta-analysis. *Perceptual & Motor Skills, 79,* 1611-1616.

Eling, P. (1983). Comparing different measures of laterality: do they relate to a single mechanism? *Journal of Clinical Neuropsychology, 5,* 135-147.

Eling, P. (1984). Broca on the relation between handedness and cerebral speech dominance. *Brain & Language, 22,* 158-159.

Elliott, D. (1985). Manual asymmetries in the performance of sequential movement by adolescents and adults with Down syndrome. *American Journal of Mental Deficiency, 91,* 90-97.

Elliott, D., Edwards, J.M. Weeks, D.J., Lindley, S. & Carnahan, H. (1987). Cerebral specialization in young adults with Down syndrome. *American Journal of Mental Deficiency, 91,* 480-485.

Elliott, D., Weeks, D.J. & Jones, R. (1986). Lateral asymmetries in finger-tapping by adolescents and young adults with Down syndrome. *American Journal of Mental Deficiency, 90,* 472-475.

Elliott, J. & Connolly, K. (1974). Hierarchical structure in skill development. In K.Connolly & J.S.Bruner (Eds). *The growth of competence.* New York: Academic Press. p.135-168.

Ellis, L. (1990). Left- and mixed-handedness and criminality: explanations for a probable relationship. In S.Coren (Ed). *Left-handedness: behavioral implications and anomalies*. Amsterdam: North Holland. p.485-507.

Ellis, L. & Ames, M.A. (1989). Delinquency, sidedness and sex. *Journal of General Psychology, 116,* 57-62.

Ellis, S.J., Ellis, P.J. & Marshall, E. (1988). Hand preference in a normal population, *Cortex, 24,* 157-163.

Enstrom, e.a. (1962). The extent of the use of the left hand in handwriting. *Journal of Educational Research, 55,* 234-235.

Entwistle, D.R. & Astone, N.M. (1994). Some practical guidelines for measuring youth's race/ethnicity and socioeconomic status. *Child Development, 65,* 1521-1540.

Era, P., Jokela, J. & Heikkinen, E. (1986). Reaction and movement times in men of different ages: a population study. *Perceptual & Motor Skills, 63,* 111-130.

Ericson, M.C. (1946). Child rearing and social status. *American Journal of Sociology, 52,* 190-192.

Ericsson, K.A., Krampe, R.T. & Heizmann, S. (1993). Can we create gifted people? In G.R.Bock & K.Ackrill (Eds). *The origins and development of high ability*. Chichester: Wiley. p.222-249.

Ericsson, K.A., Krampe, R.T. & Tesch-Romer, C. (1993). The role of deliberate practice in the acquisition of expert performance. *Psychological Review, 100,* 363-406.

Ertl, J. & Schafer, E.W.P. (1967). Cortical activity preceding speech. *Life Sciences, 6,* 473-479.

Ertl, J. & Schafer, E.W.P. (1969). Erratum. *Life Sciences, 8,* 559.

Espenschade, A. (1960). Motor development. In W.R.Johnson (Ed). *Science and medicine of exercise and sports*. New York: Harper. p.419-439.

Ettlinger, G. (1969). Apraxia considered a disorder of movements that are language dependent: evidence rom cases of brain bisection. *Cortex, 5,* 285-289.

Ettlinger, G., Jackson, C.V. & Zangwill, O.L. (1956). Cerebral dominance in sinistrals. *Brain, 79,* 569-588.

Evans-Pritchard, E.E. (1953). Nuer spear symbolism. *Anthropoliogical Quarterly, 1,* 1-19. Reprinted in R.Needham (Ed). *Right and left: essays on dual symbolic classification*. Chicago: Chicago University Press, 1973. p.92-108.

Evans-Pritchard, E.E. (1973). Foreword. In: R.Needham (Ed). *Right and left: essays on dual symbolic classification*. Chicago: University of Chicago Press.

Fabbro, F. (1994). Left and right in the Bible from a neuropsychological perspective. *Brain & Cognition, 24,* 161-183.

Fabre-Thorpe, M., Fagot, J., Lorincz, E., Levesque, F. & Vauclair, J. (1993). Laterality in cats: paw preference and performance in a visuomotor activity. *Cortex, 29,* 15-24.

Fagan, L.B. (1931). The relation of dextral training to the onset of stuttering: a report of cases. *Quarterly Journal of Speech, 17,* 73-76.

Fagan-Dubin, L. (1974). Lateral dominance and development of cerebral specialization. *Cortex, 10,* 69-74.

Faglioni, P. & Basso, A. (1985). Historical perspectives on neuroanatomical correlates of limb apraxia. In E.A.Roy (Ed). *Neuropsychological studies of apraxia and related disorders.* Amsterdam: North Holland. p.3-44.

Faglioni, P. and Scarpa, M. (1989). Skull asymmetries bear no relation to the occurrence of apraxia. A clinical and CT scan study in patients with unilateral brain damage. *Cortex, 25,* 449-459.

Fagot, B.I. (1977). Consequences of moderate cross-gender behavior in preschool children. *Child Development, 48,* 902-907.

Fagot, B.I. & Leinback, M.D. (1993). Gender-role development in young children: from discrimination to abeling. *Developmental Review, 13,* 205-224.

Fagot, J. & Vauclair, J. (1988). Handedness and bimanual coordination in the lowland gorilla. *Brain, Behavior & Evolution, 32,* 89-95.

Falzi, G., Perrone, P. & Vignolo, L.A. (1982). Right-left asymmetry in anterior speech region. *Archives of Neurology, 39,* 239-240.

Feehan, M., Stanton, W.R., McGee, R. & Silva, P.A. (1990). Is there an association between lateral preference and delinquent behavior? *Journal of Abnormal Psychology, 99,* 198-201.

Fein, D., Waterhouse, L., Lucci, D. Pennington, B. & Humes, M. (1985). Handedness and cognitive functions n pervasive developmental disorders. *Journal of Autism & Developmental Disorders, 15,* 323-334.

Fennell, E.B. (1985). Handedness in neuropsychological research. In H.J.Hanney (Ed). *Experimental techniques in human neuropsychology.* New York: Oxford University Press. p.15-44.

Fennell, E.B., Satz, P. & Morris, R. (1983). The development of handedness and dichotic ear listening asymmetries in relation to school achievement: a longitudinal study. *Journal of Experimental Child Psychology, 35,* 248-262.

Fetters, L. & Todd, J. (1987). Quantitative assessment of infant reaching movements. *Journal of Motor Behavior, 19,* 147-166.

Finch, G. (1941). Chimpanzee handedness. *Science, 94,* 117-118.

Finger, S. & Almli, C.R. (1988). Margaret Kennard and her "Principle" in historical perspective. In S.Finger, T.E.Levere, C.R.Almli and D.G.Stein (Eds). *Brain injury and recovery.* New York: Plenum Press. p.117-132.

Finlayson, M.A.J. & Reitan, R.M. (1980). Effect of lateralized lesions on ipsilateral and contralateral motor functioning. *Journal of Clinical Neuropsychology, 2,* 237-243.

Finney, T. (1955). *Football around the world.* London: The Sportsmans Book Club.

Fiske, D.W. & Maddi, S.R. (1961). *Functions of varied experience.* Homewood, Ill.: Dorsey Press.

Fitts, P.M. (1954). The information capacity of the human motor system in controlling the amplitude of movement. *Journal of Experimental Psychology, 47,* 381-391.

Fitzhugh, K.B. (1973). Some neuropsychological features of delinquent subjects. *Perceptual & Motor Skills, 36,* 494.

Flannery, K.A. & Liederman, J. (1995). Is there really a syndrome involving the co-occurrence of neurodevelopmental disorder, talent, non-right-handedness and immune disorder among children? *Cortex, 31,* 503-515.

Fleishman, E.A. (1964) *The structure and measurement of physical fitness.* Eaglewood Cliffs, N.J.: Prentice-Hall.

Fleishman, E.A. & Ellison, G.D. (1962). A factor analysis of fine manipulative tests. *Journal of Applied Psychology, 46,* 96-105.

Fleishman, E.A. & Hempel, W.E. (1954). A factor analysis of dexterity tests. *Personnel Psychology, 7,* 15-32.

Fleminger, J.J., Dalton, R. & Standage, K.F. (1977a). Handedness in psychiatric patients. *British Journal of Psychiatry, 131,* 448-452.

Fleminger, J.J., Dalton, R. & Standage, K.F. (1977b). Age as a factor in the handedness of adults. *Neuropsychologia, 15,* 471-473.

Flowers, K. (1975). Handedness and controlled movement. *British Journal of Psychology, 66,* 39-52.

Fog, E. & Fog, M. (1963). Cerebral inhibition examined by associated movements. In M.Bax & R.MacKeith (Eds). *Minimal cerebral dysfunction* London: Heinemann. p.52-57.

Formby, C., Thomas, R.G. & Halsey, J.H. (1989). Regional cerebral blood flow for singers and nonsingers while speaking, singing, and humming a rote passage. *Brain & Language, 36,* 690-698.

Forward, E., Warren, J.M. & Hara, K. (1962). The effects of unilateral lesions in sensorimotor cortex on manipulation by cats. *Journal of Comparative & Physiological Psychology, 55,* 1130-1135.

Fox, J.J. (1973). On bad death and the left hand: a study of Rotinese symbolic inversions. In R.Needham (Ed). *Right and left: essays on dual symbolic classification.* Chicago: Chicago University Press. p.342-368.

Freeman, R.B. (1984). The apraxias, purposeful motor behavior, and left hemisphere function. In W.Prinz & A.F.Sanders (Eds). *Cognition and motor processes.* Berlin: Springer-Verlag. p.29-50.

French, C.C. & Richards, A. (1990). The relationship between handedness, anxiety and questionnaire response patterns. *British Journal of Psychology, 81,* 57-61.

French, L.A. & Johnson, D.R. (1955). Observations on the motor system following cerebral hemispherectomy. *Neurology, 5,* 11-14.

Freund, H-J. (1987). Abnormalities of motor behavior after cortical lesions in humans. In V.B.Mountcastle and F.Plum (Eds). *Handbook of physiology, Section1. The nervous system. Vol. 5. Higher functions of the brain, Part2.* Bethesda: American Physiological Society. p.763-810.

Fried, I., Ojemann, G. & Fetz, E. (1981). Language-related potentials specific to human language cortex. *Science, 212,* 353-356.

Frisch, K.von (1971). *Bees; their chemical senses, vision and language.* Ithaca, New York: Cornell University Press. Revised Edition.

Fullwood, D. (1986). Australian norms for hand and finger strength of boys and girls aged 5 - 12 years. *Australian Occupational Therapy Journal, 33,* 26-36.

Gabrielli, W.F. & Mednick S.A. (1980). Sinistrality and delinquency. *Journal of Abnormal Psychology, 89,* 654-661.

Gaillard, F. & Satz, P. (1989). Handedness and reading disability: a developmental study. *Archives of Clinical Neuropsychology, 4,* 63-69.

Galaburda, A.M. (1984) Anatomical asymmetries. In N.Geschwind & A.M.Galaburda (Eds). *Cerebral dominance: the biological foundations.* Cambridge, Mass.: Harvard University Press. p.11-25.

Galaburda, A.M. (1991). Asymmetries of cerebral neuroanatomy. In J.Marsh (Ed). *Biological asymmetry and handedness.* Chichester: Wiley. p.219-233.

Galaburda, A.M. (1993). The planum temporale. *Archives of Neurology, 50,* 457.

Galaburda, A.M. (1995). Anatomic basis of cerebral dominance. In R.J.Davidson & K.Hugdahl (Eds). *Brain asymmetry.* Cambridge, Mass.: MIT Press. p.51-73.

Galaburda, A.M., Corsiglia, J., Rosen, G.D. & Sherman, G.F. (1987). Planum temporale asymmetry: reappraisal since Geschwind and Levitsky. *Neuropsychologia, 25,* 853-868.

Galaburda, A.M., LeMay, M., Kemper, T.I. & Geschwind, N. (1978) Right-left asymmetries in the brain. *Science, 199,* 852-856.

Galaburda, A.M., Sherman, G. & Geschwind, N. (1985). Cerebral lateralization: historical note on animal studies. In S.D.Glick (Ed). *Cerebral lateralization in nonhuman species*. New York: Academic Press, p.1-10.

Galin, D. & Ornstein, R. (1972). Lateral specialization of cognitive mode: an EEG study. *Psychophysiology, 9,* 412-418.

Galin, D., Ornstein, R., Herron, J. & Johnstone, J. (1982). Sex and handedness differences in EEG measures of hemisphere specialization. *Brain & Language, 16,* 19-55.

Gardner, H. (1983). *Frames of mind.* New York: Basic Books.

Gardner, J., Lewkowicz, D. & Turkewitz, G. (1977). Development of postural asymmetry in premature human infants. *Developmental Psychobiology, 10,* 471-480.

Gardner, L.P. (1941). Experimental data on the problem of motor lateral dominance in feet and hands. *Psychological Record, 5,* 2-63.

Gardner, W.J., Karnosh, L.J., McClure, C.C. & Gardner, A.K. (1955). Residual function following hemispherectomy for tumour and for infantile hemiplegia. *Brain, 78,* 487-502.

Gasser, T., Verleger, R., Bacher, P. & Sroka, L. (1988a). Development of the EEG of school-age children and adolescents. I. Analysis of band power. *Electroencephalography & Clinical Neurophysiology, 69,* 91-99.

Gasser, T., Jennen-Steinmetz, C., Sroka, L. Verleger, R. & Mocks, J. (1988b). Development of the EEG of school-age children and adolescents. II. Topography. *Electroencephalography & Clinical Neurophysiology, 69,* 100-109.

Gazzaniga, M.S. (1970). *The bisected brain.* New York: Appleton-Century-Crofts.

Gazzaniga, M.S., Bogen, J.E. & Sperry, R.W. (1967). Dyspraxia following division of cerebral commissures. *Archives of Neurology, 16,* 606-612.

Gazzaniga, M.S. & LeDoux, J.E. (1978) *The integrated mind.* New York: Plenum Press.

Gazzaniga, M.S., Le Doux, J.E. & Wilson, D.H. (1977). Language, praxis and the right hemisphere: clues to some mechanisms of consciousness. *Neurology, 27,* 1144-1147.

Gazzaniga, M.S. & Sperry, R.W. (1967). Language after section of the cerebral commissures. *Brain, 90,* 131-148.

Gazzaniga, M.S., Volpe, B.T., Smylie, C.S., Wilson, D.H. & Le Doux, J.E. (1979). Plasticity in speech organization following commissurotomy. *Brain, 102,* 805-815.

Geschwind, N. (1965). Disconnection syndromes in animals and man. *Brain, 88,* 237-294, 585-644.

Geschwind, N. (1971). Aphasia. *New England Journal of Medicine, 284,* 654-656.

Geschwind, N. (1975). The apraxias: neural mechanisms of disorders of learned movement. *American Scientist, 63,* 188-195.

Geschwind, N. (1983). Biological associations of left-handedness. *Annals of Dyslexia, 33,* 29-40.

Geschwind, N. (1984) The biology of cerebral dominance: implications for cognition. *Cognition, 17,* 193-208.

Geschwind, N. & Behan, P. (1982). Left-handedness: association with immune disease, migraine and developmental learning disorder. *Proceedings of the National Academy of Sciences, 79,* 5097-5100.

Geschwind, N. & Behan, P. (1984). Laterality, hormones and immunity. In N.Geschwind & A.M.Galaburda (Eds). *Cerebral dominance: the biological foundations.* Cambridge, Mass.: Harvard University Press. p.211-224.

Geschwind, N. & Galaburda, A.M. (1985a). Cerebral lateralization. Biological mechanisms, associations, and pathology: I. A hypothesis and a program for research. *Archives of Neurology, 42,* 428-459.

Geschwind, N. & Galaburda, A.M. (1985b). Cerebral lateralization. Biological mechanisms, associations, and pathology. II. A hypothesis and a program for research. *Archives of Neurology, 42,* 521-552.

Geschwind, N. & Galaburda, A.M. (1985c). Cerebral lateralization. Biological mechanisms, associations, and pathology. III. A hypothesis and a program for research. *Archives of Neurology, 42,* 634-654.

Geschwind, N. & Galaburda, A.M. (1987). *Cerebral lateralization: biological mechanisms, associations and pathology.* Cambridge, Mass.: M.I.T. Press.

Geschwind, N. & Kaplan, E. (1962). A human cerebral disconnection syndrome. *Neurology, 12,* 675-685.

Geschwind, N. & Levitsky, W. (1968).Human brain: left-right asymmetries in temporal speech region. *Science, 161,* 186-187.

Gevins, A. (1983). Brain potential (BP) evidence for lateralization of higher cognitive functions. In J.Hellige (Ed). *Cerebral hemispheric asymmetry.* New York: Praeger. p.335-382.

Gevins, A.S., Zeitlin, G.M., Doyle, J.C., Yingling, C.D., Schaffer, R.E., Callawy, E. & Yeager, C.L. (1979). Electroencephalogram correlates of higher cortical functions. *Science, 203,* 665-668.

Gibson, J.B. (1973). Intelligence and handedness. *Nature, 243,* 482.

Giesecke, M. (1936). The genesis of hand preference. *Monograph of the Society for Research in Child Development, 1, series No.5,* 1-102.

Gilbert, A.N. & Wysocki, C.J. (1992). Hand preference and age in the United States. *Neuropsychologia, 30,* 601-608.

Gillberg, C. (1983). Autistic children's hand preferences: results from an epidemiological study of infantile autism. *Psychiatry Research, 10,* 21030.

Gillberg, C. & Coleman, M. (1992). *The biology of the autistic syndromes, 2nd Edition.* Oxford: MacKeith Press.

Gillberg, C. & Rasmussen, P. (1982). Perceptual, motor and attentional deficits in seven-year-old children: background factors. *Developmental Medicine & Child Neurology, 24,* 752-770.

Gillberg, C. Waldenstrom, E. & Rasmussen, P. (1984). Handedness in Swedish 10 year olds. Some background and associated factors. *Journal of Child Psychology & Psychiatry, 25,* 421-432.

Glanville, A.D. & Antonitis, J.J. (1955). The relationship between occipital alpha activity and laterality. *Journal of Experimental Psychology, 49,* 294-299.

Glees, P. (1980). Functional cerebral reorganization following hemispherectomy in man and after small experimental lesions in primates. In P.Bach-y-Rita (Ed). *Recovery of function: theoretical considerations for brain injury rehabilitation.* Baltimore: University Park Press. p.106-126.

Glencross, D.J. (1970). Serial organization and timing in a motor skill. *Journal of Motor Behavior, 2,* 229-237.

Glencross, D.J. (1973). Temporal organization in a repetitive speed skill. *Ergonomics, 16,* 765-776.

Gloning, K. (1977). Handedness and aphasia. *Neuropsychologia, 15,* 355-358.

Gloning, I., Gloning, K., Hant, G. & Quatember, R. (1969). Comparison of verbal behaviour in right-handed and non right-handed patients with anatomically verified lesion of one hemisphere. *Cortex, 5,* 43-52.

Gloning, K. & Quatember, R. (1966). Statistical evidence of neuropsychological syndromes in left-handed and ambidextrous patients. *Cortex, 2,* 484-488.

Goldfarb, W. (1961). *Childhood schizophrenia.* Cambridge, Mass: Harvard University Press.

Goldman, P.S. (1976). The role of experience in recovery of function following orbital prefrontal lesions in infant monkeys. *Neuropsychologia, 14,* 401-412.

Goodall, J. (1964). Tool using and aimed throwing in a community of free-living chimpanzees. *Nature, 201,* 1264-1266.

Goodglass, H. & Quadfasel, F.A. (1954). Language laterality in left-handed aphasics. *Brain, 77,* 521-548.

Goodman, C.H. (1946). The MacQuarrie Test for Mechanical Ability. I. Selecting radio assembly operators. *Journal of Applied Psychology, 30,* 586-595.

Goodman, C.H. (1947). The MacQuarrie Test for Mechanical Ability. II. Factor analysis. *Journal of Applied Psychology, 31,* 150-154.

Goodman, R. & Graham, P. (1996). Psychiatric problems in children with hemiplegia: cross sectional epidemiological survey. *British Medical Journal, 312,* 1065-1069.

Goodman, R. & Yude, C. (1997). Do unilateral lesions of the developing brain have side-specific psychiatric consequences in childhood? *Laterality, 2,* 103-115.

Gordon, H. (1921). Left-handedness and mirror writing especially among defective children. *Brain, 43,* 313-368.

Goring, C. (1913). *The English convict: a statistical study.* London: H.M.S.O. Reprinted by Patterson Smith: Montclair, N.J., 1972.

Gotestam, K.O. (1990). Left-handedness among students of architecture and music. *Perceptual & Motor Skills, 70,* 1323-1327.

Gottfried, A.W. & Bathurst, K. (1983). Hand preference across time is related to intelligence in young girls, but not boys. *Science, 221,* 1074-1076.

Gould, S. (1977). *Ontogeny and phylogeny.* Cambridge, Ma: Harvard University Press.

Govind, C.K. (1989). Asymmetry in lobster claws. *American Scientist, 77,* 468-474.

Grabow, J.D. & Elliott, F.W. (1974). The electrophysiological assessment of hemispheric asymmetries during speech. *Journal of Speech & Hearing Research, 17,* 64-72.

Grace, W.C. (1987). Strength of handedness as an indicant of delinquent's behavior. *Journal of Clinical Psychology, 43,* 151-155.

Green, A. & Vaid, J. (1986). Methodological issues in the use of the concurrent activities paradigm. *Brain & Cognition, 5,* 465-476.

Green, M.F., Satz, P., Smith, C. & Nelson, L. (1989). Is there atypical handedness in schizophrenia? *Journal of Abnormal Psychology, 98,* 57-61.

Greenough, W.T. (1986). What's special about development? Thoughts on the bases of experience-sensitive synaptic plasticity. In: W.T.Greenough & J.M.Juraska (Eds). *Developmental Neuropsychobiology.* New York: Academic Press. p.387-417.

Greenough, W.T., Larson, J.R. & Withers, G.S. (1985). Effects of unilateral and bilateral training in a reaching task on dendritic branching of neurons in the rat motor-sensory forelimb cortex. *Behavioral & Neural Biology, 44,* 301-314.

Gregg, R.A., Mastellone, A.F. & Gersten, J.W. (1957). Cross exercise - a review of the literature and study utilizing electromyographic techniques. *American Journal of Physical Medicine, 36,* 269-280.

Gregory, R.L. (1961). The brain as an engineering problem. In W.H.Thorpe and O.L.Zangwill (Eds). *Current problems in animal behaviour.* Cambridge: Cambridge University Press. p.307-330.

Griffiths, C.H. (1931). A study of some "motor ability" tests. *Journal of Applied Psychology, 15,* 109-125.

Grogono, J.L. (1968). Children with agenesis of the corpus callosum. *Developmental Medicine & Child Neurology, 10,* 613-616.

Grozinger, B., Kornhuber, H.H. & Kriebel, J. (1975). Methodological problems in the investigation of cerebral potentials preceding speech: determining the onset and suppressing artifacts caused by speech. *Neuropsychologia, 13,* 263-270.

Grozinger, B., Kornhuber, H.H. & Kriebel, J. (1977). Human cerebral potentials preceding speech production, phonation, and movements of the mouth and tongue, with reference to respiratory and extracerebral potentials. In J.E.Desmedt (Ed). *Language and hemispheric specialization in man: cerebral event-related potentials.* Basel: Karger. p.87-103.

Grozinger, B., Kornhuber, H.H., Kriebel, J. Szirtes, J. & Westphal, K.T.P. (1980). The Bereitschaftspotential preceding the act of speaking. Also an analysis of artifacts. In Kornhuber, H.H. & Deecke, L. (Eds), *Motivation, motor and sensory processes of the brain. Electrical potentials, behavior and clinical use. Progress in Brain Research. Vol.54.* p.798-804.

Gruber, D., Waanders, R., Collins, R.L., Wolfer, D.P. & Lipp, H-P. (1991). Weak or missing paw lateralization n a mouse strain (I/LnJ) with congenital absence of the corpus callosum. *Behavioural Brain Research, 46,* 9-16.

Grunewald, G., Grunewald-Zuberbier, E., Homberg, E. & Netz, J. (1979). Cerebral potentials during smooth goal-directed hand movements in right-handed and left-handed subjects. *Pflugers Archives, 381,* 39-46.

Gubbay, S.S. (1975). *The clumsy child: a study of developmental apraxic and agnosic ataxia.* London: Saunders.

Gudmundsson, E. (1993). Lateral preference of preschool and primary school children. *Perceptual & Motor Skills, 77,* 819-823.

Guereiro, M., Castro-Caldas, A. & Martins, I.P. (1995). Aphasia following right hemisphere lesion in a woman with left hemisphere injury in childhood. *Brain & Language, 49,* 280-288

Guo, N.F.(1984). Studies of lateralization of Chinese language functions. In H.S.R.Kao & R.Hoosain (Eds). *Psychological studies of the Chinese language.* Hong Kong: Chinese Language Society of Hong Kong. p.1-9.

Gur, R.E. (1977). Motoric laterality imbalance in schizophrenia. *Archives of General Psychiatry, 34,* 33-37.

Gureje, O. (1988). Sensorimotor laterality in schizophrenia: which features transcend cultural influences? *Acta Psychiatrica Scandinavica, 77*, 188-193.

Haaland, K.Y. & Yeo, R. (1989). Neuropsychological and neuroanatomic aspects of complex motor control. n E.D.Bigler, R.A.Yeo and E.Turkheimer (Eds). *Neuropsychological function and brain imaging*. New York: Plenum Press. p.219-244.

Habib, M. (1989). Anatomical asymmetries of the human cerebral cortex. *International Journal of Neuroscience, 47*, 67-79.

Habib, M., Gayraud, D., Oliva, A., Regis, J., Salamon, G. & Khalil, R. (1991). Effects of handedness and sex on the morphology of the corpus callosum: a study with brain magnetic resonance imaging. *Brain & Cognition, 16*, 41-61.

Habib, M., Robichou, F., Levrier, O., Khalil, R. & Salamon, G. (1995). Diverging asymmetries of temporo-parietal cortical areas: a reappraisal of Geschwind/Galaburda theory. *Brain & Language, 48*, 238-258.

Haefner, R. (1929). The educational significance of left-handedness. *Contributions to Education No. 360, Teachers College, Columbia University, N.Y.*

Hall, K.R.L. (1963). Tool-using performances as indicators of behavioral adaptability. *Current Anthropology, 4*, 479-494.

Halpern, D.F. & Coren, S. (1988). Do right-handers live longer? *Nature, 333*, 213.

Halpern, D.F. & Coren, S. (1990). Laterality and longevity: is left-handedness associated with a younger age at death? In S.Coren (Ed). *Left-handedness: behavioral implications and anomalies*. Amsterdam: North Holland. p.509-545.

Halpern, D.F. & Coren, S. (1991). Handedness and life span. *New England Journal of Medicine, 324*, 998.

Halpern, D.F. & Coren, S. (1993). Left-handedness and life span: a reply to Harris. *Psychological Bulletin, 114*, 235-241.

Halsey, J., Blauenstein, U., Wilson, E. & Wills, E. (1979). Regional cerebral blood flow comparison of right and left hand movement. *Neurology, 29*, 21-28.

Halsy, J.H., Blauenstein, U.W., Wilson, E.M. & Wills, E.L. (1980). Brain activation in the presence of brain damage. *Brain & Language, 9*, 47-60.

Halverson, L.E., Robertson, M.A. & Langendorfer, S. (1982). Development of the overarm throw: movement and ball velocity changes by seventh grade. *Research Quarterly for Exercise & Sport, 53*, 198-205.

Hamilton, C.R, & Vermeire, B.A. (1988). Cognition, not handedness is lateralized in monkeys. *Behavioral & Brain Sciences, 11*, 723-725.

Hanvik, L.J. & Kaste, L.M. (1973). Mixed cerebral dominance in clinic and school populations. *Perceptual & Motor Skills, 37*, 900-902.

Harburg, E., Roeper, P., Ozgoren, G. & Feldstein, A.M. (1981). Handedness and temperament. *Perceptual & Motor Skills, 52*, 283-290.

Hardyck, C. (1977). Laterality and intellectual ability: a just not noticeable difference? *British Journal of Educational Psychology, 47*, 305-311.

Hardyck, C., Goldman, R. & Petrinovich, L. (1975). Handedness and sex, race and age. *Human Biology, 47*, 369-375.

Hardyck, C., Petrinovich, L.F. & Goldman, R.D. (1976). Left-handedness and cognitive deficit. *Cortex, 12*, 266-279.

Hare, R.D. & Forth, A.E. (1985). Psychopathy and lateral preference. *Journal of Abnormal Psychology, 94*, 541-546.

Harlow, H.F., Akert, K. & Schiltz, K.A. (1964). The effects of bilateral prefrontal lesions on learned behavior of neonatal, infant, and preadolescent monkeys. In J.M.Warren and K.Akert (Eds). *The frontal granular cortex and behavior.* New York: McGraw-Hill. p.126-148.

Harrington, AS. (1987). *Medicine, mind and the double brain.* New Jersey: Princeton University Press.

Harris, A.J. (1957). Lateral dominance, directional confusion and reading disability. *Journal of Psychology, 44, 283-294.*

Harris, L.J. (1980). Left-handedness: early theories, facts and fancies. In: J.Herron (Ed). *Neuropsychology of left-handedness.* New York: Academic Press.

Harris, L.J. (1985). Teaching the right brain: historical perspective on a comtemporary educational fad. In: C.T.Best (Ed). *Hemispheric function and collaboration in the child.* New York: Academic Press. p.231-274.

Harris, L.J. (1988) Right-brain training: Some reflections on the application of research on cerebral hemisphere specialization to education. In: D.L.Molfese & S.J.Segalowitz (Eds). *Brain lateralization in children.* New York: Guilford Press. p.207-235.

Harris, L.J. (1991). Cerebral control for speech in right-handers and left-handers: an analysis of the views of Paul Broca, his contemporaries and his successors. *Brain & Language, 40*, 1-50.

Harris, L.J. (1993a). Do left-handers die sooner than right-handers? Commentary on Coren & Halpern's (1991) "Left-handedness: a marker for decreased survival fitness" *Psychological Bulletin, 114*, 203-234.

Harris, L.J. (1993b). Reply to Halpern and Coren. *Psychological Bulletin, 114*, 242-247.

Harris, L.J. (1993c). Broca on cerebral control for speech in right-handers and left-handers: a note on translation and some further comments. *Brain & Language, 45,* 108-120.

Harris, L.J. & Carlson, D.F. (1988). Pathological left-handedness: an analysis of theories and evidence. In D.L.Molfese & S.J.Segalowitz (Eds). *Brain lateralization in children: developmental implications.* New York: Guilford Press. p.289-372.

Harrison, G.A., Tanner, J.M., Pilbeam, D.R. & Baker, P.T. (1988). *Human Biology: an introduction to human evolution, variation, growth and adaptability.* Oxford: Oxford University Press.

Harshman, R.A., Hampson, E. & Berenbaum, S.A. (1983). Individual differences in cognitive abilities and brain organization. Part 1. Sex and handedness differences in ability. *Canadian Journal of Psychology, 37,* 144-192.

Hassler, M. & Birbaumer, M. (1988). Handedness, musical abilities and dichaptic and dichotic performance in adolescents: a longitudinal study. *Developmental Neuropsychology, 4,* 129-145.

Hassler, M. & Gupta, D. (1993). Functional brain organization, handedness, and immune vulnerability in musicians and non-musicians. *Neuropsychologia, 31,* 655-660.

Hatta, T. & Nakatsuka, Z. (1976). Note on hand preference of Japanese people. *Perceptual & Motor Skills, 42,* 530.

Hauser, M., Perry, S., Manson, J.H., Ball, H., Williams, M., Pearson, E. & Berard, J. (1991). It's all in the hands of the beholder: new data on free-ranging rhesus monkeys. *Behavioral & Brain Sciences, 14,* 342-347.

Hauser, R.M. (1994). Measuring socioeconomic status in studies of child development. *Child Development, 65,* 1541-1545.

Hay, D.A., Prior, M., Collett, S. & Williams, M. (1987). Speech and language development in preschool twins. *Acta Geneticae et Gamellologica, 36* 213-223.

Head, H. (1926). *Aphasia and kindred disorders of speech.* London: Cambridge University Press.

Healey, J.M., Liederman, J. & Geschwind, N. (1986). Handedness is not a unidimensional trait. *Cortex, 22,* 33-54.

Hecaen, H. (1962). Clinical symptomatology in right and left hemisphere lesions. In V.B.Mountcastle (Ed). *Interhemispheric relations and cerebral dominance.* Baltimore: Johns Hopkins Press. p.215-243.

Hecaen, H. (1976). Acquired aphasia in children and the ontogenesis of hemispheric functional specialization. *Brain & Language, 3,* 114-134.

Hecaen, H. (1983). Acquired aphasia in children revisited. *Neuropsychologia, 21,* 581-587.

Hecaen, H. & Ajuriaguerra, J. de (1964). *Left-handedness. Manual superiority and cerebral dominance.* New York: Grune and Stratton.

Heilman, K.M., Coyle, J.M., Gonyea, E.F. & Geschwind, N. (1973). Apraxia and agraphia in a left-hander. *Brain, 96,* 21-28.

Heilman, K.M., Gonyea, E.F. & Geschwind, N. (1974). Apraxia and agraphia in a right-hander. *Cortex, 10,* 284-288.

Heim, A.W. & Watts, K.P. (1976). Handedness and cognitive bias. *Quarterly Journal of Experimental Psychology, 28,* 355-360.

Heinlein, J.H. (1929). A study of dexterity in children. *Journal of Genetic Psychology, 36,* 91-117.

Hellebrandt, F., Parrish, M. & Houtz, S.J. (1947). Cross education, the influence of unilateral exercise on the contralateral limb. *Archives of Physical Medicine, 28,* 76-85.

Hemenway, D. (1983) Bimanual dexterity in baseball players. *New England Journal of Medicine, 309,* 1587-1588.

Hempel, W.E. & Fleishman, E.A. (1955). A factor analysis of physical proficiency and manipulative skills. *Journal of Applied Psychology, 39,* 12-16.

Henderson, V.W. (1986). Paul Broca's less heralded contributions to aphasia research: historical perspective and contemporary relevance. *Archives of Neurology, 43,* 609-612.

Hendry, J. (1986). *Becoming Japanese: the world of the pre-school child.* Honolulu: The University of Honolulu Press.

Henschen, S.E. (1926). On the function of the right hemisphere of the brain in relation to the left in speech, music and calculation. *Brain, 49,* 110-123.

Hertz, R. (1909). The re-eminence of the right hand: a study in religious polarity. *Revue Philosophique, 68,* 553-580. Translated and reprinted by R. & C. Needham (1960). *Death and the right hand.* London: Cohen & West. p.89-174.

Hicks, R.A. & Dusek, C.M. (1980). The handedness distribution of gifted and non-gifted children. *Cortex, 16,* 479-481.

Hicks, R.A., Pass, K., Freeman, H., Bautista, J. & Johnson, C. (1993). Handedness and accidents with injury. *Perceptual & Motor Skills, 77,* 1119-1122.

Hicks, R.A. & Pellegrini, R.J. (1978). Handedness and anxiety, *Cortex, 14,* 119-121.

Hicks, R.E. (1975). Intrahemispheric response competition between vocal and unimanual performance in normal adult human males. *Journal of Comparative & Physiological Psychology, 89,* 50-60.

Hicks, R.E., Bradshaw, G.J., Kinsbourne, M. & Feigin, D.S. (1978). Vocal-manual trade-offs in hemisphere sharing of human performance control. *Journal of Motor Behavior, 10*, 1-6.

Hicks, R.E. & Barton, K. (1975). A note on left-handedness and severity of mental retardation. *Journal of Genetic Psychology, 127*, 323-324.

Hicks, R.E. & Beveridge, R. (1978). Handedness and intelligence. *Cortex, 14*, 304-307.

Hicks, R.E. & Kinsbourne, M. (1976). Human handedness: a partial cross-fostering study. *Science, 192*, 908-910.

Hicks, R.E., Provenzano, F.J. & Rybstein, E.D. (1975). Generalized and lateralized effects of concurrent verbal rehearsal upon performance of sequential movements of the fingers by the left and right hands. *Acta Psychologica, 39*, 119-130.

Hildreth, G. (1949). The development and training of hand dominance: I. Characteristics of handedness. *Journal of Genetic Psychology, 75*, 197-220.

Hiscock, C.K., Hiscock, M., Benjamins, D. & Hillman, S. (1989). Motor asymmetries in hemiplegic children: implications for the normal and pathological development of handedness. *Developmental Neuropsychology, 5*, 169-186.

Hiscock, M. (1982). Verbal-manual time sharing in children as a function of task priority. *Brain & Cognition, 1*, 119-131.

Hiscock, M., Antonuik, D., Prisciak, K. & von Hessert, D. (1985). Generalized and lateralized interference between concurrent tasks performed by children. Effects of age, sex and skill. *Developmental Neuropsychology, 1*, 29-48.

Hiscock, M. & Kinsbourne, M. (1978). Ontogeny of cerebral dominance: evidence from time-sharing asymmetry in children. *Developmental Psychology, 14*, 321-329.

Hiscock, M. & Kinsbourne, M. (1980). Asymmetry of verbal-manual time sharing in children: a follow-up study. *Neuropsychologia, 18*, 151-162.

Hiscock, M., Kinsbourne, M., Samuels, M. & Krause, A.E. (1985). Effects of speaking upon the rate and variability of concurrent finger tapping in children. *Journal of Experimental Child Psychology, 40*, 486-500.

Hiscock, M., Kinsbourne, M., Samuels, M. & Krause, A.E. (1987). Dual task performance in children: generalized and lateralized effects of memory encoding upon the rate and variability of concurrent finger tapping. *Brain & Cognition, 6*, 24-40.

Ho, D.Y-F. (1994). Filial piety, authoritarian moralism, and cognitive conservatism in Chinese societies. *Genetic, Social & General Psychology Monographs, 120*, 347-365.

Hodos, W. (1986). Comparative neuroanatomy and the evolution of intelligence. In: H.J.Jerison & I.Jerison (Eds). *Intelligence and evolutionary biology.* Berlin: Springer-Verlag. p.93-107.

Hodos, W. & Campbell, C.B.G. (1969). Scalae naturae: why there is no theory in comparative psychology. *Psychological Review, 76,* 337-350.

Hofsten, C. von (1993). Prospective control: a basic aspect of action development. *Human Development, 36,* 253-270.

Holding, D.H. (1989a). Skills research. In D.H.Holding (Ed). *Human Skills.* Chichester: Wiley. p.1-16.

Holding, D.H. (Ed). (1989b). *Human Skills.* Chichester: Wiley.

Holloway, R.L. (1974). The casts of fossil hominid brains. *Scientific American, 231,* 105-115.

Homzie, M.J. & Lindsay, J.S. (1984). Language and the young stutterer: a new look at old theories and findings. *Brain & Language, 22,* 232-252.

Hood, P. & Perlstein, M. (1955). Infantile spastic hemiplegia: 2, Laterality of involvement. *American Journal of Physical Medicine, 34, 457-466.*

Hoosain, R. (1990). Left-handedness and handedness switch amongst the Chinese. *Cortex, 26,* 451-454.

Hopkins, W.D. (1993). Posture and reaching in chimpanzees (Pan troglodytes) and orangutans (Pongo pygmaeus). *Journal of Comparative Psychology, 107,* 162-168.

Hopkins, W.D. (1995). Hand preferences for simple reaching in juvenile chimpanzees (Pan troglodytes): continuity in development. *Developmental Psychology, 31,* 619-625.

Hopkins, W.D. & Bard, K.A. (1993). The ontogeny of lateralized behavior in nonhuman primates with special reference to chimpanzees (Pan troglodytes). In J.P.Ward & W.D.Hopkins (Eds). *Primate laterality: current behavioral evidence of primate asymmetries.* New York: Springer-Verlag, p.251-265.

Hopkins, W.D. & Morris, R.D. (1993). Handedness in great apes: a review of findings. *International Journal of Primatology, 14,* 1-25.

House, J.F. & Naitoh, P. (1979). Lateral-frontal slow potential shifts preceding language acts in deaf and hearing adults. *Brain & Language, 8,* 287-302.

Howe, C.E.A. (1959). A comparison of motor skills of mentally retarded and normal children. *Exceptional Children, 25,* 352-354.

Hsu, F.L.K. (1955). *Americans and Chinese: passage to differences.* London: The Cresset Press.

Hugdahl, K., Satz., Mitrushina, M. & Miller, E.M. (1993). Left-handedness and old age: do left-handers die earlier? *Neuropsychologia, 31*, 325-333.

Hughes, M. & Sussman, H.N. (1983). An assessment of cerebral dominance in language-disordered children via a time-sharing paradigm. *Brain & Language, 19*, 48-64.

Hull, C.J. (1936). A study of laterality test items. *Journal of Experimental Education, 4*, 287-290.

Humphrey, D.R. (1986). Representation of movements and muscles within the primate precentral motor cortex: historical and current perspectives. *Federation Proceedings, 45*, 2687-2699.

Humphrey, M. (1951). Consistency of hand usage: a preliminary enquiry. *British Journal of Educational Psychology, 21*, 214-225.

Humphrey, M.E. & Zangwill, O.L. (1952a). Dysphasia in left-handed patients with unilateral brain lesions. *Journal of Neurology, Neurosurgery & Psychiatry, 15*, 184-193.

Humphrey, M.E. & Zangwill, O.L. (1952b). Effects of a right-sided occipito-parietal brain injury in a left-handed man. *Brain, 75*, 312-324.

Hung, C-C., Tu, Y-K., Chen, S-H. & Chen, R-C. (1985). A study on handedness and cerebral speech dominance in right-handed Chinese. *Neurolinguistics, 1*, 143-163.

Hutt, C.W. (1917). Education of the left hand of disabled sailors and soldiers. *The Lancet, (i)*, 553.

Huttenlocher, P.R. (1979). Synaptic density in human frontal cortex - developmental changes and effects of aging. *Brain Research, 163*, 195-205.

Huttenlocher, P.R. (1990). Morphometric study of human cerebral cortex development. *Neuropsychologia, 28*, 517-527.

Huttenlocher, P.R. (1992). Neural plasticity. In A.K.Astbury, G.M.McKhann & W.I.McDonald (Eds). *Diseases of the nervous system. Clinical neurobiology.* Philadelphia: W.B.Saunders (2nd Edition, Vol.1). p.63-71.

Huttenlocher, P.R. (1994). Synaptogenesis, synaptic elimination, and neural plasticity in human cerebral cortex. In C.A.Nelson (Ed). *Threats to optimal development: integrating biological, psychological, and social risk factors.* Hillsdale, NJ: Erlbaum. p.35-54.

Huttenlocher, P.R., de Courten, C., Garey, L.J. & Van der Loos, H. (1982). Synaptogenesis in human visual cortex - evidence for synapse elimination during normal development. *Neuroscience Letters, 33*, 247-252.

Huxley, J. (1951). New bottles for new wine: ideology and scientific knowledge. *Journal of the Royal Anthropological Institute, 80*, 7-23.

Ida, Y. & Bryden, M.P. (1996). A comparison of hand preference in Japan and Canada. *Canadian Journal of Experimental Psychology, 50*, 234-239.

Ingram, D. (1975). Motor asymmetries in young children. *Neuropsychologia, 13*, 95-102.

Ingvar, D.H. (1983). Serial aspects of language and speech related to prefrontal cortical activity: a selective review. *Human Neurobiology, 2*, 177-189.

Ingvar, D.H. & Risberg, J. (1967). Increase of regional blood flow during mental effort in normals and in patients with focal brain disorders. *Experimental Brain Research, 3*, 195-211.

Ingvar, D.H. & Schwartz, M.S. (1974). Blood flow patterns induced in the dominant hemisphere by speech and reading. *Brain, 97*, 273-288.

Ireland, P. & Watter, P. (1995). Development of lumbrucal control in children aged four to six years. *Australian Physiotherapy, 41*, 13-18.

Itani, J., Tokuda, K., Foruya, Y., Kano, K. & Shin, Y. (1963). The social construction of natural troops of Japanese monkeys in Takasakiyama. *Primates, 4*, 1-42.

Jaegers, S.M.H.J., Peterson, R.F., Dantuma, R., Hillen, B., Geuze, R. & Schelleken, J. (1989). Kinesiologic aspects of motor learning in dart throwing. *Journal of Human Movement Studies, 16*, 161-171.

Jancke, L., Schlaug, G. & Steinmetz, H. (1997). Hand skill asymmetry in professional musicians. *Brain & Cognition, 34*, 424-432.

Jason, G.W. (1986). Performance of manual copying tasks after focal cerebral lesions. *Neuropsychologia, 24*, 181-191.

Jay, R. (1986). *Learned pigs and fireproof women.* New York: Warner Books.

Jeannerod, M. (Ed). (1990). *Attention and performance XIII. Motor representation and control.* Hillsdale, N.J.: Lawrence Erlbaum.

Jebson, R.H., Griffith, E.R., Long, E.W. & Fowler, R. (1971). Function of "normal" hand in stroke patients. *Archives of Physical Medicine & Rehabilitation, 52*, 170-174.

Jebson, R.H., Taylor, N., Treischmann, R.B., Trotter, M.J. & Howard, L.A. (1969). An objective and standardized test of hand function. *Archives of Physical Medicine, 50*, 311-319.

Jeeves, M.A. (1986). Callosal agenesis: neuronal and developmental adaptations. In F.Lepore, M.Pitto and H.H.Jasper (Eds). *Two hemispheres - one brain: functions of the corpus callosum.* New York: Alan R. Liss nc. p.403-421.

Jeeves, M.A. (1990). Agenesis of the corpus callosum. In F.Boller and J.Grafman (Eds). *Handbook of Neuropsychology. Vol. 4.* Amsterdam: Elsevier. p.99-114.

Jeeves, M.A. & Milner, A.D. (1987). Specificity and plasticity in interhemispheric integration: evidence from callsoal agenesis. In D.Ottoson (Ed). *Duality and unity of the brain*. London: Macmillan. p.416-441.

Jeeves, M.A., Silver, P.H. & Jacobsen, I. (1988). Bimanual coordination in callosal agenesis and partial commissurotomy. *Neuropsychologia, 26*, 833-850.

Jeeves, M.A., Silver, P.H. & Milne, A.B. (1988). Role of the corpus callosum in the development of a bimanual motor skill. *Developmental Neuropsychology, 4*, 305-323.

Jenkins, W.M., Merzenich, M.M. & Recanzone, G. (1990). Neocortical dynamics in adult primates: implications for neuropsychology. *Neuropsychologia, 28*, 573-584.

Jensen, B.T. (1952a). Left-right orientation in profile drawing. *American Journal of Psychology, 65*, 80-83.

Jensen, B.T. (1952b). Reading habits and left-right orientation in profile drawing by Japanese children. *American Journal of Psychology, 65*, 306-307.

Jerison, H.J. (1973). *Evolution of the brain and intelligence*. New York: Academic Press.

Jerison, H.J. (1985). Animal intelligence as encephalization. *Philosophical Transactions of the Royal Society (London), B, 308*, 21-35.

Jerison, H.J. (1986). Evolutionary biology of intelligence: the nature of the problem. In H.J.Jerison & I.Jerison (Eds). *Intelligence and evolutionary biology*. Berlin: Springer-Verlag. p.1-11.

Jerison, H.J. (1991). *Brain size and the evolution of mind. 59th James Arthur lecture on the evolution of the human brain*. New York: American Museum of Natural History.

Joanette, Y. (1989). Aphasia in left-handers and crossed aphasia. In F.Boller and T.Grafman (Eds). *Handbook of Neuropsychology. Vol. 2*. Amsterdam: Elsevier. p.173-183.

Johnson, D. and Almli, C.R. (1978). Age, brain damage, and performance. In S.Finger (Ed). *Recovery from brain damage: research and theory*. New York: Plenum Press. p.115-134.

Johnson, J.S. & Newport, E.L. (1989). Critical period effects in second language learning: the influence of maturational state on the acquisition of English as a second language. *Cognitive Psychology, 21*, 60-99.

Johnson, J.S. & Newport, E.L. (1991). Critical period effects on universal properties of language: the status of subjacency in the acquisition of a second language. *Cognition, 39*, 215-258.

Johnson, M.S., Clarke, B. & Murray, J. (1990). The coil polymorphism in *Partula Suturalis* does not favor sympatric speciation. *Evolution, 44*, 459-464.

Johnson, O. & Harley, C. (1980). Handedness and sex differences in cognitive tests of brain laterality. *Cortex, 16,* 73-82.

Johnson, O. & Kozma. A. (1977). Effects of concurrent verbal and musical tasks on a unimanual skill. *Cortex, 13,* 11-16.

Johnson, W. & King, A. (1942). An angle board and hand usage study of stutterers and non-stutterers. *Journal of Experimental Psychology, 31,* 293-311.

Johnston, C., Prior, M. & Hay, D. (1984). Prediction of reading disability in twin boys. *Developmental Medicine & Child Neurology, 26,* 588-595.

Johnstone, J., Galin, D. & Herron, J. (1979). Choice of handedness measures in studies of hemisphere specialization. *International Journal of Neuroscience, 9,* 71-80.

Jones, B. & Bell, J. (1980). Handedness in engineering and psychology students. *Cortex, 16,* 621-625.

Jones, H.E. (1931). Dextrality as a function of age. *Journal of Experimental Psychology, 14,* 125-143.

Jones, P., Guth, C., Lewis, S. & Murray, R. (1994). Low intelligence and poor educational achievement precede early onset schizophrenic psychosis. In A.S.David & J.C.Cutting (Eds). *The neuropsychology of schizophrenia.* Hillsdale: Lawrence Erlbaum. p.131-144.

Jordan, H.E. (1922). The crime against left-handedness. *Good Health, 57(ii),* 378-383.

Jordan, J.E. (1911). The inheritance of left-handedness. *American Breeders Magazine, 2,* 19-29, 113-124.

Jordan, J.E. (1914). Hereditary left-handedness, with a note on twinning (study III). *Journal of Genetics, 4,* 67-81.

Joseph, J.A. (1988). *Central determinants of age-related declines in motor function.* New York: New York Academy of Sciences.

Jung, R. (1984). Electrophysiological cues of the language-dominant hemisphere in man: slow brain potentials during language processing and writing. In O.Creutzfeldt, R.F.Schmidt & W.D.Willis (Eds). *Sensory-motor integration in the nervous system.* Berlin: Springer-Verlag. p.430-450.

Kaas, J.H. (1991) Plasticity of sensory and motor maps in adult mammals. *Annual Review of Neuroscience, 14,* 137-167.

Kahl, J.A. & Davis, J.A. (1955). A comparison of indexes of socioeconomic status. *American Sociological Review, 20,* 317-325.

Kameyama, T., Niwa, S., Hiramatsu, K. & Saitoh, O. (1983). Hand preference and eye dominance patterns in Japanese schizophrenics. In P.Flor-Henry & J.Gruzelier (Eds). *Laterality and psychopathology.* Amsterdam: Elsevier. p.163-180.

Kann, J. (1950). A translation of Broca's original article on the location of the speech center. *Journal of Speech & Hearing Disorders, 15,* 16-20.

Kaufman, A.S., Zalma, R. & Kaufman, N.L. (1978). The relationship of hand dominance to the motor coordination, mental ability, and right-left awareness of young children. *Child Development, 49,* 885-888.

Kee, D.W., Gottfried, A. & Bathurst, K. (1991). Consistency of hand preference: predictions to intelligence and school achievement. *Brain & Cognition, 16,* 1-10.

Keele, S.W. (1982). Learning and control of coordinated motor patterns: the programming perspective. In J.A.Kelso (Ed). *Human motor behavior: an introduction.* Hillsdale, N.J.; Lawrence Erlbaum.

Keeley, L.H. (1977). The functions of paleolithic flint tools. *Scientific American, 237,* 108-127.

Keeley, L.H. & Newcomer, M.H. (1977). Microwear analysis of experimental flint tools: a test case. *Journal of Archeological Science, 4,* 29-62.

Kelley, M.P. & Coursey, R.D. (1992). Lateral preference and neuropsychological correlates of schizotypy. *Psychiatry Research, 41,* 115-135.

Kempf, H.D. (1917). The preferential use of the hands in monkeys with modification by training and retention of the new habit. *Psychological Bulletin, 141,* 297-301.

Kennard, M.A. (1936). Age and other factors in motor recovery from precentral lesions in monkeys. *American Journal of Physiology, 115,* 138-146.

Kennard, M.A. (1938). Reorganization of motor function in the cerebral cortex of monkeys deprived of motor and premotor areas in infancy. *Journal of Neurophysiology, 1,* 477-496.

Kennard, M.A. (1940). Relation of age to motor impairment in man and in subhuman primates. *Archives of Neurology & Psychiatry, 44,* 377-397.

Kennard, M.A. (1942). Cortical re-organization of motor function. *Archives of Neurology & Psychiatry, 48,* 227-240.

Kennedy, F. (1916). Stock-brainedness, the causative factor in the so-called "crossed aphasias". *American Journal of the Medical Sciences, 152,* 849-859.

Kerr, M., Mingay, R. & Elithorn, A. (1963). Cerebral dominance in reaction time responses, *British Journal of Psychology, 54,* 325-336.

Kertesz, A. (1985). Apraxia and aphasia. Anatomical and clinical relationship. In E.A.Roy (Ed). *Neuropsychological studies of apraxia and related disorders.* Amsterdam: North-Holland. p.163-178.

Kertesz, A. (1988). What do we learn from recovery from aphasia? In S.G.Waxman (Ed). *Functional recovery n neurological disease.* New York: Raven Press. p.277-292.

Kertesz, A. (1994). Localization and function: old issues revisited and new developments. In A.Kertesz (Ed). *Localization and neuroimaging in neuropsychology*. San Diego: Academic Press. p.1-33.

Kertesz, A., Black, S.E., Polk, M. & Howell, J.E. (1986). Cerebral asymmetries on magnetic resonance imaging. *Cortex, 22*, 117-127.

Kertesz, A., & Naeser, M.A. (1994). Anatomical asymmetries and cerebral lateralization. In A.Kertesz (Ed). San Diego: Academic Press. p.213-244.

Kertesz, A., Polk, M., Black, S.E. & Howell, J. (1990). Sex, handedness, and the morphometry of cerebral asymmetries on magnetic resonance imaging. *Brain Research, 530*, 40-48.

Kertesz, A., Polk, M., Black, S.A. & Howell, J. (1992). Anatomical asymmetries and functional laterality. *Brain, 115, 589-605*.

Kertesz, A., Polk, M., Howell, J. & Black, S.E. (1987). Cerebral dominance, sex and callosal size in MRI. *Neurology, 37*, 1385-1388.

Kilshaw, D. & Annett, M. (1983). Right and left hand skill: I. Effects of age, sex, and hand preference showing superior skill in left-handers. *British Journal of Psychology, 74*, 253-268.

Kim, D., Raine, A., Triphon, N. & Green, M.F. (1992). Mixed handedness and features of schizotypal personality in a nonclinical sample. *Journal of Nervous and Mental Disease, 180*, 133-135.

Kimura, D. (1961a). Some effects of temporal-lobe damage on auditory perception. *Canadian Journal of Psychology, 15*, 156-165.

Kimura, D. (1961b). Cerebral dominance and the perception of verbal stimuli. *Canadian Journal of Psychology, 15*, 166-171.

Kimura, D. (1967). Functional asymmetry of the brain in dichotic listening. *Cortex, 3*, 163-178.

Kimura, D. (1983). Sex differences in cerebral organization for speech and praxic functions. *Canadian Journal of Psychology, 37*, 19-35.

Kimura, D. (1987). Are men's and women's brains really different? *Canadian Psychology, 28*, 133-147.

Kimura, D. & Archibald, Y. (1974). Motor functions of the left hemisphere. *Brain, 97*, 337-350.

Kimura, D. & Davidson, W. (1975). Right arm superiority for tapping with distal and proximal joints. *Journal of Human Movement Studies, 1*, 199-202.

Kimura, D & Vanderwolf, C.H. (1970). The relation between hand preference and the performance of individual finger movements by left and right hands. *Brain, 93*, 769-774.

Kinsbourne, M. (1973a). Minimal brain dysfunction as a neurodevelopmental lag. *Annals of the New York Academy of Sciences, 205,* 268-273.

Kinsbourne, M. (1973b). The control of attention by interaction between the cerebral hemispheres. In S.Korblum (Ed). *Attention and performance IV.* New York: Academic Press. p.239-256.

Kinsbourne, M. (1974). Mechanisms of hemispheric interaction in man. In M.Kinsbourne and W.L.Smith (Eds). *Hemispheric disconnection and cerebral function.* Springfield: Chas. C. Thomas. p.260-285.

Kinsbourne, M. (1975). The ontogeny of cerebral dominance. *Annals of the New York Academy of Sciences, 263,* 244-250.

Kinsbourne, M. (1981). Single channel theory. In D.H.Holding (Ed). Chichester: Wiley. p.65-89.

Kinsbourne, M. & Cook, J.(1971). Generalized and lateralized effects of concurrent verbalization on a unimanual skill. *Quarterly Journal of Experimental Psychology, 23,* 341-345.

Kinsbourne, M. & Hicks, R.E. (1978). Functional cerebral space: a model for overflow, transfer and interference effects in human performance. A tutorial review. In J.Requin (Ed). *Attention and performance VII.* Hillsdale, NJ: Erlbaum. p.345-362.

Kinsbourne, M. & Hiscock, M. (1983). Asymmetries of dual-task performance. In J.B.Hellige (Ed). *Cerebral hemispheric asymmetry: methods, theory and application..* New York: Praeger Press. p.255-334.

Kinsbourne, M. & McMurray, J. (1975). The effect of cerebral dominance on time sharing between speaking and tapping by preschool children. *Child Development, 46,* 240-242.

Klemmer, E.T. (1962). Communication and human performance. *Human Factors, 4,* 75-79.

Knights, R.M. & Moule, A.D. (1967). Normative and reliability data on finger and foot tapping in children. *Perceptual & Motor Skills, 25,* 717-720.

Koff, E., Naeser, M.A., Pieniadz, J.M., Foundras, A.L. & Levine, H.L. (1986). Computed tomographic scan hemispheric asymmetries in right- and left-handed male and female subjects. *Archives of Neurology, 43,* 487-491.

Kolb, B. (1987). Recovery from early cortical damage in rats. I. Differential behavioral and anatomical effects of frontal lesions at different ages of neural maturation. *Behavioral Brain Research, 25,* 205-220.

Kolb, B. (1992). Mechanisms underlying recovery from cortical injury: reflections on progress and directions for the future. In F.D.Rose and D.A.Johnson (Eds). *Recovery from brain damage.* New York: Plenum Press. p.169-186.

Kolb, B. & Tomie, J. (1988). Recovery from early cortical damage in rats IV. Effects of hemidecortication at 1, 5, or 10 days of age on cerebral anatomy and behavior. *Behavioral Brain Research, 28,* 259-274.

Kolb, B. & Whishaw, I.Q. (1985). Can the study of praxis in animals aid in the study of apraxia in humans? In E.A.Roy (Ed). *Neuropsychological studies of apraxia and related disorders.* Amsterdam: North-Holland. p.203-223.

Komai, T. & Fukuoka, G. (1934). A study on the frequency of left-handedness and left-footedness among Japanese school children. *Human Biology, 6,* 33-42.

Konishi, Y., Kuriyama, M., Mikawa, H. & Suzuki, J. (1987). Effect of body position on later postural and functional lateralities of preterm infants. *Developmental Medicine & Child Neurology, 29,* 751-757.

Koppel, H. & Innocenti, G.M. (1983). Is there a genuine exuberancy of callosal projections in development? A quantitative electron microscope study in the cat. *Neuroscience Letters, 41,* 33-40.

Kornhuber, H.H., Deecke, L., Lang, W., Lang, M. & Kornhuber, A. (1989). Will, volitional action, attention and cerebral potentials in man: Bereitschaftspotential, performance-related potentials, directed attention potential, EEG spectrum changes. In W.A.Hershberger (Ed). *Volitional action.* Amsterdam: Elsevier. p.107-168.

Kosaka, B., Hiscock, M., Strauss, E., Wada, J.A. & Purves, S. (1993). Dual task performance by patients with left or right speech dominance as determined by carotid amytal tests. *Neuropsychologia, 31,* 127-136.

Kounin, J.S. (1938). Laterality in monkeys. *Journal of Genetic Psychology, 52,* 375-393.

Krashen, S. (1973). Lateralization, language learning and the critical period: some new evidence. *Language Learning, 23,* 63-74.

Kristeva, R. & Deecke, L. (1980). Cerebral potentials preceding right and left unilateral and bilateral finger movements in sinistrals. In H.H.Kornhuber & L.Deecke (Eds). *Motivation, motor and sensory processes of the brain: electrical potentials, behavior and clinical use.* Amsterdam: Elsevier. p.748-754.

Kristeva, R., Keller, E., Deecke, L. & Kornhuber, H.H. (1979). Cerebral potentials preceding unilateral and simultaneous bilateral finger movements. *Electroencephalography & Clinical Neurophysiology, 47,* 229-238.

Kristeva, R.G. & Tchakaroff, V. (1986). Bereitschaftspotential in children. In F.Klix & H.Hagensdorf (Eds). *Human memory and cognitive capabilities: mechanisms and performances.* Amsterdam: Elsevier. p.745-753.

Krombholtz, H. (1989). Laterality and force of handgrip during the first two years at school. *Perceptual & Motor Skills, 68,* 955-962.

Kruse, R.D. & Mathews, D.K. (1958). Bilateral effects of unilateral exercise. *Archives of Physical Medicine & Rehabilitation, 39,* 371-376.

Krynauw, R.A. (1950). Infantile hemiplegia treated by removing one cerebral hemisphere. *Journal of Neurology, Neurosurgery & Psychiatry, 13,* 243-267.

Krynicki, V.E. (1978). Cerebral dysfunction in repetitively assaultive adolescents. *Journal of Nervous & Mental Disease, 166,* 59-67.

Kurthen, M., Helmstaedter, C., Linke, D.B., Hufnagel, A., Elger, C.E. & Schramm, J. (1994). Quantitative and qualitative evaluation of patterns of cerebral language dominance. An amobarbital study. *Brain & Language, 46,* 536-564.

Kurthen, M., Helmstaedter, C., Linke, D.B., Solymosi, L., Elger, C.E. & Schramm, J. (1992). Interhemispheric dissociation of expressive and receptive language functions in patients with complex-partial seizures: an amobarbital study. *Brain & Language, 43,* 694-712.

Kutas, M. & Donchin, E. (1974). Studies of squeezing: handedness, responding hand, response force and asymmetry of readiness potential. *Science, 186,* 545-548.

Kutas, M. & Donchin, E. (1977). The effect of handedness, of responding hand and of response forces on the contralateral dominance of the readiness potential. In J.E.Desmedt (Ed). *Attention, voluntary contraction and event-related potentials.* Basel: Karger. p.189-210.

Lake, D.A. & Bryden, M.P. (1976). Handedness and sex differences in hemisphere asymmetry. *Brain & Language, 3,* 266-282.

LaMantia, A-S & Rakic, P. (1984). The number, size, myelination and regional variation of axons in the corpus callosum and anterior commissure of the developing rhesus monkey. *Society for Neuroscience Abstracts, 10,* 1081.

Lang, W., Lang, M., Uhl, F., Koska, Ch., Kornhuber, A. & Deecke, L. (1988). Negative cortical DC shifts preceding and accompanying simultaneous and sequential finger movements. *Experimental Brain Research, 71,* 579-587.

Lang, W., Obrig, H., Lindinger, G., Cheyne, D. & Deecke, L. (1990). Supplementary motor area activation while tapping bimanually different rhythms in musicians. *Experimental Brain Research, 79,* 504-514.

Lang, W., Zilch, O., Koska, C., Lindinger, C. & Deecke, L. (1989). Negative cortical DC shifts preceding and accompanying simple and complex sequential movements. *Experimental Brain Research, 74,* 99-104.

Lansky, L.M. Feinstein, H. & Peterson, J.M. (1988). Demography of handedness in two samples of randomly selected adults (N = 2,083). *Neuropsychologia, 26,* 465-477.

Laponce, J.A. (1976). The left-hander and politics. In A.Somit (Ed). *Biology and politics*. The Hague: Monton. p.45-57.

Larkin, D. & Hoare, D. (1991). *Out of step: coordinating kids' movement*. Perth, W.A.: Active Life Foundation.

Larsen, B., Skinhoj, E. & Lassen, N.A. (1978). Variations in regional cortical blood flow in the right and left hemispheres during automatic speech. *Brain, 101*, 193-209.

Larsen, B., Skinhoj, E., Soh, K., Endo, H. & Lassen, N.A. (1977). The pattern of cortical activity provoked by listening and speech revealed by rCBF measurements. *Acta Neurologica Scandinavica, 56, (Supp.64)*, 268-269.

Larson, J.R. & Greenough, W.T. (1981). Effects of handedness training on dendritic branching of neurons in forelimb area of rat motor cortex. *Society for Neuroscience Abstracts, 7*, 65.

Lashley, K.S. (1917). Modifiability of the preferential use of the hands in the rhesus monkey. *Journal of Animal Behavior, 7*, 178-186.

Lassen, N.A. & Larsen, B. (1980). Cortical activity in the left and right hemispheres during language-related brain functions. *Phonetica, 37*, 27-37.

Lassonde, M. (1986). The facilitatory influence of the corpus callosum on intrahemispheric processing. In F.Lepore, M.Ptito and H.H.Jasper (Eds). *Two hemispheres - one brain: functions of the corpus callosum*. New York: Alan R. Liss. p.385-401.

Lauritzen, M., Henriksen, L. & Lassen, B. (1981). Regional cerebral blood flow during rest and skilled hand movements measured by Xenon-133 inhalation and emission computerized tomography. *Journal of Cerebral Blood Flow & Metabolism, 1*, 385-389.

Lauterbach, C.E. (1933a). The measurement of handedness. *Journal of Genetic Psychology, 43*, 207-212.

Lauterbach, C.E. (1933b). Shall the left-handed child be transferred? *Journal of Genetic Psychology, 43*, 454-462.

Lazarus, J.C. & Todor, J.I. (1987). Age differences in the magnitude of associated movement. *Developmental Medicine and Child Neurology, 29*, 726-733.

Leakey, R.E. (1981). *The making of mankind*. London: Michael Joseph.

Leakey, R.E. (1994). *The origin of humankind*. London: Wedienfeld and Nicolson.

LeBoeuf, A. (1986). Handedness and anxiety in male and female agoraphobics. *Research Communications in Psychology, Psychiatry, & Behavior, 11*, 74-78.

Lecours, A.R., Basso, A., Moraschini, S. & Nespoulous, J-L. (1984). Where is the speech area, and who has seen it? In D.Caplan, A.R.Lecours and A Smith (Eds). *Biological perspectives on language*. Cambridge, Mass.: M.I.T. Press. p.220-246.

Le Gros Clark, W.E. (1956). *History of the primates: an introduction to the study of fossil man*. London: British Museum.

Le Gros Clark, W.E. (1971). *The antecedents of man*. Edinburgh: Edinburgh University Press.

Lehman, R.A.W. (1970). Hand preference and cerebral predominance in 24 rhesus monkeys. *Journal of the Neurological Sciences, 10*, 185-192.

Lehman, R.A.W. (1978). The handedness of rhesus monkeys. I. Distribution. *Neuropsychologia, 16*, 33-42.

Lehman, R.A.W. (1980a). The handedness of rhesus monkeys. III. Consistency within and across activities. *Cortex, 16*, 197-204.

Lehman, R.A.W. (1980b). Persistence of primate hand preference despite initial training to the contrary. *Behavioral Brain Research, 1*, 547-551.

Lehman, R.A.W. (1980c). Distribution and changes in strength of hand preferences of Cynomolgus monkeys. *Brain, Behavior & Evolution, 17*, 209-217.

Lehman, R.A.W (1989). Hand preferences of rhesus monkeys on differing tasks. *Neuropsychologia, 27*, 1193-1196.

Lehman, R.A.W. (1993). Manual preference in prosimians, monkeys and apes. In J.P.Ward & W.D.Hopkins (Eds). *Primate laterality: current behavioral evidence of primate asymmetries*. New York: Springer-Verlag, p.149-181.

LeMay, M. (1976). Morphological cerebral asymmetries of modern man, fossil man and non-human primate. *Annals of the New York Academy of Sciences, 280*, 349-366.

LeMay, M. (1977). Asymmetries of the skull and handedness. *Journal of the Neurological Sciences, 32*, 243-253.

LeMay, M. (1984). Radiological, developmental and fossil asymmetries. In N.Geschwind & A.M.Galaburda (Eds). *Cerebral dominance: the biological foundations*. Cambridge, Mass.: Harvard University Press. p.26-42.

LeMay, M. (1985). Asymmetries of the brains and skulls of nonhuman primates. In S.D.Glick (Ed). *Cerebral ateralization in nonhuman species*. New York: Academic Press. p.233-245.

LeMay. M. & Geschwind, N. (1978). Asymmetries of the human cerebral hemispheres. In A.Caramazza & E.B.Zurif (Eds). *Language acquisition and language breakdown*. Baltimore: Johns Hopkins Press. p.311-328.

Le May, M. & Kido, D.K. (1978). Asymmetries of the cerebral hemispheres in computed tomograms. *Journal of Computer Assisted Tomography, 2*, 471-476.

Lenneberg, E.H. (1967). *Biological foundations of language*. New York: Wiley.

Levander, M. & Schalling, D. (1988). Hand preference in a population of Swedish college students. *Cortex, 24,* 149-156.

Levy, J. (1969). Possible basis for the evolution of lateral specialization of the human brain. *Nature, 224,* 614-615.

Levy, J. (1974). Psychobiological implications of bilateral asymmetry. In S.J.Dimond & J.G.Beaumont (Eds). *Hemisphere function in the human brain.* London: Elek Science. p.121-183.

Levy, J. (1985). Interhemispheric collaboration: single-mindedness in the asymmetric brain. In C.T.Best (Ed). *Hemispheric function and collaboration in the child.* New York: Academic Press. p.11-30.

Levy, J., Nebes, R.D. & Sperry, R.W. (1971). Expressive language in the surgically separated minor hemisphere. *Cortex, 7,* 49-58.

Levy, R.S. (1977). The question of electrophysiological asymmetries preceding speech. In H.Whitaker & H.A.Whitaker (Eds). *Studies in Neurolinguistics, Vol.3.* New York: Academic Press. p.287-318.

Lewandowski, L. & Kohlbreuer, R. (1985). Lateralization in gifted children. *Developmental Neuropsychology, 1,* 277-282.

Lewis, S.W., Chitkara, B. & Reveley, A.M. (1989). Hand preference in psychotic twins. *Biological Psychiatry, 25,* 215-221.

Lieberman, P. (1975). *On the origins of language: an introduction to the evolution of human speech.* New York: Macmillan.

Lieberth, A.K. (1982). Functional speech therapy for the deaf child. In D.G.Sims, G.G.Walker & R.L. Whitehead (Eds). *Deafness and communication: assessment and training.* Baltimore: Williams & Wilkins. p.245-257.

Liederman, J. & Foley, L.M. (1987). A modified finger lift test reveals an asymmetry of motor overflow in adults. *Journal of Clinical & Experimental Neuropsychology, 9,* 498-510.

Liederman, J. & Healey, J.M. (1986). Independent dimensions of hand preference: reliability of the factor structure and the handedness inventory. *Archives of Clinical Neuropsychology, 1,* 371-386.

Liederman, J. & Kinsbourne, M. (1980). Rightward motor bias in newborns depends upon parental right-handedness. *Neuropsychologia, 18,* 579-584.

Lippman, H.S. (1927). Certain behavior responses in early infancy. *Journal of Genetic Psychology, 34,* 424-440.

Lippold, O.C.J. (1973). *The origin of the alpha rhythm.* Edinburgh: Churchill Livingstone.

Lishman, W.A. & McMeekan, E.R.L. (1976). Hand preference patterns in psychiatric patients. *British Journal of Psychiatry, 129,* 158-166.

Littlejohn, J. (1967). The choreography of left hand and right. *New Society*, 9 February, 198-199.

Littlejohn, J. (1973). Temne right and left: an essay on the choreography of everyday life. In R.Needham (Ed). *Right and left: essays on dual symbolic classification*. Chicago: University of Chicago Press. p.288-298.

Lomas, J. & Kimura, D. (1976). Intrahemispheric interaction between speaking and sequential manual activity. *Neuropsychologia,14*, 23-33.

Lombroso, C. (1903). Left-handedness and left-sidedness. *North American Review*, 177, *440-444*.

Longoni, A.M. & Orsini, L. (1988). Lateral preferences in preschool children: a research note. *Journal of Child Psychology & Psychiatry, 29*, 533-539.

Longstreth, L.E. (1980). Human handedness: more evidence for genetic involvement. *Journal of Genetic Psychology, 137*, 275-283.

Loo, R. & Schneider, R.(1979). An evaluation of the Briggs-Nebes modified version of Annett's handedness inventory. *Cortex, 15*, 683-686.

Loring, D.W., Meador, E.J., Lee, G.P. & King, D.W. (1992). *Amobarbital effects and lateralized brain function. The Wada test*. New York: Springer-Verlag.

Loring, D.W., Meador, K.J., Lee, G.P., Murro, A.M., Smith, J.R., Flanigin, H.F., Gallagher, B.B. & King, D.W. (1990). Cerebral language lateralization: evidence from intracarotid amobarbital testing. *Neuropsychologia, 28*, 831-838.

Low, M.D. & Fox, M. (1977). Scalp-recorded slow potential asymmetries preceding speech in man. In J.E.Desmedt (Ed). *Language and hemispheric specialization in man: cerebral event-related potentials*. Basel: Karger. p.104-111.

Low, M.D., Wada, J.A. & Fox, M. (1976). Electroencephalographic localization of conative aspects of language production in the human brain. In W.C.McCallum & J.R.Knott (Eds). *The responsive brain*. Bristol: Wright. p.165-168.

Lucas, J.A., Rosenstein, L.D. & Bigler, E.D. (1989). Handedness and language among the mentally retarded: implications for the model of pathological left-handedness and gender differences in hemispheric specialization. *Neuropsychologia, 27*, 713-723.

Lunderwold, A.J.S. (1951). Electromyographic investigations of position and manner of working in typewriting. *Acta Physiologica Scandinavica, 24*, Suppl. 84.

Luria, A.R. (1965). Neuropsychological analysis of focal brain lesions. In B.B.Wolman (Ed). *Handbook of clinical psychology*. New York: McGraw-Hill. p.689-754.

Luria, A.R. & Yudovich, F.la. (1959). *Speech and the development of mental processes in the child.* London: Staples Press.

Lytton, H., Conway, D. & Sauve, R. (1977). The impact of twinship on parent-child interaction. *Journal of Personality & Social Psychology, 35,* 97-107.

MacNeilage, P.F., Studdert-Kennedy, M.G. & Lindblom, B. (1987). Primate handedness reconsidered. *Behavioral & Brain Sciences, 10,* 247-303.

Macphail, E.M. (1982). *Brain and intelligence in vertebrates.* Oxford: Clarendon Press.

Maehara, K., Negishi, N., Tsai, A., Lizuka, R., Otsuki, N., Suzuki, S., Takahashi, T. & Sumiyoshi, Y. (1988). Handedness in the Japanese. *Developmental Neuropsychology, 4,* 117-127.

Majeres, R.L. (1975). The effect of unimanual performance on speed of verbalization. *Journal of Motor Behavior, 7,* 57-58.

Malina, R.M. (1975). Anthropometric correlates of strength and motor performance. *Exercise & Sport Sciences Reviews, 3,* 249-274.

Malpass, L.F. (1963). Motor skills in mental deficiency, In N.R.Ellis (Ed). *Handbook of mental deficiency.* New York: McGraw-Hill. p.602-631.

Manoache, D.S., Maher, B.A. & Manschreck, T.C. (1988). Left-handedness and thought disorder in the schizophrenias. *Journal of Abnormal Psychology, 97,* 97-99.

Marchant, L.F. & McGrew, W.C. (1991). Laterality of function in apes: a meta analysis of methods. *Journal of Human Evolution, 21,* 425-438.

Marchant, L.F. & Steklis, H.D. (1986). Hand preference in a captive island group of chimpanzees (Pan troglodyta). *American Journal of Primatology, 10,* 301-313.

Margerison, J.H., St.John-Loe, P. & Binnie, C.D. (1967). Electromyography. In P.H.Venables & I.Martin (Eds). *A manual of psychophysiological methods.* Amsterdam: North-Holland. p.351-402.

Marlow, N., Roberts, B.L. & Cooke, R.W.I. (1989a). Motor skills in extremely low birth weight children at the age of 6 years. *Archives of Disease in Childhood, 64,* 839-847.

Marlow, N., Roberts, B.L. & Cooke, R.W.I. (1989b). Laterality and prematurity. *Archives of Disease in Childhood, 64,* 1713-1716.

Marrion, L.V. (1986). Writing hand differences in Kwakiutls and Caucasians. *Perceptual & Motor Skills, 62,* 760-762.

Marrion, L.V. & Rosenblood, L.K. (1986). Handedness in the Kwakiutl totem poles: an exception to 50 centuries of right-handedness. *Perceptual & Motor Skills, 62,* 755-759.

Marsden, C.D. & Sheehy, M.P. (1990). Writers' cramp. *Trends in Neurological Science, 13,* 148-153.

Martenuik, R.G. (1974). Individual differences in motor performance and learning. *Exercise & Sport Sciences Reviews, 2,* 103-130.

Martin, R.D. (1973). Comparative anatomy and primate systematics. *Symposia of the Zoological Society of London, 33,* 301-337.

Martins, I.P., Castro-Caldas, A., van Dougen, H.R. & van Hout, A. (1991). *Acquired aphasia in children: acquisition and breakdown of language in the developing brain.* Dordrecht: Kluwer.

Mascie-Taylor, C.G.N. (1980). Hand preference and components of I.Q. *Annals of Human Biology, 7,* 235-248.

Mascie-Taylor, C.G.N. (1981). Hand preference and personality traits. *Cortex, 17,* 319-322.

Mateer, C.A. (1983). Functional organization of the right nondominant cortex: evidence from electrical stimulation. *Canadian Journal of Psychology, 37,* 36-58.

Mateer, C.A. & Cameron, P.A. (1989). Electrophysiological correlates of language: stimulation mapping and evoked potential studies. In H.Goodglass & A.R.Damasio (Eds). *Handbook of Neuropsychology. Vol.2.* Amsterdam: Elsevier. p.91-116.

Mateer, C.A., Polen, S.B. & Ojemann, G.A. (1982). Sexual variation in cortical localization of naming as determined by stimulation mapping. *Behavioral & Brain Sciences, 5,* 310-311.

Mathiowetz, V., Wiemer, D.M. & Federman, S.M. (1986). Grip and pinch strength: norms for 6 to 19 year-olds. *American Journal of Occupational Therapy, 40,* 705-711.

McAdam, D.W. & Whitaker, H.A. (1971). Language production: electrophysiological localization in the normal human brain. *Science, 172,* 499-502.

McCarthy, D. (1970). *McCarthy scales of children's abilities.* New York: Psychological Corporation.

McCreadie, R.G., Crorie, J., Barron, E.T. & Winslow, G.S. (1982). The Nithsdale schizophrenic survey: III. Handedness and tardive dyskinesia. *British Journal of Psychiatry, 140,* 591-594.

McFarland, K. & Anderson, J. (1980). Factor stability of the Edinburgh Inventory as a function of test-retest performance, age and sex. *British Journal of Psychology, 71,* 135-142.

McFarland, K. & Ashton, R. (1975a). The lateralized effects of concurrent cognitive activity on a unimanual skill. *Cortex, 11,* 283-290.

McFarland, K. & Ashton, R. (1975b). A developmental study of the influence of cognitive activity on an ongoing manual task. *Acta Psychologica, 39*, 447-456.

McFarland, K. & Ashton, R. (1978a). The influence of concurrent task difficulty on manual performance. *Neuropsychologia, 16*, 735-741.

McFarland, K. & Ashton, R. (1978b). The lateralized effects of concurrent cognition and motor performance. *Perception & Psychophysics, 23*, 344-349.

McFarland, K. & Ashton, R. (1978c). The influence of brain lateralization of function on a manual skill. *Cortex, 14*, 102-111.

McFie, J. (1961). The effects of hemispherectomy on intellectual functioning in cases of infantile hemiplegia. *Journal of Neurology, Neurosurgery & Psychiatry, 24*, 240-249.

McGlone, J. (1977). Sex differences in the cerebral organization of verbal functions in patients with unilateral brain lesions. *Brain, 100*, 775-793.

McGlone, J. (1978). Sex differences in functional brain asymmetry, *Cortex, 14*, 122-128.

McGlone, J. (1980). Sex differences in human brain asymmetry: a critical survey. *Behavioral & Brain Sciences, 3*, 215-263.

McGonigle, B. & Flook, J. (1978). The learning of hand preferences by squirrel monkeys. *Psychological Research, 40*, 93-98.

McKee, G., Humphrey, B. & McAdam, D.W. (1973). Scaled lateralization of alpha activity during linguistic and musical tasks. *Psychophysiology, 10*, 441-443.

McKeever, W.F. & Rich, D.A. (1990). Left-handedness and immune disorders. *Cortex, 26*, 33-40.

McLean, J.M. & Ciurczak, F.M. (1982). Bimanual dexterity in major league baseball players: a statistical study. *New England Journal of Medicine, 307*, 1278-1279.

McLeod, P. & Posner, M.I. (1985). Privileged loops from percept to act. In: H.Bouma & Bouwhuis, D.G. (Eds). *Attention and performance X, Control of language processes.* London: Erlbaum. p.55-66.

McManus, A. & Armstrong, N. (1995). Patterns of physical activity among primary schoolchildren. In F.J.Ring (Ed). *Children in sport.* Bath: University of Bath. p.17-23.

McManus, I.C. (1980a). Handedness in twins: a critical review. *Neuropsychologia, 18*, 347-355.

McManus, I.C. (1980b). Left-handedness and epilepsy. *Cortex, 16*, 487-491.

McManus, I.C. (1983). Pathologic left-handedness: does it exist? *Journal of Communication Disorders, 16*, 315-344.

McManus, I.C. (1984). Genetics of handedness in relation to language disorder. In F.C. Rose (Ed). *Progress in Aphasiology.* New York: Raven Press. p.125-138.

McManus, I.C. (1985a). Handedness, language dominance and aphasia: a genetic model. *Psychological Medicine, Monograph Supplement No.8.*

McManus, I.C. (1985b). Right- and left-hand skill: failure of the right-shift model. *British Journal of Psychology, 76,* 1-16.

McManus, I.C. (1992). Are paw preference differences in HI and LO mice the result of specific genes or of heterosis and fluctuating asymmetry? *Behavior Genetics, 22,* 435-451.

McManus, I.C. & Bryden, M.P. (1992). The genetics of handedness, cerebral dominance and lateralization. In I.Rapin & S.J.Segalowitz (Eds). *Handbook of neuropsychology, Vol. 6. Child neuropsychology.* Amsterdam: Elsevier Science. p.115-144.

McManus, I.C., Murray, B., Doyle, K. & Barron-Cohen, S. (1992). Handedness in childhood autism shows a dissociation of skill and preference. *Cortex, 28,* 373-381.

McManus, I.C., Naylor, J. & Booker, B.L. (1990). Left-handedness and myasthenia gravis. *Neuropsychologia, 28,* 947-955.

McManus, I.C., Shergill, S. & Bryden, M.P. (1993). Annett's theory that individuals heterozygous for the right shift gene are intellectually advantaged: theoretical and empirical problems. *British Journal of Psychology, 84,* 517-537.

McManus, I.C., Sik, G., Cole, D.R., Mellon, A.F., Wong, J. & Kloss, J. (1988). The development of handedness in children. *British Journal of Developmental Psychology, 6,* 257-273.

McNamara, P., Flannery, K.A., Obler, L.K. & Schacter, S. (1994). Special talents in Geschwind's theory of cerebral lateralization: an examination in a female population. *International Journal of Neuroscience, 78,* 167-176.

Mebert, C.J. & Michel, G.F. (1980). Handedness in artists. In J.Herron (Ed). *Neuropsychology of left-handedness* New York: Academic Press. p.273-279.

Merckelbach, H., de Ruiter, C. & Olff, M. (1989). Handedness and anxiety in normal and clinical populations. *Cortex, 25,* 599-606.

Merrin, E.L. (1984). Motor and sighting dominance in schizophrenic and affective disorder. *British Journal of Psychiatry, 146,* 539-544.

Methany, E. (1940). Breathing capacity and grip strength of preschool children. *University of Iowa Studies. Studies in Child Welfare, 18,* No. 2.

Meudell, P.R. & Greenhalgh, M. (1987). Age related differences in left and right hand skill and in visuo-spatial performance: their possible relationships to the hypothesis that the right hemisphere ages more rapidly than the left. *Cortex, 23,* 431-445.

Meyer, B.C. (1945). Psychosomatic aspects of stuttering. *Journal of Nervous & Mental Diseases, 101,* 127-157.

Michalewski, H.J., Weinberg, H. & Patterson, J.V. (1977). The contingent negative variation (CNV) and speech production: slow potentials and the area of Broca. *Biological Psychology, 5,* 83-96.

Middleton, J.M. (1968). Some categories of dual classification among the Lugbara of Uganda. *History of Religions, 7,* 187-208. Reprinted in R.Needham (Ed). *Right and left: essays on dual symbolic classification.* Chicago: University of Chicago Press. p.369-390.

Miklyaeva, E.I., Ioffe, M.E. & Kulikov, M.A. (1991). Innate versus learned factors determining limb preference in the rat. *Behavioral Brain Research, 46,* 103-115.

Miles, W.R. (1931). Measures of certain human abilities throughout the life span. *Proceedings of the National Academy of Sciences, 17,* 627-633.

Miller, E. (1971). Handedness and the pattern of human ability. *British Journal of Psychology, 62,* 111-112.

Milner, A.D. & Jeeves, M.A. (1979). A review of behavioral studies of agenesis of the corpus callosum. In I.S.Russell, M.W. van Hof and G.Berlucchi (Eds). *Structure and function of the cerebral commissures.* London: Macmillan. p.428-448.

Mitrushina, M., Fogel, T., D'Elia, L., Uchiyama, C. & Satz, P. (1995). Performance on motor tasks as an indication of increased behavioral asymmetry with advancing age. *Neuropsychologia, 33,* 359-364.

Mittler, P. (1970). Biological and social aspects of language development in twins. *Developmental Medicine & Child Neurology, 12,* 741-757.

Mohay, H., Burns, Y., Luke, D., Tudehope, D. & O'Callaghan, M. (1986). The effects of prenatal and postnatal twin environments on development. In C.Pratt, A.F.Garton, W.E.Turner & A.R.Neasdale (Eds). *Research issues in child development.* Sydney: Allen & Unwin. p.5-12.

Molfese, D.L. (1983). Event related potentials and language processes. In A.W.K.Gaillard & W.Ritter (Eds). *Tutorials in ERP research: endogenous components.* Amsterdam, North-Holland. p.345-367.

Molfese, V. & Betz, J. (1986). Parallels between motor and language development. In H.Whiting & M.Wade (Eds). *Themes in motor development.* Dordrecht: Nijhoff. p.329-340.

Molfese, V. & Betz, J. (1987). Language and motor development in infancy: three views with neuropsychological implications. *Developmental Neuropsychology, 3,* 255-274.

Molnar, G.E. & Alexander, J. (1983). Strength development in retarded children: a comparative study on the effect of intervention. In J.Hogg & P.J.Mittler (Eds). *Advances in Mental Handicap Research, 2,* 285-307.

Montagu, A. (1976). Tool making, hunting and the origin of language. *Annals of the New York Academy of Sciences, 280,* 266-274.

Morgan, A.H., Macdonald, H. & Hilgard, E.R. (1974). EEG alpha: lateral asymmetries related to task and hypnotizability. *Psychophysiology, 11,* 275-282.

Morgan, M.J. & McManus, I.C. (1988). The relationship between brainedness and handedness. In: F.C.Rose, R.Whurr & M.A.Wyke (Eds). *Aphasia.* London: Whurr. p.85-130.

Morrell, L.K. & Huntington, D.A. (1971). Electrocortical localization of language production. *Science, 174,* 1359-1360.

Morrell, L.K. & Huntington, D.A. (1972). Cortical potentials time-locked to speech production: evidence for probable cerebral origin. *Life Sciences, 11,* 921-929.

Morris, R.D. & Romski, M.A. (1993). Handedness distribution in a non-speaking population with mental retardation. *American Journal of Mental Retardation, 97,* 443-448.

Munch, P.A. (1964). Culture and super-culture in a displaced community: Tristan da Cunha. *Ethnology, 3,* 369-376.

Munn, N. (1965). *The evolution and growth of human behavior.* Boston: Houghton Mifflin.

Muscio, B. (1922). Motor capacity with special reference to vocational guidance. *British Journal of Psychology, 13,* 157-184.

Nachshon, I., Denno, D. & Aurand, S. (1983). Lateral preferences of hand, eye and foot: relation to cerebral dominance. *International Journal of Neuroscience, 18,* 1-10.

Nagylaki, T. & Levy, J. (1973). The sound of one paw clapping isn't sound. *Behavior Genetics, 3,* 279-292.

Napier, J.R. (1961) Prehensibility and opposability in the hands of primates. *Symposia of the Zoological Society of London, 5,* 115-132.

Napier, J.R. & Tuttle, R.H. (1993). *Hands.* Princeton, N.J.: Princeton University Press.

Nasrallah, H.A., Keeler, K., van Schroeder, C. & McCalley-Whitters, M. (1981). Motoric lateralization in schizophrenic males. *American Journal of Psychiatry, 138,* 1114-1115.

Nasrallah, H.A., McCalley-Whitters, M. & Kuperman, S. (1982a). Neurological differences between paranoid and nonparanoid schizophrenics: Part 1. Sensory-motor lateralization. *Journal of Clinical Psychiatry, 43,* 305-306.

Nasrallah, H.A. & McCalley-Whitters, M. (1982b). Motor lateralization in manic males. *British Journal of Psychiatry, 140,* 521-522.

Nebes, R.D. (1971). Handedness and the perception of part-whole relationship. *Cortex, 7,* 350-356.

Needham, R. (1967). Right and left in Nyoro symbolic classification. *Africa, 37,* 425-451. Reprinted in R.Needham (Ed). *Right and left: essays on dual symbolic classification.* Chicago: Chicago University Press, 1973. p.299-341.

Needham, R. (Ed). (1973). *Right and left; essays on dual symbolic classification.* Chicago: Chicago University Press.

Needles, W. (1942). Concerning transfer of cerebral dominance in the function of speech. *Journal of Nervous & Mental Disease, 95,* 270-277.

Nelligan, G. & Prudham, D. (1969). Norms for four standard developmental milestones by sex, social class, and place in family. *Developmental Medicine & Child Neurology, 11,* 413-422.

Nelson, L.D., Satz, P. Green, M. & Cicchetti, D. (1993). Re-examining handedness in schizophrenia: now you see it - now you don't! *Journal of Clinical & Experimental Neuropsychology, 15,* 149-158.

Newcombe, F.G., Ratcliff, G.G., Carrivick, P.J., Hiorns, R.W., Harrison, G.A. & Gibson, J.B. (1975). Hand preference and I.Q. in a group of Oxfordshire villages. *Annals of Human Biology, 2,* 235-242.

Newell, A. & Rosenbloom, P.S. (1981). Mechanism of skill acquisition and the law of practice. In J.R.Anderson (Ed). *Cognitive skills and their acquisition.* Hillsdale, N.J.: Lawrence Erlbaum. p.1-55.

Newell, K.M. (1976). Knowledge of results and motor learning. *Exercise & Sport Sciences Reviews, 4,* 195-228.

Newell, K.M. (1985). Coordination, control and skill. In D.Goodman, R.B.Wilberg & I.M.Franks (Eds). *Differing perspectives in motor learning, memory and control.* Amsterdam: North Holland. p.295-317.

Newport, E.L. (1990). Maturational constraints on language learning. *Cognitive Science, 14,* 11-28.

Newport, E.L. (1991). Contrasting conceptions of the critical period for language. In S.Carey & R.Gelman (Eds). *Epigenesis of mind: essays on biology and cognition.* Hillsdale, N.J.: Erlbaum. p.111-130.

Nichols, R.C. (1964). Parental attitudes of mothers of intelligent adolescents, and creativity of their children. *Child Development, 35,* 1041-1049.

Njiokiktjien, C., Driessen, M. & Itabraken, L. (1986). Development of supination-pronation movements in normal children. *Human Neurobiology, 5,* 199-203.

Nunez, P.L. & Katznelson, R.D. (1981). *Electric fields of the brain: the neurophysics of EEG.* Oxford: Oxford University Press.

Oakley, K.P. (1958). *Man the tool-maker.* London: British Museum.

Oakley, K.P. (1972). Skill as a human possession. In: S.L.Washburn & P.Dolhinow (Eds). *Perspectives on human evolution, Vol.2.* New York: Holt, Rinehart & Winston. p.14-50.

Oakley, K.P. (1981). Emergence of higher thought 3.0-2.0 Ma B.P. *Philosophical Transactions of the Royal Society (London) B, 252,* 205-211.

O'Boyle, M.W. & Benbow, C.P. (1990). Handedness and its relationship to ability and talent. In S.Coren (Ed). *Left-handedness: behavioral implications and anomalies.* Amsterdam: North Holland. p.343-372.

O'Boyle, M.W., Gill, H.S., Benbow, C.P. & Alexander, J.E. (1994). Concurrent finger-tapping in mathematically gifted males: evidence for enhanced right hemisphere involvement during linguistic processing. *Cortex, 30,* 519-526.

Obrzut, J.E., Hynd, G.W., Obrzut, A. & Leitget, J.L. (1980). Time-sharing and dichotic listening asymmetry in normal and learning-disabled children. *Brain & Language, 11,* 181-194.

O'Callaghan, M.J., Tudehope, D.I., Dugdale, A.E., Mohay, H., Burns, Y. & Cook, F. (1987). Handedness in children with birthweights below 1000g. *The Lancet, (i),* 1155.

O'Callaghan, M.J., Burns, Y.R., Mohay, H.A., Rogers, Y. & Tudehope, D.I. (1993a). The prevalence and origins of left hand preference in high risk infants, and its implications for intellectual, motor and behavioural performance at four and six years. *Cortex, 29,* 617-627.

O'Callaghan, M.J., Burns, Y.R., Mohay, H.A., Rogers, Y. & Tudehope, D.I. (1993b). Handedness in extremely low birth weight infants: aetiology and relationship to intellectual abilities, motor performance and behaviour at four and six years. *Cortex, 29,* 629-637.

Oddy, H.C. & Lobstein, T.J. (1972). Hand and eye dominance in schizophrenia. *British Journal of Psychiatry, 120,* 331-332.

Offord, D.R. & Cross, L.A. (1971). Adult schizophrenia with scholastic failure or low I.Q. in childhood. *Archives of General Psychiatry, 24,* 431-436.

Ojemann, G.A. (1983a). The intrahemispheric organization of human language, derived with electrical stimulation techniques. *Trends in Neurosciences, 6,* 184-189.

Ojemann, G.A. (1983b). Brain organization for language from the perspective of electrical stimulation mapping. *Behavioral & Brain Sciences, 6,* 189-230.

Ojemann, G.A. (1994). Cortical stimulation and recording in language. In A.Kertesz (Ed). *Localization and neuroimaging in neuropsychology.* San Diego: Academic Press, p.35-55.

Ojemann, G.A. & Mateer, C.A. (1979a). Human language cortex: localization of memory, syntax, and sequential motor-phoneme identification systems. *Science, 205,* 1401-1403.

Ojemann, G.A. & Mateer, C.A. (1979b). Human language cortex: Identification of common sites for sequencing motor activity and speech discrimination. *Experimental Brain Research, Supp.2.,* 205-211.

Ojemann, G.A. & Mateer, C.A. (1979c). Cortical and subcortical organization of human communication: evidence from stimulation studies. In H.D.Steklis & M.J. Raleigh (Eds). *Neurobiology of social communication in primates.* New York: Academic Press. p.111-131.

Ojemann, G., Ojemann, J., Lettich, E. & Berger, M. (1989). Cortical language lateralization in left, dominant hemisphere. *Journal of Neurosurgery, 71,* 316-327.

Ojemann, G.A. & Whitaker, H.A. (1978). Language localization and variability. *Brain & Language, 6,* 239-260.

O'Kusky, J., Strauss, E., Kosaka, B. Wada, J., Li, D., Druhan, M. & Petrie, J. (1988). The corpus callosum is larger with right hemisphere cerebral speech dominance. *Annals of Neurology, 24,* 379-383.

Oldfield, R.C. (1969). Handedness in musicians. *British Journal of Psychology, 60,* 91-99.

Oldfield, R.C. (1971). The assessment and analysis of handedness. The Edinburgh Inventory. *Neuropsychologia, 9,* 97-113.

Olesen, J. (1971). Contralateral focal increase of cerebral blood flow in man during arm work. *Brain, 94,* 635-646.

Olson, D.A., Ellis, J.E. & Nadler, R.D. (1990). Hand preferences in captive gorillas, orangutans and gibbons. *American Journal of Primatology, 20,* 83-94.

Orgogozo, J.M. & Larsen, B. (1979). Activation of the supplementary motor area during voluntary movement suggests that it works as a supramotor area. *Science, 206,* 847-850.

Orlando, C.P. (1972). Measures of handedness as indicators of language lateralization. *Bulletin of the Orton Society, 22,* 14-26.

Orme, J.E. (1970). Left-handedness, ability and emotional instability. *British Journal of Social & Clinical Psychology, 9,* 87-88.

Orsini, D.L. & Satz, P. (1986). A syndrome of pathological left-handedness: correlates of early left-hemisphere injury. *Archives of Neurology, 43,* 333-337.

Orsini, D.L., Satz, P., Soper, H.V. & Light, R.K. (1985). The role of familial sinistrality in cerebral organization. *Neuropsychologia,, 23,* 223-232.

Orton, S.T. (1928). A physiological theory of reading disability and stuttering in children. *New England Journal of Medicine, 199,* 1046-1052.

Orton, S.T. (1930). Familial occurrence of disorders in acquisition of language. *Eugenics, 3,* No.4.

Orton, S.T. (1937). *Reading, writing, and speech problems in children.* London: Chapman & Hall.

Overby, L.A. (1993). Handedness patterns of psychosis-prone college students. *Personality & Individual differences, 15,* 261-265.

Oxbury, S.M. & Oxbury, J.M. (1984). Intracarotid amytal test in the assessment of language dominance. In F.C.Rose (Ed). *Progress in Aphasiology.* New York: Raven Press. p.115-123.

Palmer, R.E. & Corballis, M.C. (1996). Predicting reading ability from handedness measures. *British Journal of Psychology, 87,* 609-620.

Papaioannou, J. (1972). Handedness in mice: some behavioral correlates. *Experientia, 28,* 273-274.

Papakostopoulos, D. (1980). The Bereitschaftspotential in left- and right-handed subjects. In H.H.Kornhuber & L.Deecke (Eds). *Motivation, motor and sensory processes of the brain.* Amsterdam: Elsevier. p.742-747.

Paquier, P. and Van Dongen, H.R. (1993). Current trends in acquired childhood aphasia: an introduction. *Aphasiology, 7,* 421-440.

Parlow, S. (1978). Differential finger movements and hand preference. *Cortex, 14,* 608-611.

Parlow, S.E. (1990). Asymmetrical movement overflow in children depends on handedness and task characteristics. *Journal of Clinical & Experimental Neuropsychology, 12,* 270-280.

Parson, B.S. (1924). *Left-handedness.* New York: Macmillan.

Passingham, R.E. (1975a). Changes in the size and organization of the brain in man and his ancestors. *Brain, Behavior & Evolution, 11,* 73-90.

Passingham, R.E. (1975b). The brain and intelligence. *Brain, Behavior & Evolution, 11, 1-15.*

Passingham, R.E. (1981). *Primate specialization in brain and intelligence.* Symposia of the Zoological Society of London, 46, *36-388.*

Passingham, R.E., Perry, V.H. and Wilkinson, F. (1983). The long-term effects of removal of sensorimotor cortex in infant and adult rhesus monkeys. *Brain, 106,* 675-705.

Payne, M. (1981a). Incidence of left-handedness for writing: a study of Nigerian primary schoolchildren. *Journal of Cross-cultural Psychology, 12,* 233-239.

Payne, M. (1981b). Attitudes towards the use of the left hand for writing: a pilot study in Nigeria. *Nigeria Education Forum, 4,* 164-172.

Payne, M. (1987). Impact of cultural pressures on self-reports of actual and approved hand use. *Neuropsychologia, 25,* 247-258.

Penfield, W. and Jasper, H. (1954). *Epilepsy and the functional anatomy of the human brain.* Boston: Little, Brown.

Penfield, W. and Rasmussen, T. (1950). *The cerebral cortex of man.* New York: Macmillan.

Penfield, W. and Roberts, L. (1959). *Speech and brain mechanisms.* Princeton: Princeton University Press.

Penfield, W. and Steelman, H. (1947). The treatment of focal epilepsy by cortical excision. *Annals of Surgery, 126,* 740-762.

Penfield, W. and Welch, K. (1951). The supplementary motor area of the cerebral cortex. *Archives of Neurology and Psychiatry, 66,* 289-317.

Pennington, B.F., Smith, S.D., Kimberling, W.J., Green, P.A. & Haith, M.M. (1987). Left-handedness and immune disorders in familial dyslexics. *Archives of Neurology, 44,* 634-639.

Perelle, I., Ehrman, L. & Manowitz, J.W. (1981). Human handedness: the influence of learning. *Perceptual & Motor Skills, 54,* 967-977.

Perello, J. (1970). Digressions on the biological foundations of language. *Journal of Communication Disorders, 3,* 140-149.

Peters, J.F., Varner, J.L. & Ellingson, R.J. (1981). Interhemispheric amplitude symmetry in the EEGs of normal full term, low risk and trisomy-21 infants. *Electroencephalography & Clinical Neurophysiology, 51,* 165-169.

Peters, M. (1976). Prolonged practice of a simple motor task by preferred and non-preferred hands. *Perceptual & Motor Skills, 43,* 447-450.

Peters, M. (1981). Handedness: effect of prolonged practice on between hand performance differences. *Neuropsychologia, 19,* 587-590.

Peters, M. (1988). Footedness: asymmetries in foot preference and skill, and neurological assessment of foot movement. *Psychological Bulletin, 103,* 179-192.

Peters, M. (1990). Interaction of vocal and manual movements. In G.R.Hammond (Ed). *Cerebral control of speech and limb movements.* Amsterdam: North-Holland. p.535-574.

Peters, M. & Durding, B.M. (1978). Handedness measured by finger tapping: a continuous variable. *Canadian Journal of Psychology, 32,* 257-261.

Peters, M. & Durding, B.M. (1979). Left-handers and right-handers compared on a motor task. *Journal of Motor Behavior, 11,* 103-111.

Peters, M. & Servos, P. (1989). Performance of subgroups of left-handers and right-handers. *Canadian Journal of Psychology, 43,* 341-358.

Petersen, S.E., Fox, P.T., Posner, M.I., Mintun, M. & Raichle, M. (1988). Positron emission tomographic studies of the cortical anatomy of single word processing. *Nature, 331,* 585-589.

Petersen, S.E., Fox, P.T., Posner, M.I., Mintun, M. & Raichle, M.E. (1989). Positron emission tomographic studies of the processing of single words. *Journal of Cognitive Science, 1,* 154-170.

Peterson, G.M. (1931). A preliminary report on right and left handedness in the rat. *Journal of Comparative & Physiological Psychology, 12,* 243-250.

Peterson, G.M. (1934). Mechanisms of handedness in the rat. *Comparative Psychology Monographs, 9,* 1-67.

Peterson, G.M. (1938). The influence of cerebral destructions upon the handedness of the rat in the latch box. *Journal of Comparative & Physiological Psychology, 26,* 445-457.

Peterson, G.M. (1951). Transfers in handedness in the rat from forced practice. *Journal of Comparative & Physiological Psychology, 44, 184-190.*

Peterson, G.M. and Barnett, P.E. (1961). The cortical destruction necessary to produce a transfer of a forced-practice function. *Journal of Comparative & Physiological Psychology, 54,* 382-385.

Peterson, G.M. & Devine, J.V. (1963). Transfers in handedness in the rat resulting from small cortical lesions after limited forced practice. *Journal of Comparative & Physiological Psychology, 56,* 752-756.

Peterson, G.M. & Fracarol, C. (1938). The relative influence of the locus and mass of destruction upon the control of handedness by the cerebral cortex. *Journal of Comparative Neurology, 68,* 173-190.

Peterson, G.M. and Gucker, D.K. (1959). Factors influencing identification of the handedness area in the cerebral cortex of the rat. *Journal of Comparative & Physiological Psychology, 52,* 279-283.

Peterson, J.M. (1979). Left-handedness: differences between student artists and scientists. *Perceptual & Motor Skills, 48,* 961-962.

Peterson, J.M. & Lansky, L.M. (1974). Left-handedness among architects: some facts and speculation. *Perceptual & Motor Skills, 38,* 547-550.

Peterson, J.M. & Lansky, L.M. (1977). Left-handedness among architects: partial replication and some new data. *Perceptual & Motor Skills, 45,* 1216-1218.

Piazza, D. (1977). Cerebral lateralization in young children as measured by dichotic listening and finger tapping tasks. *Neuropsychologia, 15,* 417-425.

Pieniadz, J.M. & Naeser, M.A. (1984). Computed tomographic scan, cerebral asymmetries and morphologic brain asymmetries. *Annals of Neurology, 41,* 403-409.

Pipe, M.E., (1988). Atypical laterality and retardation. *Psychological Bulletin, 104*, 343-347.

Pipe, M.E. (1990). Mental retardation and left-handedness: evidence and theories. In S.Coren (Ed). *Left-handedness: behavioral implications and anomalies.* Amsterdam: Elsevier. p.293-318.

Piran, N., Bigler, E.D. & Cohen, D. (1982). Motoric laterality and eye dominance suggest unique pattern of cerebral organization in schizophrenia. *Archives of General Psychiatry, 39*, 1006-1010.

Pizzamiglio, L. (1974). Handedness, ear preference and field dependence. *Perceptual & Motor Skills, 38*, 700-702.

Plato, C.C., Fox, K.M. & Garruto, R.M. (1984). Measures of lateral functional dominance: hand dominance. *Human Biology, 56*, 259-275.

Plomin, R. (1991) Behavioral genetics. In P.R.McHugh & V.A.McKusik (Eds). *Genes, brain and behavior.* New York: Raven Press. p.165-180.

Plomin, R. & De Fries, J.C. (1985). *Origins of individual differences in infancy. The Colorado Adoption Project.* New York: Academic Press.

Poeck, K. (1988). The relationship between aphasia and motor apraxia. In F.C.Rose, R.Whurr and M.Wyke (Eds). *Aphasia.* London: Whurr. p.288-301.

Policansky, D. (1982). The asymmetry of flounders. *Scientific American, 246*, 96-102.

Porac, C. (1993). Are age trends in adult hand preference best explained by developmental shifts or generational differences? *Canadian Journal of Experimental Psychology, 47*, 697-713.

Porac, C. & Coren, S. (1981). *Lateral preferences and human behavior.* New York: Springer-Verlag.

Porac, C., Coren, S. & Duncan, P. (1980a). Lateral preferences in retardates: relationships between hand, eye, foot and ear preference. *Journal of Clinical Neuropsychology, 2*, 173-187.

Porac, C., Coren, S. & Duncan, P. (1980b). Life-span age trends in laterality. *Journal of Gerontology, 35*, 715-721.

Porac, C., Coren, S., Steiger, J.H. & Duncan, P. (1980). Human laterality: a multidimensional approach. *Canadian Journal of Psychology, 34*, 91-96.

Poreh, A.M. (1994). Re-examination of mixed handedness in psychosis-prone college students. *Personality & Individual Differences, 17*, 445-448.

Porfert, A.R. & Rosenfield, D.B. (1978). Prevalence of stuttering. *Journal of Neurology, Neurosurgery & Psychiatry, 41*, 954-956.

Powell, G.E., Polkey, C.E. & Canavan, A.G.M. (1987). Lateralization of memory functions in epileptic patients by use of the sodium amytal (Wada) technique. *Journal of Neurology, Neurosurgery & Psychiatry, 50*, 665-672.

Preilowski, B. (1972). Possible contribution of the anterior forebrain commissures to bilateral motor coordination. *Neuropsychologia, 10*, 267-277.

Preilowski, B. (1975). Bilateral motor interaction: Perceptual-motor performance of partial and complete "split-brain" patients. In K.J.Zulch, O.Creutzfeldt and G.C. Galbraith (Eds). *Cerebral localization*. Berlin: Springer-Verlag. p.115-132.

Preilowski, B. (1977). Phases of motor-skills acquisition: a neuropsychological approach. *Journal of Human Movement Studies, 3*, 169-181.

Preilowski, B. (1990). Intermanual transfer, interhemispheric interaction and handedness in man and monkeys. In C.Trevarthen (Ed). *Brain circuits and functions of the mind*. Cambridge: Cambridge University Press, p.168-180.

Preilowski, B., Reger, M. & Engele, H.C. (1986). Handedness and cerebral asymmetry in nonhuman primates. In D.M.Taub and F.A.King (Eds). *Current perspectives in primate biology*. New York: Van Nostrand & Reinhold, p.270-282.

Pringle, M.L.K. (1961). The incidence of some supposedly adverse family conditions and of left-handedness in schools for maladjusted children. *British Journal of Educational Psychology, 31*, 183-193.

Prohovnik, I., Hakansson, K. & Risberg, J. (1980). Observations on the functional significance of regional cerebral blood flow in "resting" subjects. *Neuropsychologia, 18*, 203-217.

Provins, K.A. (1956). Handedness and skill. *Quarterly Journal of Experimental Psychology, 8*, 79-95.

Provins, K.A. (1958). The effect of training and handedness on the performance of two simple motor tasks. *Quarterly Journal of Experimental Psychology, 10*, 29-39.

Provins, K.A. (1967). Motor skills, handedness and behaviour. *Australian Journal of Psychology, 19*, 137-150.

Provins, K.A. (1990). Handedness and conformity in a small isolated community. *International Journal of Psychology, 25*, 343-350.

Provins, K.A. (1992). Early infant motor asymmetries and handedness: a critical evaluation of the evidence. *Developmental Neuropsychology, 8*, 325-365.

Provins, K.A. (1997a). Handedness and speech: a critical re-appraisal of the role of genetic and environmental factors in the cerebral lateralization of function. *Psychological Review, 104*, 554-571.

Provins, K.A. (1997b). The specificity of motor skill and manual asymmetry: a review of the evidence and its implications. *Journal of Motor Behavior, 29*, 183-192.

Provins, K.A. & Cunliffe, P. (1972a). The relationship between EEG activity and handedness. *Cortex, 8*, 136-146.

Provins, K.A. & Cunliffe, P. (1972b). The reliability of some motor performance tests of handedness. *Neuropsychologia, 10*, 199-206.

Provins, K.A. & Dalziel, F.R. (1969). Handedness: an unusual case of spontaneous change of writing hand. *Journal of Motor Behavior, 1*, 163-167.

Provins, K.A., Dalziel, D.F. & Higginbottom, G. (1987). Asymmetrical hand usage in infancy: an ethological approach. *Infant Behavior & Development, 10*, 165-172.

Provins, K.A. & Glencross, D.J. (1968). Handwriting, typewriting and handedness. *Quarterly Journal of Experimental Psychology, 20*, 282-289.

Provins, K.A. & Magliaro, J. (1993). The measurement of handedness by preference and performance tests. *Brain & Cognition, 22*, 171-181.

Provins, K.A., Milner, A.D. & Kerr, P. (1982). Asymmetry of manual preference and performance. *Perceptual & Motor Skills, 54*, 179-194.

Quinan, C. (1930). The principal sinistral types. *Archives of Neurology & Psychiatry, 24*, 35-47.

Raczkowski, D., Kalat, J.W. & Nebes, R. (1974). Reliability and validity of some handedness questionnaire items. *Neuropsychologia, 12*, 43-47.

Ramaley, F. (1913). Inheritance of left-handedness. *The American Naturalist, 47*, 730-738.

Ramsay, D.S. (1980). Onset of unimanual handedness in infants. *Infant Behavior & Development, 3*, 377-385.

Raney, E.T. (1939). Brain potentials and lateral dominance in identical twins. *Journal of Experimental Psychology, 24*, 21-39.

Rapin, I. & Allen, D.A. (1988). Syndromes in developmental dysphasia and adult aphasia. In: F.Plum (Ed). *Language, communication, and the brain.* New York: Raven Press. p.57-73.

Rapin, I., Tourke, L.M. & Costa, L.D. (1966). Evaluation of the Purdue pegboard as a screening test for brain damage. *Developmental Medicine & Child Neurology, 8*, 45-54.

Rarick, G.L. (1973). Motor performance of mentally retarded children. In G.L.Rarick (Ed). *Physical activity: human growth and development.* New York: Academic Press. p.225-256.

Rasmussen, G.L. (1993). Persistent mirror movements: a clinical study of 17 children, adolescents and young adults. *Developmental Medicine & Child Neurology, 35,* 699-707.

Rasmussen, T. & Milner, B. (1977). The role of early left brain injury in determining lateralization of cerebral functions. *Annals of the New York Academy of Sciences, 299,* 355-368.

Rausch, R. (1987). Psychological evaluation. In J.Engel (Ed). *Surgical treatment of the epilepsies.* New York: Raven Press. p.181-195.

Rawnsley, K. & Loudon, J.B. (1964). Epidemiology of mental disorder in a closed community. *British Journal of Psychiatry, 110,* 830-839.

Ray, W.J. & Cole, H.W. (1985). EEG alpha activity reflects attentional demands, and beta activity reflects emotional and cognitive processes. *Science, 228,* 750-752.

Record, R.G., McKeown, T. & Edwards, J.H. (1970). An investigation of the difference in measured intelligence between twins and single births. *Annals of Human Genetics, 34,* 11-20.

Records, M.A., Heimbuch, R.C. & Kidd, K.K. (1977). Handedness and stuttering: a dead horse? *Journal of Fluency Disorders, 2,* 271-282.

Reed, G.F. & Smith, A.C. (1962). A further experimental investigation of the relative speeds of left- and right-handed writers. *Journal of Genetic Psychology, 100,* 275-288.

Reed, J.C. & Reitan, R.M. (1969). Verbal and performance differences among brain-injured children with lateralized motor deficits. *Perceptual & Motor Skills, 29,* 747-752.

Reid, G. (1986). Motor performance and learning by the mentally retarded. In L.D.Zaichkowsky & C.Z.Fuchs (Eds). *The psychology of motor behavior: development, control, learning and performance.* Ithaca, N.Y.: Movement Publications.

Rey, M., Dellatolas, G., Bancaud, J. & Talairach, J. (1988). Hemispheric lateralization of motor and speech functions after early brain lesion: study of 73 epileptic patients with intracarotid amytal test. *Neuropsychologia, 26,* !67-172.

Rice, T. & Plomin, R. (1983). Hand preferences in the Colorado adoption project. *Behavior Genetics, 13,* 550.

Rice, T., Plomin, R. & De Fries, J.C. (1984). Development of hand preference in the Cororado adoption project. *Perceptual & Motor Skills, 58,* 683-689.

Richardson, J.T.E. (1978). A factor analysis of self reported handedness. *Neuropsychologia, 16,* 747-748.

Rife, D.C. (1940). Handedness with special reference to twins. *Genetics, 25,* 178-186.

Rigal, R.A. (1992). Which handedness: preference or performance? *Perceptual & Motor Skills, 75,* 851-866.

Rigby, P. (1966). Dual symbolic classification among the Gogo of central Tanzania. *Africa, 36,* 1.16. Reprinted in R.Needham (Ed) *Right and left: essays on dual symbolic classification.* Chicago: Chicago University Press. p.263-287.

Risberg, J. (1980). Regional cerebral blood flow measurement by 133Xe inhalation: methodology and application in neuropsychology and psychiatry. *Brain & Language, 9,* 9-34.

Risberg, J. (1987). Hemisphere functions evaluated by measurements of the regional cerebral blood flow. In D.Ottoson (Ed). *Duality and unity of the brain.* London: Macmillan. p.442-453.

Roberts, L. (1959). Handedness and cerebral dominance. In W.Penfield and L.Roberts. *Speech and Brain Mechanisms.* Princeton, NJ: Princeton University Press. p.89-102.

Rohrbaugh, J.W., McCallum, W.C., Gaillard, R.W.K., Simons, R.F., Birbaumer, N. & Papakostopoulos, D. (1986). ERP's associated with preparatory and movement-related processes: a review. In W.C.McCallum, R.Zappoli & F.Denoth (Eds). *Cerebral psychophysiology: studies in event-related potentials (EEG Supp.38).* Amsterdam: Elsevier. p.189-229.

Roland, P.E. (1985). Cortical organization of voluntary behavior in man. *Human Neurobiology, 4,* 155-167.

Roland, P.E. (1987). Metabolic mapping of sensorimotor integration in the human brain. In G.Bock, M.O'Connor & J.Marsh (Eds). *Motor areas of the cerebral cortex. Ciba Foundation Symposium 132.* New York: Wiley. p.251-268.

Roland, P. & Larsen, B. (1976). Focal increase of cerebral blood flow during stereognostic testing in man. *Archives of Neurology, 33,* 551-558.

Roland, P.E., Larsen, B., Lassen, N.A. & Skinhoj, E. (1980). Supplementary motor area and other cortical areas in organization of voluntary movements in man. *Journal of Neurophysiology, 43,* 118-136.

Roland, P.E., Meyer, E., Shibasaki, T., Yamamoto, Y.L. & Thompson, C.J. (1982). Regional cerebral blood flow changes in cortex and basal ganglia during voluntary movements in normal human volunteers. *Journal of Neurophysiology, 48,* 467-480.

Rosen, G.D., Galaburda, A.M. & Sherman, G.F. (1987). Mechanisms of brain asymmetry: new evidence and hypotheses. In D.Ottoson (Ed). *Duality and unity of the brain.* New York: Plenum Press. p.29-36.

Rosen, G.D., Galaburda, A.M. & Sherman, G.F. (1990). The ontogeny of anatomic asymmetry: constraints derived from basic mechanisms. In A.B.Sheibel & A.F.Wechsler (Eds). *Neurobiology of higher cognitive function.* New York: Guilford Press. p.215-237.

Rosenzweig, M.R. (1971). Effects of environment on development of brain and behavior. In: E.Tobach, L.R.Aronson & E.Shaw (Eds). *The biopsychology of development.* New York: Academic Press. p.303-342.

Ross, G., Lipper, E. & Auld, P. (1987). Hand preference of four-year-old children: its relationship to premature birth and neurodevelopmental outcome. *Developmental Medicine & Child Neurology, 29,* 625-622.

Ross, G., Lipper, E. & Auld, P. (1992). Hand preference, prematurity and developmental outcome at school age. *Neuropsychologia, 30,* 483-494.

Rossi, G.F. & Rosadini, G. (1967). Experimental analysis of cerebral dominance in man. In C.H.Milliken and F.L.Darley (Eds). *Brain mechanisms underlying speech and language.* New York: Grune and Stratton. p.167-184.

Roszkowski, M.J., Snelbecker, G.E. & Sacks, R. (1980). Is age correlated with right-handedness? *Perceptual & Motor Skills, 51,* 862.

Roszkowski, M.J., Snelbecker, G.E. & Sacks, R. (1981). Children's, adolescent's and adult's reports on hand preference: homogeneity and discriminating power of selected tasks. *Journal of Clinical Neuropsychology, 3,* 199-214.

Roy, E.A. & Elliott, D. (1986). Manual asymmetries in visually directed aiming. *Canadian Journal of Psychology, 40,* 109-121.

Roy, E.A. & Square-Storer, P.A. (1990). Evidence for common expressions of apraxia. In G.E.Hammond (Ed). *Cerebral control of speech and limb movements.* Amsterdam: Elsevier. p.477-502.

Rubens, A.B. (1977). Anatomical asymmetries of the human cerebral cortex. In S.Harnad, R.W.Doty, J.Jaynes, L.Goldstein & G.Krauthamer (Eds). *Lateralization in the nervous system.* New York: Academic Press. p.503-516.

Rugg, M.D. (1983). The relationship between evoked potentials and lateral asymmetries of processing. In A.W.K.Gaillard & W.Ritter (Eds). *Tutorials in ERP research: endogenous components.* Amsterdam: North-Holland. p.369-384.

Rugg, M.D. & Dickens, A.M.J. (1982). Dissociation of alpha and theta activity as a function of verbal and visuospatial tasks. *Electroencephalography & Clinical Neurophysiology, 53,* 201-207.

Rugg, M., Kok, A., Barrett, G. & Fischler, I. (1986). ERP's associated with language and hemispheric specialization: a review. In W.C.McCallum, R.Zappoli & F.Denoth (Eds). *Cerebral psychophysiology: studies in event-related potentials (EEG Supp.38).* Amsterdam: Elsevier. p.273-300.

Rumelhart, D.E. & Norman, D.A. (1982). Simulating a skilled typist: a study of skilled cognitive motor performance. *Cognitive Science, 6,* 1-36.

Ruoff, P., Doer, R.H., Fuller, P., Martin, D. & Ruoff, L.O. (1981). Motor and cognitive interactions during lateralized cerebral functions in children: an EEG study. *Cortex, 17,* 5-18.

Rutter, M. (1984a) Introduction. Concepts of brain dysfunction syndromes. In M.Rutter (Ed). *Developmental Neuropsychiatry.* Edinburgh: Churchill Livingstone. p.1-14.

Rutter, M. (1984b). Issues and prospects in developmental neuropsychiatry. In M.Rutter (Ed). *Developmental Neuropsychiatry.* Edinburgh: Churchill Livingstone. p.577-598.

Rutter, M., Graham, P. & Yule, W. (1970). *A neuropsychiatric study in childhood.* London: Heinemann Medical.

Rutter, M. & Yule, W. (1970). Neurological aspects of intellectual retardation and specific reading retardation. In M.Rutter, J.Tizzard & K.Whitmore (Eds). *Education, health and behaviour.* London: Longman. p.54-74.

Ryding, E., Bradvik, B. & Ingvar, D.H. (1987). Changes of regional cerebral blood flow measured simultaneously in the right and left hemisphere during automatic speech and humming. *Brain, 110,* 1345-1358.

Rymar, K., Kameyama, T., Niwa, S., Hiramatsu, K. & Saitoh, O. (1984). Hand and eye preference patterns in elementary and junior high school students. *Cortex, 20,* 441-446.

Saigal, S., Rosenbaum, P., Szatmari, P. & Hoult, L. (1992). Non-right handedness among ELBW and term children at eight years in relation to cognitive function and school performance. *Developmental Medicine & Child Neurology, 324,* 425-433.

Salcedo, J.R., Spiegler, B.J., Gibson, E. & Magilavy, D.B. (1985). The autoimmune disease systemic lupus erythematosus is not associated with left-handedness. *Cortex, 21,* 645-647.

Salmaso, D. & Longoni, A.M. (1985). Problems in the assessment of hand preference. *Cortex, 21,* 533-549.

Salmoni, A.W., Schmidt, R.A. & Walter, C.B. (1984). Knowledge of results and motor learning: a review and critical appraisal. *Psychological Bulletin, 95* 355-386.

Salthouse, T.A. (1984). Effects of age and skill in typing. *Journal of Experimental Psychology: General, 113,* 345-371.

Sanders, B., Wilson, J.R. & Vandenberg, S.G. (1982). Handedness and spatial ability. *Cortex, 18,* 79-90.

Sandino, K. & McManus, I.C. (1998). Handedness, footedness, eyedness and earedness in the Colorado Adoption Project. *British Journal of Developmental Psychology, 16*, 167-174.

Sappington, J.T. (1980). Measures of lateral dominance: interrelationships and temporal stability. *Perceptual & Motor Skills, 50*, 783-790.

Satz. P. (1991). Symptom pattern and recovery outcome in childhood aphasia: a methodological and theoretical critique. In I.P.Martins, A.Castro-Caldas, H.R.van Dongen and A.van Hout (Eds). *Acquired aphasia in children: acquisition and breakdown of language in the developing brain.* Dordrecht: Kluver. p.95-114.

Satz, P., Baymur, L. & Van der Vlugt, H. (1979). Pathological left-handedness: cross-cultural tests of a model. *Neuropsychologia, 17*, 77-81.

Satz, P. & Bullard-Bates, C. (1981). Acquired aphasia in children. In M.T.Sarno (Ed). *Acquired aphasia.* New York: Academic Press. p.199-426.

Satz, P. & Fletcher, J.M. (1987). Left-handedness and dyslexia: an old myth revisited. *Journal of Pediatric Psychology, 12*, 291-298.

Satz, P., Miller, E.N., Selnes, O., Van Gorp. W., D'Elia, L.F. & Visscher, B. (1991). Hand preference in homosexual men. *Cortex, 27*, 295-306.

Satz, P. Y Soper, H.V. (1986). Left-handedness, dyslexia and auto-immune disorder: a critique. *Journal of Clinical & Experimental Neuropsychology, 8*, 453-458.

Satz, P., Strauss, E., Wada, J. & Orsini, D.L. (1988). Some correlates of intra and interhemispheric speech organization after left focal brain injury. *Neuropsychologia, 26*, 345-350.

Sauerwein, H.C. & Lassonde, M. (1983). Intra and interhemispheric processing of visual information in callosal agenesis. *Neuropsychologia, 21*, 167-171.

Sauerwein, H.C., Lassonde, M.C., Cardu, B. & Geoffroy, G. (1981). Interhemispheric integration of sensory and motor functions in agenesis of the corpus callosum. *Neuropsychologia, 19*, 445-454.

Sauerwein, H.C., Nolin, P. & Lassonde, M. (1994). Cognitive functioning in callosal agenesis. In M.Lassonde & M.A.Jeeves (Eds). *Callosal agenesis: a natural split-brain?* New York: Plenum Press. p.221-233.

Saul, R.E. & Sperry, R.W. (1968). Absence of commissurotomy symptoms with agenesis of the corpus callosum. *Neurology, 18*, 307.

Scarre, C. (1988). *Past Worlds: The Times atlas of archaeology.* London: Times Books.

Schacter, S.C. & Ransil, B.J. (1996). Handedness distributions in nine professional groups. *Perceptual & Motor Skills, 82*, 51-63.

Schaefer, E.S. & Bell, R.Q. (1958). Development of a parental attitude research instrument. *Child Development, 29*, 339-361.

Scheibel, A.B. (1984). A dendritic correlate of human speech. In N.Geschwind & A.M.Galaburda (Eds). *Cerebral dominance: the biological foundations.* Cambridge, MA: Harvard University Press. p.43-52.

Scheibel, A.B., Paul, L.A., Fried, I., Forsyth, A.B., Tomiyasu, U., Wechsler, A., Kao, A. & Slotnick, J. (1985). Dendritic organization of the anterior speech area. *Experimental Neurology, 87,* 109-117.

Scheidemann, N.V. (1930). A study of the handedness of some left-handed writers. *Journal of Genetic Psychology, 38,* 510-516.

Scheidemann, N.V. & Colyer, H. (1931a). A study in reversing the handedness of some left-handed writers. *Journal of Educational Psychology, 22,* 191-196.

Scheidemann, N.V. & Colyer, H. (1931b). A study of causes of left-hand writing preferences of some right-handed children. *Psychological Clinic, 20,* 116-119.

Schick, K.D. & Toth, N. (1993). *Making silent stones speak: human evolution and the dawn of technology.* New York: Simon & Schuster.

Schiers, J.G.M. (1990). Relationships between the direction of movements and handedness in children. *Neuropsychologia, 28,* 743-748.

Schlichting, C.L. (1982). Handedness in navy and student populations. *Perceptual & Motor Skills, 55,* 699-702.

Schmidt, R.A. (1976). The schema as a solution to some persistent problems in motor learning theory. In G.E.Stelmach (Ed) *Motor control: issues and trends.* New York: Academic Press. p.41-65.

Schmidt, R.A., Zelaznik, H.N. & Frank, J.S. (1978). Sources of inaccuracy in rapid movement. In G.E.Stelmach (Ed). *Information processing in motor control and learning.* New York: Academic Press. p.183-203.

Schmidt, R.T. & Toews, J.V. (1970). Grip strength as measured by the Jamar dynamometer. *Archives of Physical Medicine & Rehabilitation, 51,* 321-327.

Schneider, G.E. (1979). Is it really better to have your brain lesion early? A revision of the "Kennard Principle". *Neuropsychologia, 17,* 557-583.

Schott, G.D. (1980). Mirror movements of the left arm following peripheral damage to the preferred right arm. *Journal of Neurology, Neurosurgery & Psychiatry, 43,* 768-773.

Schreiber, H., Lang, M., Lang, W., Kornhuber, A., Heise, B., Keidel, M., Deecke, L. & Kornhuber, H.H. (1983). Frontal hemisphere differences in the Bereitschaftspotential associated with writing and drawing. *Human Neurobiology, 2,* 197-202.

Schultz, A.H. (1969). *The life of primates.* London: Weidenfeld & Nicolson.

Schur, P.H. (1986). Handedness in systemic lupus erythenatosus. *Arthritis Y Rheumatism, 29,* 419-420.

Scott, J.P., Bradt, D. & Collins, R.L. (1986). Fighting in female mice in lines selected for laterality. *Aggressive Behavior, 12*, 41-44.

Searleman. A. & Fugagli, A.K. (1987). Suspected autoimmune disorders and left-handedness: evidence from individuals with diabetes, Crohn's disease and ulcerative colitis. *Neuropsychologia, 25*, 367-374.

Searleman, A. & Fugagli, A.K. (1988). Left-handedness and suspected autoimmune disorders: a reply to Persson and Ahlbom. *Neuropsychologia, 26*, 739-740.

Searleman, Herrmann, D.J. & Coventry, A.K. (1984). Cognitive abilities and left-handedness: an interaction between familial sinistrality and strength of handedness. *Intelligence, 8*, 295-304.

Seashore, R.H. (1930). Individual differences in motor skills. *Journal of General Psychology, 3*, 38-66.

Seashore, R.H. (1951). Work and motor performance. In S.S.Stevens (Ed). *Handbook of experimental psychology.* New York: Wiley. p.1341-1362.

Seltzer, C., Forsythe, C. & Ward, J.P. (1990). Multiple measures of motor lateralization in human primates (Homo sapiens). *Journal of Comparative Psychology, 104*, 159-166.

Semenov, S.A. (1964). *Prehistoric technology,* New York: Barnes & Noble.

Shafer, D.D. (1987). Patterns of hand preference among captive gorillas. Unpublished Masters thesis quoted by W.D.Hopkins & R.D.Morris (1993). Handedness in great apes: a review of findings. *International Journal of Primatology, 14*, 1=25.

Shallice, T., McLeod, P. & Lewis, K. (1985). Isolating cognitive modules with the dual-task paradigm: are speech perception and production separate processes? *Quarterly Journal of Experimental Psychology, 37A*, 507-532.

Shanon, B. (1979). Graphological patterns as a function of handedness and culture. *Neuropsychologia, 17*, 457-465.

Shatz, C.J. (1992). The developing brain. *Scientific American, 267*, 35-41.

Sheehan, E.P. & Smith, H.V. (1986). Cerebral lateralization and handedness and their effect on verbal and spatial reasoning. *Neuropsychologia, 24*, 531-540.

Sheehy, M.P. & Marsden, C.D. (1982). Writer's cramp - a focal dystonia. *Brain, 105*, 461-480.

Sheehy, M.P., Rothwell, J.C. & Marsden, C.D. (1988). Writer's cramp. *Advances in Neurology, 50*, 457-472.

Sheridan, E.M. (1976). *Sex differences and reading. An annotated bibliography.* Newark, Del.: International Reading Association.

Shettel-Neuber, J. & O'Reilly, J. (1983). Handedness and career choice: another look at supposed left/right differences. *Perceptual & Motor Skills, 57,* 391-397.

Shimizu, A. & Endo, M. (1983). Handedness and familial sinistrality in a Japanese student population. *Cortex, 19,* 265-272.

Sidtis, J.J., Volpe, B.T., Wilson, D.H., Rayport, M. & Gazzaniga, M.S. (1981). Variability in right hemisphere language function after callosal section: evidence for a continuum of generative capacity. *Journal of Neuroscience, 1,* 323-331.

Signore, P., Nosten-Bertrand, M., Chaoui, M., Roubertoux, P.L., Marchaland, C. & Perez-Diaz, F. (1991a). An assessment of handedness in mice. *Physiology & Behavior, 49,* 701-704.

Signore, P., Chaoui, M., Nosten-Bertrand, M., Perez-Diaz, F. & Marchaland, C. (1991b). Handedness in mice: comparison across eleven inbred strains. *Behavior Genetics, 21,* 421-429.

Silverberg, R., Obler, L.K. & Gordon, H.W. (1979). Handedness in Israel. *Neuropsychologia, 17,* 83-87.

Silverman, A.J., Adevai, G. & McGough, W.E. (1966). Some relationships between handedness and perception. *Journal of Psychosomatic Research, 10,* 151-158.

Simon, J.R. (1964). Steadiness, handedness and hand preference. *Perceptual & Motor Skills, 18, 203-206.*

Simon, T.J. & Sussman, H.M. (1987). The dual task paradigm: speech dominance or manual dominance? *Neuropsychologia, 25,* 559-569.

Simonds, R.J. & Scheibel, A.B. (1989). The postnatal development of the motor speech area: a preliminary study. *Brain & Language, 37,* 42-58.

Sinclair, D. (1957). *An introduction to functional anatomy.* Oxford: Blackwell.

Slater-Hammel, A.T. (1948). Action current study of contraction-movement relationships in the golf stroke. *Research Quarterly, 19,* 164-177.

Slater-Hammel, A.T. (1949). An action current study of contraction-movement relationships in the tennis stroke. *Research Quarterly, 20,* 424-431.

Slater-Hammel, A.T. (1950). Bilateral effects of muscle activity. *Research Quarterly, 21,* 203-209.

Sloan, W. (1951). Motor proficiency and intelligence. *American Journal of Mental Deficiency, 55,* 394-406.

Smart, J.L., Jeffery, C. & Richards, B.A. (1980). A retrospective study of the relationship between birth history and handedness at six years. *Early Human Development, 4,* 79-88.

Smith,   A.   (1966).   Speech   and   other   functions   after   left   (dominant) hemispherectomy. *Journal of Neurology, Neurosurgery & Psychiatry, 29*, 467-471.

Smith,   A.   (1974).   Dominant   and   nondominant   hemispherectomy.   In M.Kinsbourne and W.L.Smith (Eds). *Hemisphere disconnection and cerebral function.* Springfield: Thomas. p.5-33.

Smith, A. & Burkland, C.W. (1966). Dominant hemispherectomy: preliminary report on neuropsychological sequelae. *Science, 153*, 1280-1282.

Smith, A.C. & Reed, G.F. (1959). An experimental investigation of the relative speeds of left- and right-handed writers. *Journal of Genetic Psychology, 94*, 67-76.

Smith, J. (1987). Left-handedness: its association with allergic disease. *Neuropsychologia, 25*, 665-674.

Smith, L.G. (1917). A brief survey of right- and left-handedness. *Pedagogical Seminary, 24*, 19-35.

Snyder, P.J., Novelly, R.A. & Harris, L.J. (1990). Mixed speech dominance in the intracarotid sodium amytal procedure: validity and criteria issues. *Journal of Clinical & Experimental Neuropsychology, 12*, 629-643.

Soper, H.V., Satz, P., Orsini, D.L., Henry, R.H., Zvi, J.C. & Schulman, M. (1986). Handedness patterns in autism suggest subtypes. *Journal of Autism & Developmental Disorders, 16*, 155-167.

Soper, H.V., Satz, P., Orsini, D.L., Van Gorp, W.G. & Green, M.F. (1987). Handedness distribution in a residential population with severe and profound mental retardation. *American Journal of Mental Deficiency, 92*, 94-102.

Special Correspondent (1972) Armless need not spell helpless. *Australian Women's Weekly*, August 9, p.7.

Spenneman, D.R. (1984a). Handedness data on the European neolithic. *Neuropsychologia, 22*, 613-615.

Spenneman, D.R. (1984b). Right- and left-handedness in early south-east Asia: the graphic evidence of the Borobudur. *Bijdragen tot de Taal-Land en Volkenkunde, 140*, 163-166.

Spenneman, D.R. (1985). On the origins and development of handedness in humans: some remarks on past and current theories. *Homo: Zeitschrift für die Vergleichender Forschung am Menschen, 36*, 121-141.

Sperry, R.W. (1952). Neurology and the mind-brain problem. *American Scientist, 40*, 291-312.

Sperry, R.W. (1958). Developmental basis of behavior. In: A.Roe & G.G.Simpson (Eds). *Behavior and evolution.* New Haven: Yale University Press. p.128-139.

Sperry, R.W. (1961). Cerebral organization and behavior. *Science, 133*, 1749-1757.

Sperry, R.W. (1967). Mental unity following surgical disconnection of the cerebral hemispheres. *Harvey Lectures, 62,* 293-323.

Sperry, R.W. (1968). Plasticity of neural maturation. *Developmental Biology, Suppl. 2,* 306-327.

Sperry, R.W. (1970). Cerebral dominance in perception. In F.A.Young and D.B.Lindsley (Eds). *Early experience and visual information processing in perceptual and reading disorders.* Washington, DC: National Academy of Science. p.167-178.

Sperry, R.W. (1974). Lateral specialization in the surgically separated hemispheres. In: F.O.Schmitt & F.G.Warden (Eds). *The neurosciences third study program.* Cambridge, Mass.: M.I.T. Press. p.5-19.

Spinelli, D.N. (1990). Plasticity triggering experience, nature, and the dual genesis of brain structure and function. *Clinics in Perinatology, 17,* 77-82.

Spinelli, D.N. & Jensen, F.E. (1979). Plasticity: the mirror of experience. *Science, 203,* 75-78.

Spinelli, D.N. & Jensen, F.E. (1982). Plasticity, experience and resource allocation in motor cortex and hypothalamus. In: C.D.Wood (Ed). *Conditioning.* New York: Plenum Press. p.651-661.

Spinelli, D.N., Jensen, F.E. & Viana di Prisco, G. (1980). Early experience effect on dendritic branching in normally reared kittens. *Experimental Neurology, 68,* 1-11.

Sporns, O. & Edelman, G.M. (1993). Solving Bernstein's problem: a proposal for the development of coordinated movement by selection. *Child Development, 64,* 960-981.

Spreen, O. & Gaddes, W.H. (1969). Developmental norms for 15 neuropsychological tests age 6 to 15. *Cortex, 5,* 170-191.

Square-Storer, P. & Roy, E.A. (1989). The apraxias: commonalities and distinctions. In P.Square-Storer (Ed). *Acquired apraxia of speech in aphasic adults.* London: Taylor and Francis. p.20-63.

Square-Storer, P., Roy, E.A. & Hogg, S.C. (1990). The dissociation of aphasia from apraxia of speech, ideomotor limb, and buccofacial apraxia. In G.E.Hammond (Ed). *Cerebral control of speech and limb movements.* Amsterdam: Elsevier. p.451-476.

St James Roberts, I. (1981). A reinterpretation of hemispherectomy data without functional plasticity of the brain. 1. Intellectual function. *Brain & Language, 13,* 31-53.

Standage, K.F. (1983). Observations on the handedness preferences of patients with personality disorders. *British Journal of Psychiatry, 142,* 575-578.

Stanton, W.R.. Feehan, M., Silva, P.A. & Sears, M.R. (1991). Handedness and allergic disorders in a New Zealand cohort. *Cortex, 27,* 131-135.

Steenhuis, R.E. & Bryden, M.P. (1989). Different dimensions of hand preference that relate to skilled and unskilled activities. *Cortex, 25,* 289-304.

Steenhuis, R.E., Bryden, M.P. & Schroeder, D.H. (1993). Gender, laterality, learning difficulties and health problems. *Neuropsychologia, 31,* 1243-1254.

Stein, D.G. & Glasier, M.M. (1992). An overview of developments in research on recovery from brain injury. In F.D.Rose and D.A.Johnson (Eds). *Recovery from brain damage.* New York: Plenum Press. p.1-22.

Stein, J.F. (1988). Physiological differences between left and right. In F.C.Rose, R.Whurr & M.A.Wyke (Eds). *Aphasia.* London: Whurr.

Steingruber, H.J. (1975). Handedness as a function of test complexity. *Perceptual & Motor Skills, 40,* 263-266.

Steinmetz, H., Volkmann, J., Jancke, L. & Freund, H-J. (1991). Anatomical left-right asymmetry of language related temporal cortex is different in left- and right-handers. *Annals of Neurology, 29,* 315-319.

Steklis, H.D. & Marchant, L.F. (1987). Primate handedness: reaching and grasping for straws? *Behavioral & Brain Sciences, 10,* 284-286.

Stern, J.A., Gold, S., Hoine, H. & Barocas, V.S. (1976). Toward a more refined analysis of the "overflow" or "associated movement" phenomenon. In D.V.S.Sankar (Ed). *Mental health in children. Vol II.* Westbury, N.Y.: PJD Publications. p.113-128.

Sternberg, R.J. (1985). Cognitive approaches to intelligence. In B.Wolman (Ed). *Handbook of intelligence.* New York: Wiley. p.59-118).

Sternberg, R.J. (1988). *The nature of creativity: contemporary psychological perspectives.* Cambridge: Cambridge University Press.

Strauss, E., Kosaka, B. & Wada, J. (1983). The neurobiological basis of lateralized cerebral function: a review. *Human Neurobiology, 2,* 115-127.

Strauss, E. & Wada, J. (1983). Lateral preferences and cerebral speech dominance. *Cortex, 19,* 165-177.

Strauss, E., Wada, J. & Goldwater, B. (1992). Sex differences in interhemispheric reorganization of speech. *Neuropsychologia, 30,* 353-359.

Strien, J.W.van & Bouma, A. (1988). Cerebral organization of verbal and motor functions in left-handed and right-handed adults: effects of concurrent verbal tasks on bimanual tapping performance. *Journal of Clinical & Experimental Neuropsychology, 10,* 139-156.

Subirana, A. (1958). The prognosis in aphasia in relation to cerebral dominance and handedness. *Brain, 81,* 415-425.

Subirana, A. (1964). The relationship between handedness and language function. *International Journal of Neurology, 4,* 215-234.

Summers, J.J. (1989). Motor programs. In D.H.Holding (Ed). *Human skills.* New York: Wiley. p.49-69.

Summers, J.J. (1990). Temporal constraints on concurrent task performance. In G.E.Hammond (Ed). *Cerebral control of speech and limb movements.* Amsterdam: North-Holland. p.661-680.

Super, C.M. (1976). Environmental effects on motor development: the case of African infant precocity. *Developmental Medicine & Child Neurology, 18,* 561-567.

Sussman, H.M. (1984). A reply to Kee's comments on Hughes & Sussman's time-sharing study of cerebral aterality in language-disordered and normal children. *Brain & Language, 22,* 357-358.

Szeszko, P.R., Madden, G.M. & Piro, J.M. (1997). Factor analyses of handedness items in left- and right-handed intellectually gifted and non-gifted children. *Cortex, 33,* 579-584.

Szirtes, J. & Vaughan, H.G. (1977a). Characteristics of cranial and facial potentials associated with speech production. *Electroencephalography & Clinical Neurophysiology, 43,* 386-396.

Szirtes, J. & Vaughan, H.G. (1977b). Characteristics of cranial and facial potentials associated with speech production. In J.E.Desmedt (Ed). *Language and hemispheric specialization in man: cerebral event-related potentials.* Progress in clinical neurobiology, Vol.3. *Basel: Karger. p.112-126.*

Szmodis, I., Szabo, T., Rendi, M., Temesi, Z. & Meszaros, J. (1984). Performance in plate-tapping and simple serial reaction time of children aged 5 - 14 years. In J.Ilmarinen & I.Valimaki (Eds). *Children and sport.* Berlin: Springer-Verlag. p.42-45.

Takeda, S. & Endo, A. (1993). Paw preference in mice: a reappraisal. *Physiology & Behavior, 53,* 727-730.

Tamas, L.B. & Shibasaki, H. (1985). Cortical potentials associated with movement: a review. *Journal of Clinical Neurophysiology, 2,* 157-171.

Tambs, K., Magnus, P. & Berg, K. (1987). Left-handedness in twin families: support of an environmental hypothesis. *Perceptual & Motor Skills, 64,* 155-170.

Tan, L.E. (1983). Handedness in two generations. *Perceptual & Motor Skills, 56,* 867-874.

Tan, L.E. (1985). Laterality and motor skills in four year olds. *Child Development, 56,* 119-124.

Tan, U. (1988). The distribution of hand preference in normal men and women. *International Journal of Neuroscience, 41,* 35-55.

Tan, U. & Kutlu, N. (1991). The distribution of paw preference in right-, left-, and mixed-pawed male and female cats: The role of a female right-shift factor in handedness. *International Journal of Neuroscience, 59,* 219-229.

Tan, U., Yaprak, M. & Kutlu, N. (1990). Paw preference in cats: distribution and sex differences. *International Journal of Neuroscience, 50,* 195-208.

Tapley, S.M. & Bryden, M.P. (1985). A group test for the assessment of performance between hands. *Neuropsychologia, 23,* 215-221.

Tarkka, I.M. & Hallett, M. (1990). Cortical topography of premotor and motor potentials preceding self-paced, voluntary movement of dominant and non-dominant hands. *Electroencephalography & Clinical Neurophysiology, 75,* 36-43.

Tauber, M.A. (1979). Parental socialization techniques and sex differences in children's play. *Child Development, 50,* 225-234.

Taylor, P.J., Brown, R. & Gunn, J. (1983). Violence psychosis and handedness. In P.Flor-Henry & J.Gruzelier (Eds). *Laterality and psychopathology.* Amsterdam: Elsevier. p.181-194.

Taylor, P.J., Dalton, R. & Fleminger, J.J. (1980. Handedness in schizophrenia. *British Journal of Psychiatry, 136,* 375-383.

Taylor, P.J., Dalton, R. & Fleminger, J.J. (1982). Handedness and schizophrenic symptoms. *British Journal of Medical Psychology, 55,* 287-291.

Temple, C.M. (1990). Academic discipline, handedness and immune disorders. *Neuropsychologia, 28,* 303-308.

Teng, E.L., Lee, P-H., Yang, K-S. & Chang, P.C. (1976). Handedness in a Chinese population: biological, social and pathological factors. *Science, 193,* 1148-1150.

Teng, E.L., Lee, P-H., Yang, K-S. & Chang, P.C. (1979). Lateral preferences for hand, foot, and eye, and heir lack of association with scholastic achievement in 4,143 Chinese. *Neuropsychologia, 17,* 41-48.

Terzuolo, C.A. & Viviani, P. (1979). The central representation of learned motor patterns. In R.E.Talbott & D.R.Humphrey (Eds). *Posture and movement.* New York: Raven Press.

Teuber, H-L. (1974). Why two brains? In: F.O.Schmitt & F.G.Warden (Eds). *The neurosciences third study program.* Cambridge, Mass.: M.I.T. Press. p.71-74.

Thatcher, R.W., Walker, R.A. & Giudice, S. (1987). Human cerebral hemispheres develop at different rates and ages. *Science, 236,* 1110-1113.

Thelen, E. (1995). Motor development: a new synthesis. *American Psychologist, 50,* 79-95.

Thelen, E., Corbetta, D., Kamm, K. & Spencer, J.P. (1993). The transition to reaching: mapping intention and intrinsic dynamics. *Child Development, 64,,* 1058-1098.

Thomas, J. & French, K. (1985). Gender differences across age in motor performance: a meta-analysis. *Psychological Bulletin, 98,* 260-282.

Thompson, A.L. & Marsh, J.F. (1976). Probability sampling of manual asymmetry. *Neuropsychologia, 14,* 217-223.

Thorndike, E.L. (1927). The law of effect. *American Journal of Psychology, 39,* 212-222.

Thorndike, E.L., Bregman, E.O., Tilton, J.W. & Woodyard, E. (1928) *Adult learning.* New York: Macmillan.

Tierney, I., Smith, L., Axworthy, D. & Ratcliffe, S.G. (1984). The McCarthy scales of children's abilities - sex and handedness effects in 128 Scottish five-year-olds. *British Journal of Educational Psychology, 54,* 101-105.

Tobias, P.V. (1981). The emergence of man in Africa and beyond. *Philosophical Transactions of the Royal Society (London) B, 292,* 43-56.

Todor, J. & Kyprie, P.M. (1980). Hand difference in the rate and variability of rapid tapping. *Journal of Motor Behavior, 12,* 57-62.

Todor, J., Kyprie, P.M. & Price, H.L. (1982). Lateral asymmetries in arm, wrist and finger movements. *Cortex, 18,* 515-523.

Todor, J.I. & Lazarus, J.C. (1986a). Exertion level and the intensity of associated movements. *Developmental Medicine & Child Neurology, 28,* 205-212.

Todor, J.I. & Lazarus, J.C. (1986b). Inhibitory influences on the emergence of motor competence in childhood. In L.D.Zaichowsky & C.Z.Fuchs (Eds). *Psychology of motor behavior: development, control, learning and performance.* Ithaca, N.Y.: Movement Publications. p.239-255.

Todor, J.I. & Lazarus, J.C. (1986c). Inhibitory mechanisms in children's skill development. In H.T.A.Whiting & M.B.Wade (Eds). *Themes in motor development.* Dordrecht: Nijhoff. p.237-246.

Toews, J.V. (1964). A grip-strength study among steelworkers. *Archives of Physical Medicine, 45,* 413-417.

Tokuda, K. (1969). On the handedness of Japanese monkeys. *Primates, 10,* 41-46.

Toth, N. (1985). Archeological evidence for preferential right-handedness in the lower and middle pleistocene, and its possible implications. *Journal of Human Evolution, 14,* 607-614.

Toth, N. (1987). The first technology. *Scientific American, 256,* 104-113.

Toth, N., Clark, D. & Ligabue, G. (1992). The last stone axe makers. *Scientific American, 267*, 66-71.

Travis, L.E. (1931). *Speech pathology*. New York: Appleton.

Travis, L.E. & Johnson, W. (1934). Stuttering and the concept of handedness. *Psychological Review, 41*, 534-562.

Treves, T., Goldschmidt, I. & Korczyn, A.D. (1983). Development of manual laterality and language function. In G.Young, S.J.Segalowitz, S.J.Corter & S.E.Trehub (Eds). *Manual specialization and the developing brain*. New York: Academic Press. p.395-400.

Troster, H. & Brambring, M. (1993). Early motor development in blind infants. *Journal of Applied Developmental Psychology, 14*, 83-106.

Trowbridge, M.H. & Cason, H. (1932). An experimental study of Thorndike's theory of learning. *Journal of General Psychology, 7*, 245-258.

Tsai, L.S. & Maurer, S. (1930). Righthandedness in white rats. *Science, 72*, 436-438.

Tsai, L.Y. (1982). Handedness in autistic children and their families. *Journal of Autism & Developmental Disorders, 12*, 421-423.

Tylor, E.B. (1903) *Primitive culture: researches into the development of mythology, philosophy, religion, art and custom*. London: John Murray. 4th Edition.

Uhrbrock, R.S. (1973). Laterality in art. *Journal of Aesthetics & Art Criticism, 32*, 27-35.

Uvebrant, P. (1988). Hemiplegic cerebral palsy: aetiology and outcome. *Acta Paediatrica Scandinavica, Suppl. 345*.

Van Hout, A., Evrard, P.L. & Lyon. G. (1985). On the positive semiology of acquired aphasia in children. *Developmental Medicine & Child Neurology, 27*, 231-241.

VanSant, A.F. & Williams, K. (1986). Incidence of associated reactions in young boys and reliability of observations. *Physical & Occupational Therapy in Pediatrics, 6(2)*, 41-53.

Van Wagenen, W.P. & Herren, R.Y. (1940). Surgical division of commissural pathways in the corpus callosum: relation to spread of an epileptic attack. *Archives of Neurology & Psychiatry, 44*, 740-759.

Varga-Khadem, F. & Polkey, C.E. (1992). A review of cognitive outcome after hemidecortication in humans. In F.D.Rose and D.A.Johnson (Eds). *Recovery from brain damage*. New York: Plenum Press. p.137-150.

Varner, J.L., Ellingson, R.J., Danaby, T. & Nelson, B. (1977). Interhemispheric amplitude symmetry in the EEG's of full-term newborns. *Electroencephalography & Clinical Neurophysiology, 43*, 846-852.

Verhaegen, P. & Ntumba, A. (1964). Note on the frequency of left-handedness in African children. *Journal of Educational Psychology, 55,* 89-90.

Viana di Prisco, G. & Spinelli, D.N. (1981). Observations on cortical plasticity using horseradish peroxidase. *Experimental Neurology, 74,* 935-939.

Villablanca, J.R., Burgess, J.W. & Olmstead, C.E. (1986). Recovery of function after neonatal or adult hemispherectomy in cats: I. Time course, movement, posture and sensorimotor tests. *Behavioral Brain Research, 19,* 205-226.

Volkmar, F.R., Burack, J.A. & Cohen, D.J. (1990). Deviance and developmental approaches to the study of autism. In R.M.Hodapp, J.A.Burack & E.Zigler (Eds). *Issues in the developmental approach to mental retardation.* Cambridge: Cambridge University Press. p.246-271.

Wada, J.A., Clarke, R. & Hamm, A. (1975). Cerebral hemispheric asymmetry in humans. *Archives of Neurology, 32,* 239-246.

Wada, J. & Rasmussen, T. (1960). Intra-carotid injection of sodium amytal for the lateralization of cerebral speech dominance. *Journal of Neurosurgery, 17,* 266-282.

Wahl, O.F. (1976). Handedness in schizophrenia. *Perceptual & Motor Skills, 42,* 944-946.

Walker, E., Davis, D. & Baum, K. (1993). Social withdrawal. In C.G.Costello (Ed). *Symptoms of schizophrenia.* New York: Wiley. p.227-260.

Walker, H.A. & Birch, H.G. (1970). Lateral preference and right-left awareness in schizophrenic children. *Journal of Nervous & Mental Disease, 151,* 341-351.

Walker, S.F. (1980). Lateralization of functions in the vertebrate brain: a review. *British Journal of Psychology, 71,* 329-367.

Wall, J.T. (1988). Variable organization in cortical maps of the skin as an indication of the lifelong adaptation capacities of circuits in the mammalian brain. *Trends in Neuroscience, 11,* 649-557.

Walshe, F.M.R. (1935). On the "syndrome of the premotor cortex" (Fulton) and the definition of the terms "premotor" and "motor": with a consideration of Jackson's views on the cortical representation of movements. *Brain, 58,* 49-80.

Walshe, F.M.R. (1947). On the role of the pyramidal system in willed movements. *Brain, 70,* 329-354.

Walton, J.H. (1977). *Brain's diseases of the nervous system.* Oxford: Oxford University Press. 8th Edn.

Walton, J.N., Ellis, E. & Court, S.D.M. (1962). Clumsy children: a study of developmental apraxia and agnosia. *Brain, 85,* 603-612.

Ward, J.P. & Hopkins, W.D. (1993). *Primate laterality: current behavioral evidence of primate asymmetries.* New York: Springer-Verlag.

Warren, C. & Karrer, R. (1984a). Movement-related potentials during development: a replication and extension of relationships to age, motor control, mental status and IQ. *International Journal of Neuroscience, 24,* 81-96.

Warren, C. & Karrer, R. (1984b). Movement-related potentials in children: a replication of waveforms and their relationships to age, performance and cognitive development. In R.Karrer, J.Cohen & P.Tueting (Eds). *Brain and information: event-related potentials. Annals of the New York Academy of Sciences, 425,* 489-495.

Warren, J.M. (1953). Handedness in the rhesus monkey. *Science, 18,* 622-623.

Warren, J.M. (1958). The development of paw preferences in cats and monkeys. *Journal of Genetic Psychology, 93,* 229-236.

Warren, J.M. (1973). Learning in vertebrates. In: DF.A.Dewsbury & D.A.Rethlingshafer (Eds). *Comparative Psychology: a modern survey.* New York: McGraw-Hill. p.471-509.

Warren, J.M. (1974). Possibly unique characteristics of learning in primates. *Journal of Human Evolution, 3,* 445-454.

Warren, J.M. (1977a). Functional lateralization of the brain. In S.J.Dimond & D.A.Blizzard (Eds). *Evolution and lateralization of the brain.* New York: New York Academy of Sciences, p.273-280.

Warren, J.M. (1977b). Handedness and cerebral dominance in monkeys. In S.Harnad, R.W.Doty, L.Goldstein, J.Jaynes & G.Krauthamer (Eds). *Lateralization in the nervous system..* New York: Academic Press, p.151-172.

Warren, J.M. (1980). Handedness and laterality in humans and other animals. *Physiological Psychology, 8,* 351-359.

Warren, J.M., Ablanalp, J.M. & Warren, H.B. (1967). The development of handedness in cats and rhesus monkeys. In H.W.Stevenson (Ed). *Early behavior: comparative and developmental approaches.* New York: Wiley. p.73-101.

Warren, J.M., Cornwell, P.R., Webster, W.G. & Pubols, B.H. (1972). Unilateral cortical lesions and paw preferences in cats. *Journal of Comparative & Physiological Psychology, 81,* 410-422.

Warren, J.M. & Nonneman, A.J. (1976). The search for cerebral dominance in monkeys. *Annals of the New York Academy of Sciences, 280,* 732-744.

Watanabe, K. & Kawai, M. (1993). Lateralized hand use in the precultural behavior of the Koshima monkeys (Macaca fuscata). In J.P.Ward & W.D.Hopkins (Eds). *Primate laterality.* New York: Springer-Verlag, p.183-192.

Waters, N.S. & Denenberg, V.H.A. (1991). A measure of lateral paw preference in the mouse. *Physiology & Behavior, 50*, 853-856.

Watson, N.V. & Kimura, D. (1989). Right hand superiority for throwing but not for intercepting. *Neuropsychologia, 27*, 1399-1414.

Watt, N.F. & Lubensky, A.W. (1976). Childhood roots of schizophrenia. *Journal of Consulting & Clinical Psychology, 44*, 363-375.

Watt, N.F., Stolorow, R.D., Lubensky, A.W. & McClelland, D.C. (1970). School adjustment and behavior of children hospitalized for schizophrenia as adults. *American Journal of Orthopsychiatry, 40*, 637-657.

Watter, P. & Bullock, M.I. (1989). Minimal cerebral dysfunction in adults. *Australian Journal of Physiotherapy, 35*, 239-244.

Weaver, L.A. & Ravaris, C.L. (1972). Psychomotor performance of mental retardates. *Journal of Mental Deficiency Research,, 16*, 76-83.

Webster, W.G. & Poulos, M. (1987). Handedness distribution among adults who stutter. *Cortex, 23*, 705-708.

Wechsler, I.S. (1958). *A textbook of clinical neurology: with an introduction to the history of neurology*. Philadelphia: Saunders.

Weinrich, A.M., Wells, P.A. & McManus, C. (1982). Handedness, anxiety and sex differences. *British Journal of Psychology, 73*, 69-72.

Weinstein, S., Sersen, E.A. & Vetter, R.J. (1964). Phantoms and somatic sensation in cases of congenital aplasia. *Cortex, 1*, 276-290.

Weisenberg, T.H. & McBride, K.E. (1935). *Aphasia: a clinical and psychological study*. New York: Commonwealth Fund.

Weissler, K. & Landau, E. (1993). Characteristics of families with no, one or more than one gifted child. *Journal of Psychology, 127*, 143-152.

Welford, A.T. (1952). The "psychological refractory period" and the timing of high speed performance: a review and a theory. *British Journal of Psychology, 43*, 2-19.

Welford, A.T (1959). Evidence of a single channel decision mechanism limiting performance in a serial reaction task. *Quarterly Journal of Experimental Psychology, 11*, 193-210.

Welford, A.T. (1968). *Fundamentals of skill*. London: Methuen.

Weller, M.P.I. & Latimer-Sayer, D.T. (1985). Increasing right hand dominance with age on a motor skill task. *Psychological Medicine, 15*, 867-872.

Wentworth, K.L. (1938). The effect of early reaches on handedness in the rat: a preliminary study. *Journal of Genetic Psychology, 52*, 429-432.

Wentworth, K.L. (1942). Some factors determining handedness in the white rat. *Genetic Psychology Monographs, 26*, 57-117.

West, D.J. & Farrington, D.P. (1977). *The delinquent way of life.* London: Heinemann.

Westermarck, E. (1926). *Ritual and belief in Morocco.* London: Macmillan.

Wetmore, R.G. & Estabrooks, G.H. (1929). The relation of left-handedness to psycho-neurotic traits. *Journal of Educational Psychology, 20,* 628-629.

Whipple, G.M. (1911). The left-handed child. *Journal of Educational Psychology, 2,* 574-575.

Whishaw, I.Q. & Kolb, B. (1984). Behavioral and anatomical studies of rats with complete or partial decortication in infancy. In C.R.Almli and S.Finger (Eds). *Early brain damage.* New York: Academic Press. p.117-138.

Whishaw, I.Q. & Kolb, B. (1988). Sparing of skilled forelimb reaching and corticospinal projections after neonatal motor cortex removal or hemidecortication in the rat: support for the Kennard doctrine. *Brain Research, 451,* 97-114.

White, K. & Ashton, R. (1976). Handedness assessment inventory. *Neuropsychologia, 14,* 261-264.

White, N. & Kinsbourne, M. (1980). Does speech output control lateralize over time? Evidence from verbal-manual time-sharing tasks. *Brain & Language, 10,* 215-223.

Whiting, B.B. & Edwards, C. (1973). A cross cultural anaysis of sex differences in the behavior of children aged three through eleven, *Journal of Social Psychology, 91,* 171-188.

Whittington, J.E. & Richards, P.N. (1987). The stability of children's laterality prevalences and their relationship to measures of performance. *British Journal of Educational Psychology, 57,* 45-55.

Wieschoff, H.A. (1938). Concept of right and left in African cultures. *Journal of the American Oriental Society, 58,* 202-217. Reprinted in R.Needham (Ed). *Right and left: essays on dual symbolic classification.* Chicago: Chiocago University Press. p.59-73.

Wile, I.S. (1934). *Handedness right and left.* Boston: Lothrop, Lee & Shepard.

Wiley, J. & Goldstein, D. (1991). Sex, handedness and allergy: are they related to academic giftedness? *Journal for the Education of the Gifted, 14,* 412-422.

Williams, J.R. & Scott, R.B. (1953). Growth and development of Negro infants IV. Motor development and its relationship to child-rearing practices in two groups of Negro infants. *Child Development, 24,* 103-121.

Williams, S.M. (1986). Factor analysis of the Edinburgh Handedness Inventory. *Cortex, 22,* 325-326.

Wilson, D. (1891). *The right hand: left-handedness.* London: Macmillan.

Wilson, M.O. & Dolan, L.O. (1931). Handedness and ability, *American Journal of Psychology, 43,* 261-268.

Wilson, P.J.E. (1970). Cerebral hemispherectomy for infantile hemiplegia. A report of 50 cases. *Brain, 93,* 147-180.

Wilson, R.S. (1975). Twins: patterns of cognitive development as measured on the Webster Preschool and Primary Scale of Intelligence. *Developmental Psychology, 11,* 126-134.

Winstein, C.J. & Schmidt, R.A. (1989). Sensorimotor feedback. In D.H.Holding (Ed). *Human skills.* Chichester: Wiley. p.17-47.

Wise, R., Chollet, F., Hadar, U., Friston, K., Hoffner, E. & Frackowiak, R. (1991). Distribution of cortical neural networks involved in word comprehension and word retrieval. *Brain, 114,* 1803-1817.

Witelson, S.F. (1977). Anatomic asymmetry in the temporal lobes: its documentation, phylogenesis and relationship to functional asymmetry. In S.J.Dimond & D.A.Blizard (Eds). *Evolution and lateralization of the brain. Annals of the New York Academy of Sciences, 299,* 328-354.

Witelson, S.F. (1980). Neuroanatomical asymmetry in left-handers: a review and implications for functional asymmetry. In J.Herron (Ed). *Neuropsychology of left-handedness.* New York: Academic Press. p.79-113.

Witelson, S.F. (1983). Bumps on the brain: right-left asymmetry as the key to functional lateralization. In S.Segalowitz (Ed). *Language functions and brain organization.* New York: Academic Press. p.117-144.

Witelson, S.F. (1985). The brain connection: the corpus callosum is larger in left-handers. *Science, 229,* 665-668.

Witelson, S.F. (1989). Hand and sex differences in the isthmus and genu of the human corpus callosum: a postmortem morphological study. *Brain, 112,* 799-835.

Witelson, S.F. & Goldsmith, C.H. (1991). The relationship of hand preference to anatomy of the corpus callosum in men. *Brain Research, 545,* 175-182.

Witelson, S.F. & Kigar, D.L. (1988a). Asymmetry in brain function follows asymmetry in anatomical form: gross, microscopic, postmortem and imaging studies. In F.Boller & J.Grafman (Eds). *Handbook of Neuropsychology, Vol.1.* Amsterdam, Elsevier. p.111-142.

Witelson, S.F. & Kigar, D.L. (1988b). Anatomical development of the corpus callosum in humans: a review with reference to sex and cognition. In D.L.Molfese & S.J.Segalowitz (Eds). *Brain lateralization in children: developmental implications.* New York: Guilford Press. p.35-57.

Witelson, S.F. & Kigar, D.L. (1992). Sylvian fissure morphology and asymmetry in men and women: bilateral differences in relation to handedness in men. *Journal of Comparative Neurology, 323*, 326-340.

Witelson, S.F. & Nowakowski, R.S. (1991). Left out axons make men right: a hypothesis for the origin of handedness and functional asymmetry. *Neuropsychologia, 29*, 327-333.

Witelson, S.F. & Pallie, W. (1973). Left hemisphere specialization for language in the newborn. *Brain, 96*, 641-646.

Wolff, P.H., Gunnoe, C.E. & Cohen, C. (1983). Associated movements as a measure of developmental age. *Developmental Medicine & Child Neurology, 25*, 417-429.

Wong, S-L. (1970). Social change and parent-child relations in Hong Kong. In R.Hill & R.Konig (Eds). *Families in East and West: socialization process and kinship ties.* The Hague: Monton. p.167-174.

Woo, T.L. & Pearson, K. (1927). Dextrality and sinistrality of hand and eye. *Biometrika, 19*, 165-199.

Wood, C.J. & Aggleton, J.P. (1989). Handedness in "fast ball" sports: do left-handers have an innate advantage? *British Journal of Psychology, 80*, 227-240.

Wood, C.J. & Aggleton, J.P. (1991). Occupation and handedness: an examination of architects and mail survey biases. *Canadian Journal of Psychology, 45*, 395-404.

Woods, B.T. & Carey, S. (1979). Language deficits after apparent recovery from childhood aphasias. *Annals of Neurology, 6*, 405-409.

Woods, B.T. & Teuber, H-L. (1978a). Changing patterns of childhood aphasia. *Annals of Neurology, 3*, 273-280.

Woods, B.T. & Teuber, H-L. (1978b). Mirror movements after childhood hemiparesis. *Neurology, 28*, 1152-1158.

Woods, R.P., Dodrill, C.B. & Ojemann, G.A. (1988). Brain injury, handedness and speech lateralization in a series of amobarbital studies. *Annals of Neurology, 23*, 510-518.

Woolsey, C.N. (1958). Organization of somatic sensory and motor areas of the cerebral cortex. In: H.F.Harlow & C.N.Woolsey (Eds). *Biological and biochemical bases of behavior.* Madison: University of Wisconsin Press. p.63-81.

Yakovlev, P.I. & Lecours, A. (1967). The myelogenetic cycles of regional maturation of the brain. In A.Minkowski (Ed). *Regional development of the brain in early life.* London: Blackwell. p.3-70.

Yan, S-M., Flor-Henry, P., Chen, D., Li, T., Qi, S. & Ma, Z. (1985). Imbalance of hemispheric functions in the major psychoses: a study of handedness in the People's Republic of China. *Biological Psychiatry, 20*, 906-917.

Yen, W.M. (1975). Independence of hand preference and sex-linked genetic effects on spatial performance. *Perceptual & Motor Skills, 41,* 311-318.

Yeudall, L.T., Fromm-Auch, D. & Davies, P. (1982). Neuropsychological impairment of persistent delinquency. *Journal of Nervous & Mental Disease, 170,* 257-265.

Yingling, C.D. (1980). Cognition, action and mechanisms of EEG asymmetry. In G.Pfurtscheller, P.Buser, F.L.da Silva & H.Petsche (Eds). *Rhythmic EEG activities and cortical functioning.* Amsterdam: Elsevier. p.79-90.

Yokoe, K. (1970). Historical trends in home discipline. In R.Hill & R.Konig (Eds). *Families in East and West: socialization process and kinship ties.* The Hague: Monton. p.175-186.

Yoshioka, J.G. (1930a). Handedness in rats. *Psychological Bulletin, 27,* 656-657.

Yoshioka, J.G. (1930b). Handedness in rats. *Journal of Genetic Psychology, 38,* 471-474.

Young, D.E. & Schmidt, R.A. (1990). Units of motor behavior: modifications with practice and feedback. In M.Jeannerod (Ed). *Attention and performance Xiii.* Hillsdale, N.J.: Lawrence Erlbaum. p.163-195.

Zaidel, D. & Sperry, R.W. (1977). Some long-term motor effects of cerebral commissurotomy in man. *Neuropsychologia, 15,* 193-204.

Zaidel, E. (1985). Language in the right hemisphere. In D.F.Benson and E.Zaidel (Eds). *The duel brain: hemispheric specialization in humans.* New York: Guilford Press. p.205-231.

Zaidel, E. (1990). Language functions in the two hemispheres following complete cerebral commissurotomy and hemispherectomy. In R.D.Nebes and S.Corkin (Eds). *Handbook of Neuropsychology, Vol. 4.* Amsterdam: Elsevier. p.115-150.

Zangwill, O.L. (1955). Speech and handedness. *Advancement of Science (London), 12,* 55-59.

Zangwill. O.L. (1960a). Speech. In: J.Field, H.W.Magoun & V.E.Hall (Eds). *Handbook of Physiology, Vol.3. Neurophysiology.* American Physiological Society. p.1709-1722.

Zangwill, O.L. (1960b). *Cerebral dominance and its relation to psychological function.* Edinburgh: Oliver and Boyd.

Zangwill, O.L. (1964). The current status of cerebral dominance. In: D.M.Rioch & E.A.Weinstein (Eds). *Disorders of communication. Research Publication No.XL11 of the Association for Research in Nervous and Mental Disease.* Baltimore: Williams & Wilkins. p.103-118.

Zangwill, O.L. (1975). Ontogeny of cerebral dominance in man. In E.H.Lenneberg and E.Lenneberg (Eds). *Foundations of language development.* New York: Academic Press. p.137-147.

Zatorre, R.J. (1989). Perceptual asymmetry on the dichotic fused words test and cerebral speech lateralization determined by the carotid sodium amytal test. *Neuropsychologia, 27,* 1207-1219.

Zelazo, N.A., Zelazo, P.R., Cohen, K.M. & Zelazo, P.D. (1993). Specificity of practice effects on elementary neuromotor patterns. *Developmental Psychology, 29,* 686-691.

Ziviani, J. (1984). Some elaborations on handwriting speed in 7 to 14 year olds. *Perceptual & Motor Skills, 58,* 535-539.

Zuckerman, M., Ribback, B.B., Monashkin, I. & Norton, J.A. (1958). Normative data and factor analysis on the Parental Attitude Research Instrument. *Journal of Consulting Psychology, 22,* 165-171.